Cognitive Coaching

Cognitive Coaching

Developing Self-Directed Leaders and Learners

Arthur L. Costa and Robert J. Garmston
With Carolee Hayes and Jane Ellison

ROWMAN & LITTLEFIELD
Lanham • Boulder • New York • London

Executive Editor: Susanne Canavan
Associate Editor: Carlie Wall
Marketing Manager: Karin Cholak
Production Editor: Laura Reiter
Interior Designer: Rhonda Baker
Cover Designer: Meredith Nelson

Credits and acknowledgments of sources for material or information used with permission appear on the appropriate page within the text or in the credit section on page XXX.

Published by Rowman & Littlefield
A wholly owned subsidiary of The Rowman & Littlefield Publishing Group, Inc.
4501 Forbes Boulevard, Suite 200, Lanham, Maryland 20706
www.rowman.com

Unit A, Whitacre Mews, 26-34 Stannary Street, London SE11 4AB

Copyright © 2016 by Rowman & Littlefield

British Library Cataloguing in Publication Information Available

Library of Congress Cataloging-in-Publication Data

Costa, Arthur L.
  Cognitive coaching : developing self-directed leaders and learners / Aurthur L. Costa and Robert J. Garmston with Carolee Hayes and Jane Ellison.
     pages cm
  Includes bibliographical references and index.
  ISBN 978-1-4422-2365-3 (cloth : alk. paper) — ISBN 978-1-4422-2366-0 (electronic)
1. Thought and thinking—Study and teaching. 2. Cognition in children. 3. Student-centered learning. I. Title.
  LB1590.3.C6794 2015
  370.15'2—dc23

                                                                                            2015005701

∞™ The paper used in this publication meets the minimum requirements of American National Standard for Information Sciences—Permanence of Paper for Printed Library Materials, ANSI/NISO Z39.48-1992.

Printed in the United States of America

Contents

APPENDICES

Tables and Figures

FIGURES

TABLES

Foreword

Cognitive Coaching: Developing Self-Directed Learners and Leaders, 3rd Edition

Michael Fullan and Andy Hargreaves

We are honored to write the foreword to this new edition of *Cognitive Coaching: Developing Self-Directed Learners and Leaders* by Art Costa, Bob Garmston, and their colleagues, Jane Ellison and Carolee Hayes. The first edition was published over 20 years ago and was titled *Cognitive Coaching: A Foundation for Renaissance Schools*. This 1994 book put coaching on the map. Cognitive coaching is a common accompaniment to and ingredient of change implementation now: a true mark of these authors' influence. Though not always grasped or used with the sophistication and depth of these authors' intentions, their work has had a massive influence on how we now think about the best ways to improve classroom practice—with other teachers, embedded in practice, rather than in the big ballroom sessions of staff development or in the isolated classrooms where individual teachers struggle by themselves. This interest in cognitive coaching has been carefully built by Costa, Garmston, and their growing band of followers and colleagues.

The United States goes through churns of system upheaval that never seem to land: No Child Left Behind (2002), Race to the Top (2008), and now Common Core State Standards. Cognitive coaching is, by contrast, one of the constants of improvement: learning how to be more effective and reflective by working with experienced and expert peers.

The timing and nature of the third edition is perfect. Piecemeal approaches can alleviate the worst of change but can never make the best of change. This is Costa and Garmston's conclusion. Their third edition goes deeper with greater precision. It identifies the skills, and now the mind-sets, that must underpin them. It is heavily action based, taking into account the diversity of humans and their contexts, but it does this in relation to larger maps of action. It carefully examines issues of impact that in turn reverberate back on the skills, mind-sets, and actions of those who want to help. Above all, this book treats the solution as a system of interactions, not a checklist of competencies.

What the third edition adds is the guidance to convert Cognitive Coaches to true teacher leaders and instructional leaders. Both of us have just published books on the "new leadership" for system change. New leaders are people who are lead learners, who participate as learners with others moving things forward, who operate in relatively flat organization structures and modes, who develop leaders in others and thereby make a greater short-term impact, and who pave the way for sustainable leadership, with meaning and depth. This new leadership is not the instructional leadership of the individual and overwhelmed principal who must take the instructional responsibilities of the school and its teachers onto his or her own personal shoulders. Nor is it the so-called distributed leadership of benevolent dictatorships that gets teachers to pick up more of the workload, involve them a bit more in decision making, and cultivate more pleasant staff relationships. It is engaged, embedded, and integrated leadership, exercised with a sense of collective professional responsibility and combined expertise in the service of all students' success.

In *The Principal: Three Keys for Maximizing Impact*, one of us (Fullan 2014) set out to rescue principals from being led down the path of micromanaging instruction toward becoming lead learners who help develop, in very specific ways, the collective efficacy of the group. These principals have the skills of coaches cum leaders and operate as "system players" contributing to the betterment of other schools, in the district and beyond, as well as benefiting from these wider developments.

In the work that another of us coauthored on extraordinarily successful organizations in the three sectors of business, education, and sports (Hargreaves, Boyle, and Harris 2014), we found that *uplifting leaders* are not just executives and official sports "coaches" but also inspirational players who lift other members of the team to excellent performance through example and explicit assistance. The coach doesn't have to be called the coach or to have the job description of one in order to coach effectively.

There is a convergence here. Note the new subtitle of the third edition: *Developing Self-Directed Learners and Leaders*. Cognitive Coaches become leaders, and leaders become Cognitive Coaches. More than this: We *all* become leaders and coaches—for ourselves, and always for others. This is the true renaissance: deep spiritual mindsets and meaning for humankind, mediation of collective consciousness, action that engages diversity, and exquisite awareness of our impact and its perennial potential.

The first-ever coach was a vehicle constructed five hundred years ago in Hungary to convey people from one place to another. The first coaching took place when tutors accompanied their students in later coaches as they learned and traveled at the same time. Cognitive coaching has come a long way since then—and in this magnificent and comprehensive book, it has truly come of age.

Michael Fullan and Andy Hargreaves, winners of the
Grawemeyer Award in Education, 2015
March 2015

Preface

In this third edition, we not only acquaint new readers with Cognitive Coaching but also offer experienced practitioners the benefits of recent developments, including information about advanced coaching skills, neuroscientific applications to coaching, and a clear focus on one's identity as a mediator. New understandings have emerged about the function and importance of personal identity in guiding beliefs, values, and practices. This book offers a coaching map not revealed before, a map for "calibrating conversations" dedicated to the improvement of instruction. Calibrating conversations are designed to assist a staff member to calibrate his or her progress against a standard by understanding and engaging with a locally adopted and agreed-upon standard and by determining where his or her skill level falls. These conversations with an administrator, mentor, or peer coach lead to self-directed, targeted improvements in teaching practices. Because pictures add dimensions to comprehension that words do not, readers will find links to videos showing various conversations, interactions, maps, and skills central to understanding Cognitive Coaching. (See appendix G for a list of videos, their length, and the URLs to access the videos.)

Also new are expanded understandings of the process of mediation as provided by the initiator of this term, Reuven Feuerstein, the Israeli psychologist, in his last published work. Readers will learn how mediation, the primary tactic of Cognitive Coaching, not only changes the structure of behavior of the individual, not only changes the amount and quality of his repertoire, but also—according to increasingly powerful sources of evidence from fields of neurophysiology and biochemistry— changes the structure and functioning of the brain itself in very meaningful ways.

Cognitive shift, the "Aha!" moment in coaching, describes not only what is happening inside the brain but also how to recognize its occurrence and how to react and present new material. Also, we have added the important contributions that race, y, and gender make to mental processing and coaching relationships. on and congruent with the foundations that were built before.

A HISTORICAL PERSPECTIVE ON COGNITIVE COACHING

Cognitive Coaching has its source in the authors' professional experiences. We both began our educational careers in the late 1950s, a time of great ferment in American education. In the early 1970s, Nabuo Watanabe, then director of Curriculum Services of the Office of the Contra Costa County Superintendent of Schools, convened a group of California educators to develop a strategy for helping school administrators understand and apply humanistic principles of teacher evaluation. Art was a member of that group. Using the clinical supervision model of Cogan and Goldhammer, the group outlined the basic structure of the pre- and post-conference. They also identified three goals of evaluation: trust, learning, and autonomy. These goals and processes foreshadowed key concepts in Cognitive Coaching.

At about the same time, Bob was a consultant and principal for the Arabian American Oil Company Schools in Saudi Arabia. He was implementing a system-wide innovation in computer-assisted individualized instruction that cast the teacher in the roles of facilitator and mediator for student learning. Simultaneously, he and his colleagues were applying the pioneer work in clinical supervision developed by Cogan, Goldhammer, and Anderson at the Harvard Graduate School of Education. Additionally, Bob was teaching communication courses in Parent and Teacher Effectiveness Training, developed by psychologist Thomas Gordon, a forerunner to some of the nonjudgmental verbal skills found in Cognitive Coaching today.

In his early career, Art was highly influenced by leaders in education and cognitive development. Art's doctoral work at the University of California, Berkeley, emphasized curriculum, instruction, and psychology. He took courses based on curriculum and instructional theories of Hilda Taba, Jerome Bruner, and Reuven Feuerstein. Bob's early pedagogical mentors included an exceptional group of professors at San Francisco State University, Santa Rosa Center, who operated as an interdisciplinary team, and the national leaders in humanistic psychology, including Carl Rogers, Fritz Perls, and Abraham Maslow. Studies with Caleb Gattegno from the University of London and J. Richard Suchman, father of inquiry training, informed his approaches to teaching and learning.

After 20 years in the roles of teacher, principal, director of instruction, and acting superintendent, Bob completed a doctorate at the University of Southern California with an emphasis on educational administration, sociology, and staff development. About the same time that the emphases of cognition, instruction, and supervision were beginning to coalesce for Art, Bob was piecing together principles of counseling practices and strategies of group dynamics for school improvement. He joined the faculty at California State University as professor of educational administration, where he taught courses in curriculum development, school improvement, supervision, and neurolinguistics. Art was also teaching at California State University, and he and Bob were assigned to the same office. There they developed the first formal expression of Cognitive Coaching, and that December tested their ideas with staff developers at a statewide conference. The enthusiastic reception led to further conceptual work and publications, as well as invitations to present six-day "trainings" to educators.

By the summer of 1985, interest in Cognitive Coaching sparked formation of the Institute for Intelligent Behavior, an association of persons dedicated to enabling educational and corporate agencies to support their members' growth toward the five high-performance states of mind articulated by the authors. Seven senior associates helped refine and provide seminar programs in Cognitive Coaching to interested school districts and private-sector organizations throughout North America and internationally.

As more and more people and agencies became intrigued with the power of Cognitive Coaching, numerous requests were received for a book to help explain Cognitive Coaching. In 1994, the first edition of *Cognitive Coaching: A Foundation for Renaissance Schools* was published. The book came to be used as a basic text in university courses and as a reference to accompany the burgeoning seminars and leadership training.

To date, well over 200 organizational leaders have been trained to serve as resources to their own educational agencies to infuse, support, and study the effects of implementing the skills, maps, and principles of Cognitive Coaching. Additionally, more than 50 consultants have been trained to provide Foundation Seminars to educational organizations such as schools, districts, departments of education, and professional organizations as well as to business and corporate agencies.

Coordinated by Carolee Hayes and Jane Ellison, the Center for Cognitive Coaching was established in 2000 to provide leadership training and to serve as a resource to schools and districts desiring Cognitive Coaching services, information, and products.

In the 20 years since the first book was published, there have been many changes in education. Pedagogical and political ferment has been the norm in education, yet interest in and the practice of Cognitive Coaching has been sustained and is growing.

Against this background of seesawing political and policy ferment has been the steady progress in unlocking the secrets of the mind's role in learning; much of the findings supported the practices of Cognitive Coaching, which may explain, at least in part, the sustaining presence of this model. It was not until the end of 1970 that that term *cognitive neuroscience* entered the lexicon, coined by George Miller (of the "7 plus or minus 2" memory fame) and Michael Gazzaniga. Neuroscience itself was not established as a unified discipline until 1971 and it was very late in the 20th century that technologies like *transcranial magnetic stimulation* (TMS) and functional magnetic resonance imaging (fMRI) became mainstays in understanding brain functioning. Cognitive neuroscience addresses questions of how psychological functions are affected by neural circuitry.

Today integrative neuroscience attempts to consolidate data in databases and form unified descriptive models from various fields and scales: biology, psychology, anatomy, and clinical practice (Squire 2009).

In 2013, in recognition of the common focus of thinking and expanding human potential individually and in groups, Carolee Hayes and Jane Ellison, directors of the Center for Cognitive Coaching, and Michael Dolcemascolo and Carolyn McKanders, directors of the Center for Adaptive Schools, coalesced to form Thinking Collaborative, housing both Cognitive Coaching and Adaptive Schools study, leadership development, and seminars.

With the current emphasis on standards for students, teachers, and administrators; with the political desire for "high-stakes accountability"; and with increased knowledge about human learning and emphasis on 21st-century skills, it is time for a reexamination of the ideals, premises, and practices of Cognitive Coaching and how new understandings are influencing the direction of this work. The major purposes of this book, therefore, are to acquaint new readers with the promises Cognitive Coaching provides and to reaffirm and extend knowledge about Cognitive Coaching practices for readers visiting this territory again. All of this is offered toward the ultimate goal of developing teachers and students as self-directed learners capable of coping with and living productively and harmoniously in an ambiguous, technological, and global future.

OUR BELIEFS

We are pragmatic idealists. Significant advancements in educational knowledge and practice have occurred since the first edition of this book in 1994. In this third edition, we report about the edges of our own learning and how new neuroscientific research and practice can support individuals and schools in reaching higher, more satisfying, and more holistic performance.

Through the work at the Center for Cognitive Coaching and the tireless efforts of many training associates, our knowledge of Cognitive Coaching has expanded over the past 30 years. We have learned more about mediating learning. We have learned more about releasing the electrochemical energies of resourcefulness and what happens emotionally and cognitively as teachers creatively meet the daily challenges of classroom realities. We know more about how people can teach the mediational skills of Cognitive Coaching more efficiently and effectively. We also know more about supporting new teachers and applying Cognitive Coaching to work with students, parents, and groups of educators.

The enduring thread that has held us together over the course of our careers has been our desire to develop schools as homes for the mind. Throughout the past years we have come to a deep understanding about what this means. This book is an explicit reflection on what we have learned about developing thoughtful professionals and agile schools that foster self-directedness. In fact, it was through the current emphasis on teacher evaluation and external standards that we finally recognized that any attempt to enact change from the outside, with directives, rubrics, judgments, or new evaluation tools, violated one of our core values. If we have learned one lesson in our professional lives it is this: For change to happen, it needs to be understood, practiced, and ultimately owned by those who will act on the knowledge—and this is much easier said than done.

GETTING THE MOST OUT OF USING THIS BOOK

As you begin to explore this book, we suggest that you take an inventory of your own personal inner dialogue. What questions do you ask over and over? Where are your points of passion and excellence? What magic do you have that connects with your students, your colleagues, and their learning? How does this internal knowledge show up in your practices?

We invite readers to use this book as a personal resource, going to chapters of immediate interest. Some groups may wish to use the book as a study guide or text for professional development and/or university work. This book also will be a valuable reference for those participating in Cognitive Coaching seminars and training.

ABOUT THE LANGUAGE IN THIS BOOK

To the extent possible, we have worked to eliminate gender bias in this text. For example, we alternate use of "he" and "she" in our examples, or we eliminate any need for gender reference by writing sentences with plural nouns (using a construction with "coaches/their" instead of "coach/he" or "coach/she"). Occasionally we use the form of his/her. However, some gender bias is unavoidable, as with the word "craftsmanship." We searched for years for another word that would capture so precisely the attributes of this state of mind, but we could find no better alternative. "Craftspersonship" seems artificial and awkward, so we resigned ourselves to "craftsmanship." We hope readers will understand the dilemmas produced by the current limitations of the English language.

Also, we believe it is important to recognize the value of Cognitive Coaching in a variety of professions, positions, life roles, and careers. However, for the purposes of this book, we have chosen to cast the coaching process in educational settings. Thus, most of our vignettes, descriptions, samples, and examples illustrate educational settings (especially with teachers and coaches). This is despite the fact that we know of research being conducted with police officers, churches, business groups, and others. We hope readers will read beyond this narrow focus to apply the mediative process to contexts beyond education: home, community, workplace, and church.

We hope that this book supports the reader in mediating self-directed learning in others—and in themselves—as they integrate principles and values of Cognitive Coaching throughout school programs and practice. Together, we invite you to join us in this journey, the study of what it means to grow cognitively and to build schools that invest in the mind, body, and spirit.

Arthur L. Costa   Robert J. Garmston
Granite Bay, California   El Dorado Hills, California

Acknowledgments

We are indebted to scholars, practitioners, and researchers in many fields whose work has been instrumental in our own development and that has inspired and informed us, raised provocative questions, and guided our search.

The "gene pool" from which Cognitive Coaching (CC) emerges includes the work of Robert Anderson, Morris Cogan, and Robert Goldhammer (clinical supervision); Gregory Bateson (systems theory and psychology); Richard Bandler, Robert Dilts, and John Grinder (neurolinguistic programming); David Berliner (teacher "executive processes"); Reuven Feuerstein (mediated learning); Perc Marland (teacher decision making); Tom Sergiovanni (supervision theory); Richard Shavelson (the basic behaviors of teaching); Norman Sprinthall and Lois Thies-Sprinthall (teacher as adult learner); Arthur Koestler (holonomy); Robert Keagan (adult development); Robert Sternberg (metacognition and intelligence); and Antonio Damasio and Candace Pert (neurosciences). For their pioneering work we are indeed grateful. Most notably we would like to thank Carl Glickman (developmental supervision), formerly at the University of Georgia, for his scholarship, inspiration, modeling, encouragement, and friendship.

Cognitive Coaching was conceived in the mid-1980s. From the beginning it was further enhanced by dedicated educators who shaped early forms of coaching practices and seminar design. We, as well as countless educators, will always be extremely grateful to these senior associates of the Institute for Intelligent Behavior for their influence on Cognitive Coaching. We are thankful for Diane Zimmerman, who from the beginning and over the years has helped us refine and test ideas and appreciate the realities of implementation. We are appreciative, too, of the energies and rich contributions of many others associated with the Institute for Intelligent Behavior, especially members of the Institute's "Leather Apron Club." These pioneers include Bill Baker, John Dyer, Laura Lipton, Bruce Wellman, and Diane Zimmerman. Of special mention are Jenny Edwards for her research contributions, and Jane Ellison and Carolee Hayes for their integrity and leadership as co-directors of the Center for Cognitive Coaching, who 15 years ago began a concerted program providing trainers and developing leadership in this discipline. We also celebrate all the training associates who continue to expand and apply these ideas through their work with the Center for Cognitive Coaching.

We are deeply indebted to Carolee Hayes and Jane Ellison for their editorial assistance in this book and their devotion to furthering Cognitive Coaching. They have not only kept us more precise and accurate in our writing, they have also caused us to write with greater consistency and integrity through their exquisite modeling, training, and development of others. They have devoted years to the development, coordination, and refinement of the model of Cognitive Coaching, brought coherence to the training process, and developed vast materials to support an ever-increasing number of skillful Cognitive Coaches. We give special thanks to Michael Dolcemascolo and Carolyn McKanders, initially responsible for refining and developing the training activities and materials for the Adaptive Schools, who joined Carolee and Jane in their leadership of the Cognitive Coaching community, contributing to additional refinements in the work. Ultimately the four of them coalesced into a new entity, the Thinking Collaborative, sustaining the work of both programs.

We thank our international ambassadors who have introduced CC throughout the world: Bill and Ochan Powell, Gavin Grift, Father Luis Gonzales, Jenny Edwards, Luci Garza, Bridget Doogan, and David Chojnacki, executive director of Near East South East Asian International Schools.

We also wish to acknowledge and celebrate the growing numbers of dedicated people from whom we are learning as they infuse the principles and practices of Cognitive Coaching in their classrooms, schools, districts, and communities.

Sue Canavan, our editor from Rowman and Littlefield, has worked with and encouraged us and provided sound advice for over 30 years throughout both earlier editions. We are grateful for her patience and skills.

Undoubtedly, the two persons who have been the most patient with us, and have continuously extended their support and love, are our wives, Nancy and Sue.

Arthur L. Costa   Robert J. Garmston
Granite Bay, California   El Dorado Hills, California

Cognitive Coaching Resources

To learn more about Cognitive Coaching and Adaptive Schools, please go to the website www.thinkingcollaborative.com. The website contains a variety of resources, including seminars (descriptions, dates, and schedules), products for purchase, weekly information, videos, and networking opportunities. The website is designed to support learning about, practice, and implementation of the concepts, maps, and tools of both Cognitive Coaching and Adaptive Schools.

Cognitive Coaching Seminars® offerings are:

Foundation Seminar Part 1 (4 days)
Foundation Seminar Part 2 (4 days)
Advanced Seminar Part 1 (3 days)
Advanced Seminar Part 2 (3 days)
Proficiency Modules on a Variety of Topics (1 day each)
The Calibrating Conversation (1 day)
Group Coaching (1 or 2 days)
Self-Directed Evaluation Conversations (2 days)

A Cognitive Coaching[SM] app is available for Apple and Android devices.

Adaptive Schools Seminars offering are:

Foundation Seminar (4 days)
Advanced Seminar (5 days)
Presenters' Forum (4 days)
Presentation Skills (1–3 days)
Facilitation Skills (1–2 days)

Polarity Management (1–2 days)

Proficiency Modules on Norms of Collaboration

An Adaptive Schools app is available for Apple and Android devices.
For information on classroom applications, visit www.instituteforhabitsofmind.com.
For further information about Thinking Collaborative, contact:

Lisa Joseph, Office Manager
P. O. Box 630128
Highlands Ranch, CO 80163
ccclj@aol.com
303-683-6146 Voice
303-791-1772 Fax

Part 1

EXPLORING THE MEANINGS
OF COGNITIVE COACHING

In this introductory section, we explore the unique attributes of Cognitive Coaching. Although many models of coaching focus on the behaviors of teaching, here you will also find rationale and processes for enhancing teachers' cognitive capacities that relate directly to professional performance.

Cognitive Coaching is about producing self-directed learners and leaders with the disposition for continuous, lifelong learning, perhaps at no time more necessary than at the present in an increasingly complex and interconnected world. The model is described as a nonjudgmental, developmental, reflective model. It is informed by current work in brain research and focuses on the practitioner's cognitive development, which by association cannot ignore parallel learning in the affective and social domains.

Cognitive Coaches develop identities as mediators, committed to serving others in acquiring the dispositions and capacities of self-directed learners. The verbal and nonverbal skills sets designed to help transform and empower the cognitive functioning of others are introduced. Although these skills are presented in the context of Cognitive Coaching, they are useful in any dialogue between two human beings desiring to grow together, to plan, to reflect, to solve problems together, and to create deeper meaning and understanding. Drawing on the most recent work of Reuven Feuerstein, new information on mediation, not present in previous editions, is introduced.

The concepts in these chapters invite a shift from present paradigms. For many educators, a dominant sense of satisfaction has come from their expertise as problem solvers. The shift to a meditational identity creates a feeling of being rewarded by facilitating others to solve their own problems. The shift is from teaching others to helping others learn from situations; from holding power to empowering others; from telling to inquiring; and from finding strength in holding on to finding strength in letting go. A mediator refrains from giving advice, believing in what Cicero stated: "Nobody can give you wiser advice than yourself." Changing one's identity requires patience, stamina, and courage

For some, the behaviors described in these chapters come naturally. For others, it will take time and practice to become proficient and comfortable using these tools. We recommend that coaches isolate certain skills and consciously practice them rather than attempt to learn them all at once.

1

Discovering the Meanings of Cognitive Coaching

"**H**ypercomplexity—our world is changing at an unprecedented rate, becoming more complex and globally interactive. Clear-cut, unambiguous understandings are no longer an option" (Stelter 2014, 52). Thus, the capacity for self-directed and continuous learning has become the most important capacity needed for survival in the future. Fortunately the brain permits our ability to change, to grow, and to continuously develop our intellectual capacities throughout a lifetime. And that's what Cognitive Coaching is all about: producing self-directed learners and leaders with the disposition for continuous, lifelong learning.

COGNITIVE COACHING: A REVELATION

After observing a demonstration of Cognitive Coaching for the first time, educators often make comments like these:

- The coach didn't make any value judgments; the other person had to judge for himself.
- There was a lot of silence after the coach's questions; she made the person think.
- The coach gave no advice or recommendations; instead, she asked her partner to suggest what should be done.
- The coach seemed to have a strategy in mind, like he knew where he was going.
- The coach listened; she reflected back what the person said and clarified a lot.
- Cognitive Coaching seems to be like a Socratic dialogue in a form of inquiry.

People who have undergone Cognitive Coaching for the first time often reflect on the experience with comments such as these:

- I became much clearer about my plans and how to achieve them.
- I felt that the coach understood my problem and the goals I had in mind. I have a better handle on my problem now.
- The coach really made me think. She made my brain "sweat"!
- I want to know how I can learn to use this process in my work!

Although we agree with all of these comments, the essence of Cognitive Coaching is much more than these observations capture. In this chapter, we offer an overview of the model to serve as a basis for the detailed discussions of Cognitive Coaching theory and practice that follow.

A SNAPSHOT OF COGNITIVE COACHING

Research demonstrates that teachers with higher conceptual levels are more adaptive and flexible in their teaching style. They act in accordance with a disciplined commitment to human values (Sprinthall and Theis-Sprinthall 1982), and they produce higher-achieving students who are more cooperative and involved in their work (Harvey 1967). Borg (2009), citing studies on teacher cognition, notes that interest in this field faded after the 1980s, when more behaviorist conceptions of teaching became popular. Cognitive Coaching increases the capacities for sound decision making and self-directedness, which helps to achieve goals like these (Edwards 2015).

At one level, Cognitive Coaching is a simple model for conversations about planning, reflecting, or problem resolving. At deeper levels, Cognitive Coaching serves as the nucleus for professional communities that honor autonomy, encourage interdependence, and produce high achievement. Cognitive Coaching is a form of dialogue that provides a space for self-reflection, for revising and refining positions and self-concepts, where a colleague is invited to see him/herself in a new light. Cognitive Coaching is a nonjudgmental, developmental, reflective model derived from a blend of the psychological orientations of cognitive theorists and the interpersonal bonding of humanists such as Carl Rogers and Abraham Maslow. The model is informed by current work in brain research, constructivist learning theory, and practices that best promote learning (information about which is expanded in chapter 9).

Fundamental to the model is the focus on a practitioner's cognitive development. This focus is based on the belief that growth is achieved through the development of intellectual functioning. Therefore, the coaching interaction focuses on mediating a practitioner's thinking, perceptions, beliefs, and assumptions toward the goals of self-directed learning and increased complexity of cognitive processing.

Cognitive Coaching strengthens professional performance by enhancing one's ability to examine familiar patterns of practice and consider underlying assumptions that guide and direct action. Cognitive Coaching's unique contribution is that it influences another person's thought processes. Cognitive Coaching is systematic, rigorous, and data based. The purpose of this model is to enhance an individual's capacity for self-directed learning through self-management, self-monitoring, and self-modification.

We find a useful metaphor for the essence of Cognitive Coaching in the story of a boy watching a butterfly emerge from its chrysalis:

> The boy observed the chrysalis closely each day until the casing broke away and a small opening appeared. The boy could see the butterfly's head, and, as the butterfly began to emerge, an antenna appeared, then one leg.
>
> The boy watched the butterfly for several hours as it struggled to force its body through the small aperture. Then the butterfly seemed to stop making any progress. It appeared as if the butterfly could go no farther. So the boy decided to help, not realizing that it was the struggle itself that gave the butterfly strength. He scratched away the remaining scales of the confining cocoon with his thumbnail, and the butterfly easily emerged. However, it had a swollen body, small, shriveled wings, and bent legs.

So, too, the challenges and ultimate achievements of Cognitive Coaching make educators stronger and better equipped to fulfill their roles in schools today.

WHY LEARN TO COGNITIVELY COACH?

> Children grow into the intellectual life around them.
>
> —Lev Vygotsky

The answer to this question is fairly simple and direct: to enhance student learning! Enhanced student learning, however, depends on the expertise of the teacher (Hattie 2003). The expertise of the teacher, however, depends largely on the school community in which that teacher operates (Hargreaves and Fullan 2012). And the power and effectiveness of the school community depend largely on the skills and dispositions of its educational leaders (Leithwood and Seashore-Lewis 2011). Cognitive Coaches, therefore, dedicate themselves to the cascading effects of enhancing others, who, in turn, ultimately produce maximum growth in students' acquisition of desired outcomes.

COMPONENTS OF THE COGNITIVE COACHING MODEL

Cognitive Coaching comprises a set of: 1) skills, 2) capabilities, 3) mental maps, 4) beliefs, 5) values, and 6) commitments, all of which are practiced, tested over time, and assimilated into a person's day-to-day interactions. They also become part of the coach's identity as a mediator of self-directed learning. Ultimately, Cognitive Coaching's values and beliefs become an outlook on life.

Skills

Cognitive Coaches are skilled at constructing and posing questions with the intention of engaging and transforming thought. They employ nonjudgmental response behaviors to establish and maintain trust and intellectual engagement. They use nonverbal behaviors to establish and maintain rapport.

Capabilities

Cognitive Coaches know their own intentions and choose congruent behaviors. They set aside unproductive patterns of listening, responding, and inquiring. They adjust their own style preferences, and they navigate within and between several mental maps to guide their interactions.

Mental Maps

Cognitive Coaches hold in their head four maps as guides to the conversations with coaches:

1. The *Planning Conversation* occurs before a colleague conducts or participates in an event or attempts a task. The coach may or may not be present during the event being planned or available for a follow-up conversation.
2. The *Reflecting Conversation* occurs after a colleague conducts or participates in an event or completes a task. The coach may or may not have been present at or participated in the event.
3. The *Problem-Resolving Conversation* occurs when a colleague feels stuck, helpless, unclear, or lacking in resourcefulness; experiences a crisis; or requests external assistance from a mediator.
4. The *Calibrating Conversation* engages teachers in assessing their own performance related to a set of standards or external criteria, so that they share responsibility for self-assessing, and then self-prescribing personal responses to the assessment.

Beliefs

Cognitive Coaches believe that all behavior is determined by a person's perceptions and that a change in perception and thought is a prerequisite to a change in behavior. They also believe that human beings construct their own meaning through reflecting on experience and through interactions with others. They have optimism that all human beings have the capacity to continue developing their intellect throughout their lifetimes.

Values

Cognitive Coaches value self-directed learning. They delight in assisting others in becoming more self-managing, self-monitoring, and self-modifying. They cherish and work to enhance individual differences in styles, beliefs, modality preferences, developmental levels, culture, and gender.

Commitments

Cognitive Coaches are committed to learning. They continually resist complacency, and they share both the humility and the pride of admitting that there is more to learn. They dedicate themselves to serving others, and they set aside their ego needs, devoting their energies to enhancing others' resourcefulness. They commit their time and energies to make a difference by enhancing interdependence, illuminating situations from varied perspectives, and striving to bring consciousness to intentions, thoughts, feelings, and behaviors and their effect on others and the environment.

FOUNDATIONS OF COGNITIVE COACHING

Humanistic psychological orientations such as empathy, unconditional positive regard, and personal congruence guide the coach's work with relationship building. Certain neurolinguistic principles (Zoller and Landry 2010) are also applied to achieve rapport and access thinking. Rapport is essential to mediate cognitive processes, affecting brain chemistry and access to the neocortex. Thus, the Cognitive Coaching model links physiology and cognition, which increases the importance of deep relationship skills and nuance within interactions. Rapport is an important tool for building and maintaining trust in the moment, particularly when there is tension, miscommunication, or anticipated difficulty.

Cognitive Coaches are always alert to in-the-moment opportunities. Thus, they draw upon their tools, maps, and capabilities in many different situations. These include brief corridor conversations, casual planning conversations, more structured reflecting and problem-solving conversations, and formal planning or reflecting conversations, built around the observation of a lesson or event. For example, as two teachers are on the way to their classrooms, one asks, "How was the conference that you went to last week?" The response is, "It was very interesting." The first teacher asks, "So what was it about the conference that interested you?" Her colleague responds, "I found many practical strategies that I could use." The first teacher then asks, "So as you anticipate your classes this week, what insights do you plan to apply?"

HISTORICAL BACKGROUND OF THE MODEL

In the late 1960s, Morris Cogan, Robert Goldhammer, Robert Anderson, and a group of supervisors working in Harvard's master of arts in teaching program found that traditional supervision placed the supervisor in an "expert" role superior to that of the teacher. Supervisors told the teacher what should be changed and how to do it. Supervisors offered solutions to problems that concerned them, which were not necessarily the problems that concerned teachers. All efforts to change the conference style in which the supervisor did the talking and the teacher did the listening had failed. The work of Cogan and his colleagues on this dilemma was the foundation for the clinical supervision model, and it is important to understand some of Cogan's work to appreciate the distinctive attributes of Cognitive Coaching.

Cogan (1973) envisioned the purpose of clinical supervision as the development of a professionally responsible teacher who is analytical of his own performance, open to help from others, and self-directing. Clinical supervision demanded a role change in which the teacher and the supervisor worked as colleagues, respecting each other's contributions. The intent of the process was to cultivate teacher self-appraisal, self-direction, and self-supervision. Cogan's clinical supervision was conceived as a cyclic, eight-phase process organized around planning and conferring with a teacher before instruction, observing the lesson itself, and conducting a follow-up conversation after the lesson. Cogan and his colleagues believed that the act of teaching is a collection of behaviors that can be understood and controlled. These behaviors can be observed singly and in interaction. Instructional improvement could be achieved by changing or modifying these instructional behaviors.

Cognitive Coaching also draws on the model of the mediated learning experience developed by the Israeli psychologist Reuven Feuerstein et al. (2015) in which the teacher or coach has the intention of modifying a colleague's cognitive capacity to modify himself or herself.

Cognitive Coaching is a modern expression of these orientations, built on the foundation laid by Cogan and Goldhammer. There are, however, significant departures. Although the traditional model of clinical supervision addresses overt teaching behaviors, we believe that these behaviors are the product and artifacts of inner thought processes and intellectual functions.

The authors and a powerful team of associates developed Cognitive Coaching in 1984. Over the years, the forms and uses of Cognitive Coaching have evolved to include applications in the business and corporate world, peer coaching, mentor services for teachers and administrators, peer assistance and review programs, parent-teacher conferences, instructional coaches, and classrooms with students of all ages.

- Cognitive Coaching is based upon four assumptions: (1) the nonroutine and complex nature of teaching requires constant contextual decision making; (2) all behavior is directed by our individual and subjective perceptions; (3) to skillfully change behavior requires a change in perception; and (4) effective coaching mediates the perceptual changes and capacities for reasoning and decision making that promote behavioral changes toward more effective practice.
- The purpose of supervision is to increase the capacity for self-modification (Garmston, Lipton, and Kaiser 1998). Changing the overt behaviors of instruction requires the alteration and rearrangement of inner, invisible cognitive processes. This relates to the application of specific knowledge about teacher cognition, neurosciences related to conversations, and specifically coaching and psycholinguistics.

The diagram of a tree and its roots in figure 1.1 displays the relationships and connections of the various philosophical, psychological, physiological, neuroscientific, and historical concepts on which Cognitive Coaching is based.

DISTINCTIONS AMONG SUPPORT FUNCTIONS

The many terms that are used in education to describe support services intended to improve instruction are often confusing: *consulting*, *mentoring*, *peer assistance*, *catalyst*, *supervision*, *coaching*, *evaluation*, *collaborating*, and so forth.

We distinguish four functions intended to support teacher development: evaluating, collaborating, consulting, and mediating (Lipton, Wellman, and Humbard 2001). Table 1.1 elaborates on each of these.

Three of these functions—coaching, collaborating, and consulting—interact to improve instructional practice. For beginning teachers, the consulting, and collaborating features are often beginning stages of a conversation. The skillful, sensitive coach will ultimately default to Cognitive Coaching as it is most likely to support self-directed learning.

Evaluation

Evaluation is the assessment and judgment of performance based on clearly defined external criteria or standards. In most systems, personnel authorized by their position

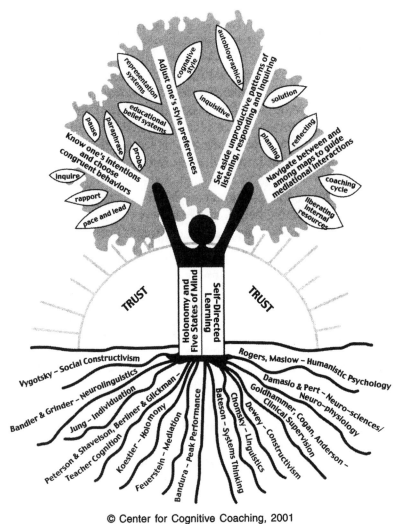

© Center for Cognitive Coaching, 2001

Figure 1.1.   The Roots of Cognitive Coaching

as administrator, supervisor, or department chairperson conduct evaluations. Evaluation serves the following purposes:

- To provide and operate with a uniform set of explicit criteria for all personnel to judge themselves and know the criteria for the district's evaluation program
- Recognize and help to reinforce outstanding service (some districts award bonuses and merit pay)
- Screen ineffective teachers
- Provide constructive feedback
- Provide direction for staff development practices
- Unify teachers and administrators in their collective efforts to educate students

When evaluation is an act of one person judging the performance of another, it is inconsistent with the goal of self-directed learning. However, "integrating Cognitive Coaching into the evaluation process brings a growth producing dimension often missing from traditional evaluation processes" (Dolcemascolo, Miori-Merola, and Ellison 2014).

Consulting

To consult is to "inform regarding processes and protocols, advise based on well-developed expertise, or advocate for particular choices and actions" (Lipton and Wellman 2011, 20). In most school districts, skillful teachers have been designated as consultants, mentors, or peer coaches. These role titles do not necessarily describe how they do their work, for they may employ consulting, collaborating, and coaching to achieve their aims. Mentoring in the educational setting is usually thought of as a relationship between a beginning teacher and a more experienced colleague; additionally, experienced principals sometimes mentor new principals.

Consultants serve as information specialists about or advocates for content or processes based upon their greater experience, broader knowledge, and wider repertoire. A supporting teacher, working as a consultant, provides technical information to a more novice teacher or to peers about the content or skills being taught, the curriculum, teaching strategies, and child growth and development. As a process advocate, a consultant informs a teacher about alternative strategies and consequences associated with different choices of methodology and content. For beginning teachers, a consultant may also provide information about school policies, assessments, procedures for obtaining special resources, protocols for parent conferences, and the like.

Consulting skills include clarifying goals, modeling expert thinking and problem-solving processes, providing data, drawing on research about best practices, making suggestions based on experience, offering advice, and advocating. The consultant is always working to support the teacher in becoming self-directed and is working from an intention of gradual release.

To be successful, the consultant must have permission from the teacher to consult, which requires a high degree of credibility and trust. The consultant also must hold commonly defined goals and the client's desired outcomes in mind.

The true test of a consulting relationship is the transfer of skills, behaviors, and increased "coachability" over time. The support person who needs to be needed can trap the teacher and him/herself into a dependency relationship. Likewise, the support person whose identity is primarily about being an expert may also trap him/herself into a dependency relationship with the teacher. Within the context of Cognitive Coaching, effective consulting functions consistently lead toward the ultimate goal of self-directedness.

es from "co-labor." Collaboration involves people with different

together as equals to achieve goals. Thus, teacher and support

provider plan, reflect, or problem solve together. Both are learners, offering ideas, listening deeply to one another, and creating new approaches toward student-centered outcomes. Both bring information to the interaction. Goals may come from a coaching question or from the expert perspective of a consulting voice. Because the purposes of collaboration are to learn with and from one another, to create new practices and ideas, and to experiment and share results, collaboration is consistent with the goal of self-directed learning.

Coaching

Cognitive Coaching is the nonjudgmental mediation of thinking. A Cognitive Coach can be anyone who is skillful in using the tools, maps, beliefs, principles, and values of mediation described earlier in this chapter. Many of the tools of Cognitive Coaching can be used in the support functions of consulting, calibrating, and collaboration. However, the greatest distinction of Cognitive Coaching is its focus on cognitive processes, on liberating internal resources, and on accessing the five states of mind as the wellsprings of constructive thought and action. (We discuss these in detail in chapter 6.) Cognitive Coaching has the expressed goal of supporting self-directedness in others. An educator committed to self-directed learning, when entering a conversation as an evaluator, consultant, or a collaborator, always exits as a Cognitive Coach.

The next section of this chapter deals specifically with an expanded definition of Cognitive Coaching.

HOW COGNITIVE COACHING IS UNIQUE

Anyone planning to use Cognitive Coaching must be able to clearly distinguish the four functions of evaluation, consulting, collaboration, and Cognitive Coaching described above. These various forms of support are summarized in table 1.1. No one cognitively coaches all the time, and it is important for a Cognitive Coach to know when it is appropriate and how each function differs from the others. Chapter 13 elaborates this dimension of a supervisor's or support provider's decision making.

Cognitive Coaching is a unique form of coaching. It differs from other forms of mentoring, supervision, and peer review in that it mediates invisible, internal mental resources and intellectual functions, as represented in figure 1.2. These resources and functions include perceptions, cognitive processes, values, and internal

Table 1.1.   Distinctions among Four Support Services

Attribute	Cognitive Coaching	Collaborating	Consulting	Evaluating
Conversations focus on:	Metacognition, decision-making processes, perceptions, values, mental models.	Generating information, co-planning, co-teaching, problem solving, and action research.	Policies, procedures, behaviors, strategies, techniques, and events.	Professional criteria, expectations, standards, and rubics.
The intention is:	To transform the effectiveness of decision making, mental models, thoughts, and perceptions and habituate reflection.	To form ideas, approaches, solutions, and focus for inquiry.	To inform regarding student needs, pedagogy, cirriculum, policies, and procedures and to provide technical assistance. To apply teaching standards.	To conform to a set of standards and criteria adopted by the organization.
The purposes are:	To enhance and habituate self-directed learning: self-managing, self-monitoring, self modifying.	To solve instructional problems, to apply and test shared ideas, to learn together.	To increase pedagogical and content knowledge and skills. To institutionalize, accepted practices and policies.	To judge and rate performance according to understood externally produced standards.
The conversations are characterized by:	Mediation, listening, questioning, pausing, paraphrasing, probing, withholding advice, judgments or interpretations.	Mutual brainstorming, clarifying, advocating, deciding, testing, assessing.	Rationale, advice, suggestions, demonstrations.	Judgments, encouragements, advice, direction, goal setting.
	"What might be some ways to approach this?"	"How should we approach this?"	"Here are several ways to approach this."	"Your approach to this was good. Here is why."
The support person's identity in relation to the teacher is:	Mediator of thinking.	Colleague.	Expert.	Boss.

(continued)

Table 1.1. *Continued*

Attribute	Cognitive Coaching	Collaborating	Consulting	Evaluating
The source of empowerment to perform this function stems from:	Trust. Competence in the maps, tools, and values of Cognitive Coaching.	Trust. Competence in forming partnerships. Knowledge and skills in the areas being explored.	Trust. Competence in consulting skills. Expertise in relevant areas.	Policy. Authority is by position, licensed, authorized by law, or a negotiated agreement to evaluate. Evaluators are held accountable for judgments and actions regarding work quality.
The source(s) of criteria and judgments about performance is (are):	The teacher.	The teacher and colleague.	The consultant.	The evaluator in reference to established criteria.
	"How will you know that you are successful?"	"How will we know that we are successful?"	"Here's how you'll know that you are successful."	"Here's how I'll know that you are successful."

*We are grateful to Laura Lipton and Bruce Wellman for sharing their thinking about distinctions across the support services of coaching, collaborating, and consulting. Like Lipton and Wellman, we regard these as being listed in order of most to least effective in transforming teacher work and self-directedness; yet each, at times, contributes to this aim. Lipton, L. and Wellman, B. with C. Humbard. (2001) *Mentoring Matters: A Practical Guide to Learning-Focused Relationships.* Sherman, CT: MiraVia—www.miravia.com.

resources. Other forms of "coaching" may focus on the behaviors, the problem, the lesson, the topic, the meeting, or the activity. We regard these forms of assistance as consulting, not coaching.

The underlying premise of Cognitive Coaching is that a person's actions are influenced by internal forces rather than overt behaviors. Therefore, Cognitive Coaches focus on the thought processes, values, identities, and beliefs that motivate, guide, influence, and give rise to the overt behaviors, as represented in figure 1.3.

Figure 1-2

Other forms of interaction focus on event or behavior.   **A cognitive coach is concerned with the mental processes.**

Figure 1.2.   Other Forms of Interaction Focus on Event or Behavior

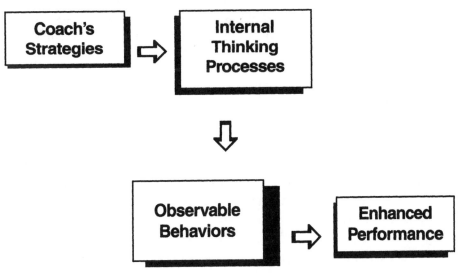

Figure 1.3.   Definition

Although only those charged with legal responsibilities can evaluate, a principal, peer, mentor, department chairperson, curriculum specialist, or resource teacher may serve as a Cognitive Coach. A Cognitive Coach helps another person take action toward his or her goals while simultaneously helping that person develop expertise in planning, reflecting, problem solving, and decision making. These are the invisible skills of being a professional, and they are the source of all teachers' choices and behaviors. The successful Cognitive Coach takes a nonjudgmental stance and uses tools of pausing, paraphrasing, and posing questions. The successful coach focuses on the other person's perceptions, thinking, and decision-making processes to mediate resources for self-directed learning.

The Cognitive Coaching model, then, is predicated on a set of values, maps, and tools that, when combined with nonjudgmental ways of being and working with others, invites the shaping and reshaping of thinking and problem-solving capacities. This shift happens for the person being coached and for the coach as well. Reciprocal learning is a critical component of the model.

Integral to Cognitive Coaching is the ability to work effectively with oneself and others across style differences and philosophical preferences. One fundamental element of a coach's effectiveness is the capacity to work within a colleague's worldview. As a result, the viewpoints of both participants are widened, which enables an exploration of new ideas along with existing beliefs and values.

THE MISSION

The mission of Cognitive Coaching is to produce self-directed persons with the cognitive capacity for excellence both independently and as members of a community. This ability to be self-directed in both independent and interdependent settings is related to holonomy—the study of interacting parts within a whole. These terms are defined below:

Self-Directedness

A self-directed person can be described as:

- *Self-Managing*: Knowing the significance of and being inclined to approach tasks with outcomes clearly in mind, a strategic plan, and necessary data, and then drawing from past experiences, anticipating success indicators, and creating alternatives for accomplishment.
- *Self-Monitoring*: Having sufficient self-knowledge about what works and establishing conscious metacognitive strategies to alert the perceptions for in-the-moment indicators of whether the strategic plan is working and to assist in the decision-making processes of altering the plan and choosing the right actions and strategies.
- *Self-Modifying*: Reflecting on, evaluating, analyzing, and constructing meaning from experience and making a commitment to apply the learning to future activities, tasks, and challenges.

Self-directed people are resourceful. They tend to engage in causal thinking, spend energy on tasks, set challenging goals for themselves, persevere in the face of barriers and occasional failure, and accurately forecast future performances. They proactively locate resources when perplexed. Seeking constant improvement, they are positive and flexible in their perspectives and are optimistic and confident with self-knowledge. They feel good about themselves, control performance anxiety, translate concepts into action, and are committed to continuous lifelong learning. (Mediating self-directedness is elaborated in chapter 4.)

HOLONOMY: TRANSCENDING AUTONOMY AND INTERDEPENDENCE

As we develop soul in our work we need to recognize our dual identity: we are both individuals and members of a group. Indeed, finding the soul of work involves the balance and integration of apparent opposites, such as head and heart, intellect and intuition, and self and group. This process is not so much based on the "shoulds" but upon "what is." It is my belief that as we attend to the soul of work we will find we feel more complete.

—Daryl Paulson 1995, 18–20

Because the mission of Cognitive Coaching is to produce self-directed persons who function well (have the cognitive capacity for excellence) individually and in groups, the concept of both—and represented by the word *holonomy*—remains key to appreciating the well-developed individual. Holonomy is a combination of two Greek words: *holos* meaning whole, and *nomy* meaning arrangements or distribution (as in eco-nomy—the study of the distribution of resources, or astro-nomy, the study of the distribution of celestial bodies). Holonomy conveys the notion that an entity is both an autonomous unit *and* a member of a larger whole simultaneously. Holonomy may be oxymoronic since it implies a combination of opposites: being both a part and a whole; acting autonomously and, at the same time, working interdependently. Because all life forces are simultaneously independent and interdependent, self-assertive and integrative—whole unto themselves yet always a part of systems larger than themselves—holonomy, therefore, is paradoxical.

When Arthur Koestler (1972) coined the word *holon*, he sought to describe something that has the characteristics of being both a part and a whole at the same time. Holonomy is the science or study of parts/whole relationships. As such, holonomy considers both our integrative tendencies and our autonomous aspects.

"It takes two to know one," Gregory Bateson was fond of saying. All beings exist within holonomous systems. That is, each person is part of several greater systems (e.g., families, teams, schools) yet maintains a unique identity and palette of choices, both as an independent agent and as the member of a group. Each system influences the individual, and, to a lesser degree, the individual influences the system.

Holonomous persons have an awareness of themselves in this somewhat oxymoronic, conflicting state of being an independent entity while also part of being and responsive to a larger system. They have the cognitive capacity to exercise flexible, responsible self-directedness in both arenas, being alert to both environmental and internal cues that inform their knowing when and how to act autonomously as well as when and how to act interdependently. Holonomous persons understand that as we transcend the self and become part of the whole, we do not lose our individuality but rather our egocentricity.

In a school, for example, teachers are autonomous decision makers. They are, however, part of a larger culture—the school—which influences and shapes their decisions. In turn, the school is an autonomous unit interacting within the influence of the district and community.

This dichotomous relationship, in which every human being exists, often gives rise to certain tensions, conflicts, and challenges. These stem from the internal drive for self-assertiveness, which conflicts with a yearning to be in harmony with others and the surrounding environment. A holonomous person, therefore, is one who possesses the capabilities to transcend this dichotomous relationship, maintaining self-directedness while acting both independently and interdependently. Holonomous people recognize their capacities to self-regulate and to be informed by the norms, values, and concerns of the larger system. Of equal importance, they recognize their capacity to influence the values, norms, and practices of the entire system.

> We provide both irritation and inspiration for each other—the grist for each other's pearl making.
>
> —Stephen Nachmonovitch

A holonomous person continually accesses and develops resources for further growth. One's goal is to become an integrated whole, capable of knowing and supporting the purposes and processes of the groups to which one belongs. A holonomous person is one who:

- Explores choices between self-assertion and integration
- Draws from prior knowledge, sensory data, and intuition to guide, hone, and refine actions
- Pursues ambiguities and possibilities to create new meanings
- Seeks balance between solitude and togetherness, action and reflection, and personal and professional goals
- Seeks perspectives beyond him/herself and others to generate resourceful responses.

Two Goals

Cognitive Coaches work to achieve their mission by supporting people in becoming self-directed, autonomous agents and self-directed members of a group. Toward this end, Cognitive Coaches regard all interactions as learning opportunities focused on self-directedness.

The goal of learning Cognitive Coaching is to develop the capacities and identity of coaches as mediators, who can in turn help to develop the capacities for self-directedness in others. The skillful Cognitive Coach:

- Establishes and maintains trust in him/herself, relationships, processes, and the environment
- Interacts with the intention of producing self-directed learning
- Envisions, assesses, and mediates for states of mind
- Generates and applies a repertoire of strategies to enhance states of mind
- Maintains optimism in the ability to mediate his own and others' capacity for continued growth

Metaphors for Coaching

> You don't see something until you have the right metaphor to let you perceive it.

—Thomas Kuhn

Think of the term *coaching*, and you may envision an athletic coach. We like to use quite a different metaphor. To us, coaching is a means of conveyance. Kocs is a village in Hungary from which the word *coach* is derived; they are pronounced the same way. The village is the birthplace of the horse-drawn vehicle called a coach in the 15th century. "To coach means to convey a valued colleague from where he or she is to where he or she wants to be" (Evered and Selman 1989, 116–32). Skillful Cognitive Coaches apply specific strategies to enhance another person's perceptions, decisions, and intellectual functions. The ultimate purpose is to enhance this person's self-directedness: the ability to be self-managing, self-monitoring, and self-modifying. Within this metaphor, the act of coaching itself, not the coach, is the conveyance.

Why Coaching?

In a time when many schools are pressed for time and money along with implementation of federal and state mandates, why is coaching so important? We have identified several compelling reasons.

1. Teachers need and want support. Earlier studies tracked the implementation of state legislative mandates in 26 national sites. Among the most significant findings was the importance of the support teacher. A cluster of studies found the support teacher-mentor to be the most powerful and cost-effective intervention in induction programs (Oja and Reiman 1998). More recently it was found that teachers with a high level of engagement in a large-scale mentoring program (California Formative Assessment and Support System for Teachers) improved both teaching practices and student achievement, producing an effect size equivalent to half a year's growth (Thompson, Goe, Paek, and Ponte 2004). Schools and classrooms today are busy, active places where teachers and students are pressured by high-stakes testing to teach and learn faster and to be held accountable for demonstrating to others their achievement of specified standards and mastery of rigorous, complex content. For that reason, classrooms are much more present- and future-oriented systems than they were in the past. Often it is easier to discard what has happened and simply move on.

Increasing evidence supports the link between student learning and staff learning. This means that as staff members learn and improve practices, the students benefit and show learning increases (Yoon, Duncan, Lee, Scarloss, and Shapley 2007). Cognitive Coaching requires a deliberate pause, a purposeful slowing down of this fast-paced life for contemplation and reflection. Coaching serves as a foundation for continuous learning by mediating another's capacity to reflect before, during, and after practice. Kahn (1992) emphasized the importance of "psychological presence" as a requisite for individual learning and high-quality performance.

2. Cognitive Coaching enhances the intellectual capacities of teachers, which in turn produces greater intellectual achievement in students. Professional development is a better predictor than age when measuring growth in adult cognitive and conceptual development. Research shows that teachers with higher conceptual levels are more adaptive and flexible in their teaching styles, and they have a greater ability to empathize, to symbolize human experience, and to act in accordance with a disciplined commitment to human values. These teachers choose new practices when classroom problems appear, vary their use of instructional strategies, elicit more conceptual responses from students (Hunt 1980), give more corrective and positive feedback to students, and produce higher-achieving students who are more cooperative and involved in their work (Calhoun 1985).

Witherall and Erickson (1978) found that teachers at the highest levels of ego development demonstrated greater complexity and commitment to the individual student; greater generation and use of data in teaching; and greater understanding of practices related to rules, authority, and moral development than their counterparts. Teachers at higher stages of intellectual functioning demonstrate more flexibility, toleration for stress, and adaptability. They take multiple perspectives, use a variety of coping behaviors, and draw from a broader repertoire of teaching models (Hunt 1977–1978). High-concept teachers are more effective with a wider range of students, including students from diverse cultural backgrounds.

We know that adults continue to move through stages of cognitive, conceptual, and ego development and that their developmental levels have a direct relationship to student behavior and student performance. Supportive organizations with a norm for growth and change promote increased levels of intellectual, social, moral, and ego states for members. The complex challenge for coaches, of course, is to understand the diverse stages in which each staff member is currently operating; to assist people in understanding their own and others' differences and stages of development; to accept staff members at their present moral, social, cognitive, and ego state; and to act in a nonjudgmental manner.

3. Few educational innovations achieve their full impact without a coaching component. Conventional approaches to staff development—workshops, lectures, and demonstrations—show little evidence of transfer to ongoing classroom practice. Several seminal studies by Bruce Joyce and Beverly Showers (2002) reveal that the level of classroom application remains very low even after high-quality training that integrates theory and demonstration. When staff development includes time for practice and nonjudgmental feedback and when the curriculum is adapted for the innovation, the implementation is higher. When staff development includes coaching

in the training design, however, the level of application increases greatly. With periodic review of both the teaching model and the coaching skills—and with continued coaching—classroom application of innovations remains very high.

4. Feedback is the energy source of goal accomplishment, growth, and self-renewal (Hattie 2003, 7; Costa and Kallick 2008, 103). However, feedback will improve practice only when it is given in a skillful way. Research by Carol Sanford (2011) has shown that value judgments or advice from others reduce the capacity for accurate self-assessment. Feedback that is data driven, value free, necessary, requested, and relevant, however, activates self-evaluation, self-analysis, and self-modification (Wiggins 2014).

5. Beginning teachers need mentors who employ Cognitive Coaching. Those entering a new profession need help to get through the struggles and quandaries of their first years. Cognitive Coaching offers a valuable initiation into the education profession by providing a model of intellectual engagement and learning that promotes self-directed learning. Cognitive Coaching also provides a leadership identity that endures in the minds of new teachers as they assume leadership roles throughout the organization. After three or four years of service, beginning teachers mentored with Cognitive Coaching in California schools gradually assumed significant teacher-leader roles (Riley 2000). Teachers often report the value of coaching over consulting. A fourth-year Arizona teacher in a Cognitive Coaching Seminar explained that she had asked for and received a great deal of consulting in her first years of teaching. She added that classroom practice in her fourth year, however, is based on thinking about the ideas she received from Cognitive Coaching (Ellison 2010).

6. Working effectively as a team member requires coaching. A harmonious collegial effort needs coordination. Consider a symphony orchestra. Its members are diversely talented individuals: an outstanding pianist, a virtuoso violinist, or an exquisite cellist. The musicians play, rehearse together, and come to a common vision of the entire score. Each musician understands how the part he or she plays contributes to the whole. They do not all play at the same time, but they do support each other in a coordinated effort. Together, they work diligently toward a common goal: producing beautiful music.

Likewise, each member of a school staff is an extremely talented professional. Together, they work to produce a positive learning environment, challenging educational experiences, and self-actualized students. In the same way, members of the school community should support each other in creating and achieving the organization's vision. Teachers neither teach the same subjects at the same time nor do they approach them in the same way. Cognitive Coaching provides a safe format for professional dialogue and develops the skills for reflection on practice, both of which are necessary for productive collaboration (Costa and O'Leary 2013).

7. Coaching develops positive interpersonal relationships that are the energy sources for adaptive school cultures and productive organizations. The pattern of

adult interactions in a school strongly influences the climate of the learning environment and the instructional outcomes for students. Integral to the Cognitive Coaching model is the recognition that human beings operate with a rich variety of cultural, personal, and cognitive style differences that can be resources for learning. Cognitive Coaching builds a knowledge of and appreciation for diversity. It also provides frameworks, skills, and tools for coaches to work with other adults and students in open and resourceful ways. Cognitive Coaching promotes cohesive school cultures in which norms of experimentation and open, honest communication enable everyone to work together in healthy, respectful ways.

Work by Susan Rosenholtz, Karen Seashore Louis, Milbrey McLaughlin, and others document that workplace culture (i.e., the shared values, quality of relationships, and collaborative norms of the workplace) has a greater influence on what people do than the knowledge, skills, or personal histories of either workers or supervisors (Garmston and Lipton 1998). Research by Gregory Moncada (1998) revealed that conversation alters the nature of the social construction of reality. His work also revealed that change in schools is determined less by environmental influences than by modifications in the social construction of reality brought about by conversation.

8. Coaching supports and makes more successful school renewal programs. Numerous organizations that have adopted Cognitive Coaching find that self-directed learning becomes central to the organization's aims; curriculum standards become broadened and organized around self-directed learning; and evaluation of teachers, students, and organizational effectiveness changes from performance reviews to mechanisms for continuous spirals of growth and learning. Schools that have adopted Cognitive Coaching have skills and values with which to accelerate growth into collaborative learning communities. Furthermore, reflection becomes habituated throughout the organization (York-Barr, Sommers, Ghere, and Montie 2001).

Schools and agencies that have adopted Cognitive Coaching have begun to profit from its effects not only in the United States but also increasingly throughout the world. Ultimately, we believe, Cognitive Coaching will significantly contribute to the creation of agile schools worldwide. Principles and examples of an agile organization are explored in chapter 15.

CONCLUSION

In personal correspondence from Bill Powell, former director of the American School of Tanganyika in Dar es Salaam, Tanzania, he stated the following:

> We interviewed internal candidates for the High School Vice-Principal's position. During the course of the interviews, I asked each of them what was the most important learning experience of their lives. Without hesitation, one candidate responded, "Cogni-

tive Coaching! It changed the way I think about teaching, about learning . . . in fact, it changed the way I think about myself."

All who wish to continually improve their craft—be they teachers, entrepreneurs, athletes, ballerinas, musicians, auto mechanics, or potters—never lose the need to be coached. Cognitive Coaching is a model of interaction that helps others take action toward goals that are important to them while simultaneously developing their capacities for self-directedness.

Cognitive Coaching is rooted in dispositions, beliefs, and values that honor the human drive for continuous learning and the spirit of collaboration. Cognitive Coaching is not giving advice or solving other people's problems, as with the boy and the butterfly chrysalis. Cognitive Coaching is a nonjudgmental process of mediation applied to those human life encounters, events, and circumstances that can be seized as opportunities to enhance one's own and another's resourcefulness.

2

Identity

> Where Cognitive Coaching provides a cognitive framework for our work as educators, the tools of Cultural Proficiency provide a moral framework. The culturally proficient educator recognizes that the disparities that exist among demographic groups are human-made and human-maintained. Recognizing these human-made disparities, the culturally proficient educator sees it as her moral obligation to create and influence change both within herself and her school (Lindsey, Martinez, and Lindsey 2007, 34).

Identity is the mental model each of us constructs of who we are as a unique self. This is an important concept because identity informs decisions and behaviors. The most sustainable way to change behaviors is to change identity. The self, writes neuroscientist Antonio Damasio, is fluid. It is not a thing; rather, it is a process (Damasio 2010). One's identity is in a constant and imperceptible gradual state of transformation. We create meaning from our interactions with others and with the environment. Identity emerges from the web of those interactions. It is not only how I see myself, but also the meanings I make from how others see me as a result of my interactions over time. There is no identity in isolation. In *A Simpler Way*, Wheatley and Kellner-Rogers observe,

> Every living thing acts to develop and preserve itself. Identity is the filter that every organism or system uses to make sense of the world. New information, new relationships, changing environments—all are interpreted through a sense of self. This tendency toward self-creation is so strong that it creates a seeming paradox. An organism will change to maintain its identity. (1999, 14)

In this chapter, we explain three forms of identity and how identity is formed. We explore a conceptual model related to changing identity and the responsive affects in other domains of human learning. We relate all this to the identity of a Cognitive Coach as a mediator of another's journey in self-directed learning.

THREE FORMS OF IDENTITY

Many theorists, cognitive and humanistic psychologists, biologists, anthropologists, and systems theorists have studied the development and effects of identity. Among these are Eric Erickson, Gregory Bateson, Robert Dilts, Milton Ericson, Carol Dweck, William James, Antonio Damasio, Abraham Maslow, Morris Rosenberg, and Oliver Sacks. Three ways of thinking about identity emerge: personal, social, and role identity.

Personal identity is a person's conception and expression of his/her individuality or group affiliations. Cognitive Coaching concerns itself with one's individuality and how that sense shapes and informs perceptions, values, beliefs, and behaviors. Psychologists refer to this as *personal identity or autobiographical self*, in which the self interacts with others and in which the relationships between self and others are organized in fairly stable representations over time (Damasio 2010). The thoughts, feelings, inferences, and interpretations of the experiences available to the autobiographical self constitute one's identity. Identity is the story we tell ourselves of who we are.

A second form is *social identities* in which a person has not one but several selves that correspond to group identifications such as ethnicity, passport country, age cohort, gender, sexual orientation, and political associations. This is a portion of self-concept related to being a member of a relevant social group. One might hold multiple social identities as being an Asian American Democrat with membership in the Sierra Club. Social identities can influence one's behavior, especially when with a group with which one is identifying.

Finally, *role identity* is concerned with the tendency for human behaviors to form characteristic patterns in different contexts (Biddle 2000). Role identity is relational since people typically interact with each other via their own role identities. At different times, in our role identities, we are husband or wife, grandchild or sibling, boss or employee, expert or student, depending on both the context in which we find ourselves and the other persons with whom we interact. These can be said to be "role identities," temporary and situational personas. The archetype of each of these roles carries certain presuppositions, orientations, and goals, which are manifested verbally and with our entire being. How we carry out each role is influenced by the identity we have developed for ourselves.

What begins as *role identity* might grow to be *personal identity* (*autobiographical self*) in which the beliefs and behaviors associated with this identity would be active in a multitude of relationships. This has been found to be true for those who develop a sense of self as mediator, the identity associated with being a Cognitive Coach. A mediator's mission is fostering a person's cognitive capacity for excellence as individuals and as members of a community. A mediator is a co-learner, engaging another person with the intent of transforming his/her capacities to become more self-directed: self-managing, self-monitoring, and self-modifying.

HOW IDENTITY IS FORMED

The mechanisms of forming identity occur in the upper brain stem and in a set of nuclei in the thalamus and regions of the cerebral cortex (Damasio 2010). People construct a set of essential characteristics to define their self-concepts. They interpret experiences and choose behaviors that are intended to maintain the continuity of those self-concepts over time. Identity formation is dependent on the state of mind of consciousness. Damasio (1999, 26) calls the simplest kind of consciousness *core consciousness*. "It provides the person with a sense of self about one moment—now and about one place—here." There is also an *extended consciousness* that provides the person with a sense of self and places "me" into a historical frame, richly aware of the lived past, the anticipated future, and the world beside oneself, the present. Damasio explains extended consciousness as a complex biological phenomenon that depends on conventional memory, working memory, reasoning, and language. The sense of self that emerges in core consciousness is a transient entity, ceaselessly re-created for each and every object with which the brain interacts. Our traditional sense of self, however, corresponds to a nontransient collection of unique facts and ways of being that characterize a person. Damasio calls this the autobiographical self.

SYSTEMS OF LEARNING

Anthropologist and cyberneticist Gregory Bateson (1972) first formulated the notion of relating systems of learning to human growth. He identified different and complementary systems of change, each developed through different systems of learning. Envisioning them as ordered on a ladder, he found that higher levels influence changes on lower levels, something like a cascade effect. Changes brought about on the lower levels are not always lasting, nor are they a guarantee that they will affect the higher ones. So it is economical to make a greater effort for change at the higher levels, without neglecting, Bateson says, to intervene at the other levels as well. And if possible, intervene at all of the levels, thereby aligning them and making a comprehensive intervention throughout the system. Bateson's system of learning is described in the four points below.

1. Any system of activity is a subsystem embedded inside another system. This system is also embedded in an even larger system and so on.
2. Learning in one subsystem produces a type of learning relative to the system in which you are operating.
3. The effect of each level of learning is to organize and control the information on the level below it.
4. Learning something on an upper level will change things on lower levels, but learning something on a lower level may or may not inform and influence levels above it.

Our reading of Bateson (1972, 2000), Tosey on Bateson (2006), and Dilts (1994, 2003, 2014) and work with thousands of people in seminars over a 30-year period lead us to place identity in a dominant position in a system of human learning. Thus identity informs and influences one's values and beliefs, capabilities or mental maps and process, and behaviors, the use of which are conditioned by the environment. So the Cognitive Coach, having an identity as mediator, forms and applies values (fostering the intellectual development of others), beliefs (resources for growth lie within a person being coached), capabilities (mental maps to guide conversations and knowing when and how to use certain skills), and finally the application of coaching behaviors (developing rapport, pausing, paraphrasing, posing questions), the use of which is conditioned by environment.

The overarching orientation guiding Cognitive Coaching work is the movement from a person's present state to a desired state. The work of Robert Dilts (2014) aligns with this. He suggests that ultimately the goal of personal coaching is to be deeply connected with ourselves and to live from a place of centeredness, presence, and fullness. We agree. Dilts originally conceived these elements as having a cascading affect, each higher level affecting the one below it.

When you were a child, you thought of yourself in certain ways. As an adult, you think of yourself differently. The root of your identity comes from your genetics, culture, and your experiences and the meaning you make of them. Your identity will change over time. It is important to know that you can modify yourself to be more in alignment with who you want to be.

ELEMENTS OF IDENTITY

Identity

This discussion focuses on *personal identity* or *autobiographical self,* which, as we noted above, may sometimes start as *role identity* but develop into a more consistent expression of self. Identity is about who I sense myself to be. Who I am relates to what mission I perceive is mine and how I go about achieving it. *Personal identity* is an individual's sense of itself, a conception and expression of one's individuality. It embodies beliefs and values. This level answers the question "Who am I?" This framework is formed and sustained through social interaction and life experiences. Identity is what makes a person a person; it is the consistently traceable thread that is "me" over time and that distinguishes me from other people. For a mediator, identity is a framework for understanding oneself as one who supports the development of thinking in others. All our perceptions of ourselves, others, and the environment flow from this sense of identity. A mediator is a facilitator of self-directed learning and is *not* the solver of the problems of others. This person will enlist his/her cognitive and emotional resources to persevere in mediative behaviors.

And, importantly, identity relates to beliefs about self. Jim Roussin, a strategic change consultant and leadership coach, executive director of Generative Learning, poses this question: What if identity and beliefs were held as one dimension in this model, reciprocally informing and reinforcing one another? We comment on this later.

Values/Beliefs

Values are our internal guidelines for what is good or bad, worthy or unworthy. Beliefs are ideas or concepts that we hold to be true. Values are absolute or relative ethical beliefs, the assumption of which can be the basis for ethical action. Values can be defined as broad preferences concerning appropriate courses of action or outcomes. As such, values reflect an individual's sense of right and wrong or what "ought" to be. Cognitive Coaches have faith that all human beings have the capacity to continue developing their intellect throughout their lifetimes. This and "We are all autonomous individuals and simultaneously interdependent members of communities" are representative of values of a Cognitive Coach.

Beliefs are what an individual has emotional investment in, based on assumptions and convictions that the person holds to be true regarding people, concepts, or things. Teachers who believe that they are more effective are likely to be more effective. Mediators who believe that behaviors are motivated by positive intentions will set aside judgmental attitudes and language. Cognitive Coaches believe that the resources for self-development lie within an individual and will behave accordingly, asking instead of telling, inquiring instead of showing. The mediator possesses a belief in his/her capacity to serve as an empowering catalyst for the growth of others.

Capabilities

These are the metacognitive maps, models, and internal skills that individuals use to guide behavioral choices: when to use, how to use, or not use specific skills sets (Dilts

2003). For example, a prime mental ability for Cognitive Coaches is to set aside unproductive patterns of listening. Relating what the coachee is saying to one's own story directs attention away from the coachee and is unproductive. The richer the mental maps and models, the more understanding about under what condition to use a behavior, ways of assessing the behavior's effectiveness, and processes for self-monitoring. To elegantly perform a behavior requires accessing the cognitive, affective, and psychomotor domains. Mental processes drive behavior through the selection of personal strategies and skills. A person operating at this level of capabilities is thinking about how to achieve something, including what skills she might need to develop to do so. As described in chapter 4, Cognitive Coaches employ four essential internal capabilities to guide and give direction to their behaviors: 1) know one's intentions and choose congruent behaviors; 2) set aside unproductive patterns of listening, responding, and inquiring; 3) adjust one's own style preferences; and 4) navigate between and within maps and support functions to guide meditational interactions.

Behaviors

Behaviors are what individuals do. They represent the application of skills and knowledge into actions. For Cognitive Coaches, they can be singular actions like paraphrasing or bundles of skills, like listening that involves being present, paraphrasing, and making meaning. Another coaching skill is constructing and posing questions with the intention of engaging and transforming thought. Cognitive Coaches employ nonjudgmental response behaviors to establish and maintain rapport, trust, and intellectual engagement. Nodding communicates receipt of information, silence infers attentiveness, paraphrasing reflects a speaker's thoughts, common filler words like *ah*, *okay*, and *right* signal attention. None of these are intended as judgmental and in the vast majority of cases are interpreted as interest rather than judgment.

Environment

This can be summarized by "where, when, and with whom." It represents the medium or the context in which we operate, in other words "what is perceived outside me," for example, work, home, social relationships, as well as political and economic contexts. Environment can be either tangible, as in things and places, or intangible, as in morale within a place. Environment influences what people do. School morale has a more than minor liberating or depressing effect on teacher thinking, behaviors, and indeed identity (Frymeir 1987; Seashore-Lewis, Marks, and Kruse 1996; Hoy, Sweetland, and Smith 2002). Modifying the environment can allow for behavioral change such as providing time and space for coaching.

Each coachee is a unique product of his/her culture, experiences, concepts, interactions, learnings, perceptions, values, skills, and behaviors. These are further affected by the school culture in which she/he works. Because each coachee brings his/her identity to interactions, the coaching process may be viewed as a collaborative interaction of identities. Each brings his/her unique perceptions into play during the dialogue. If the organizational identity is one of direct and command, it will provide the coachee with limited support for independent thought. In some situations people may be motivated to behave in ways that express or support the organization's identity rather than their own (Ellemers, De Gilder, and Haslam 2004). So when and where a Cognitive Coach uses his/her skills determines the opportunities as well as constraints.

As we shall see, all models are incomplete, including ones relating identity to other aspects of human functioning. Informed by Bateson's work (1972), Robert Dilts developed a stepladder model of learning in which elements higher on the ladder affected those below. While we have found great value in applying Dilts's model to Cognitive Coaching practices, we are recognizing new application possibilities from recent examinations of his work, including his own observations of his more recent model (Dilts 2014).

NOT LINEAR BUT NONLINEAR

Dilts presented the elements of identity, beliefs, values, capabilities, behavior, and environment as a hierarchal model with a cascading effect. In his early work, he arranged them in ladder form with elements higher on the ladder informing the lower regions. Identity rests on the top, informing beliefs and values that in turn inform capabilities and so on. Others have conceptualized these not as a ladder but as a set of nested building blocks, as shown in figure 2.1, with identity the largest holding all the other elements.

In our current practices, we are finding it useful to regard Dilts's model not as a hierarchal, linear map that assumes clocklike operation in which the same inputs will consistently produce the same result. Rather, we are seeing it as describing a complex system made of many moving parts operating within environmental systems that are constantly changing. Complex systems (also called dynamical systems) are those in which everything affects everything else, tiny inputs cause major disturbances, more data are not a greater predicator of results, and both things and energy matter. Larry Cuban (2010) likens complex systems to raising a child or getting children to succeed in school. Complicated systems, he says, have many interacting parts, like the engine in your car, but use linear logic to operate and repair. Figure 2.2 shows the Dilts elements as conceptualized by Garmston as operating within a dynamical system.

In Garmston's view (figure 2.3) of the elements of learning, we see the elements as dynamical in which everything affects everything else. Yet there are two dominant players within this system. Identity, we believe, is the major influence on all the elements within the system, informing beliefs, values, capabilities, and behavior. Next,

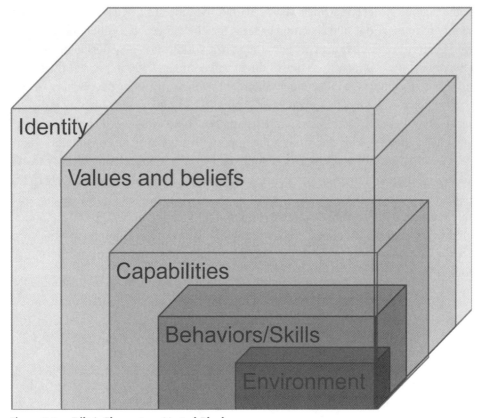

Figure 2.1.   Dilts's Elements as Nested Blocks

environment can modify the *expressions* of identity. To show this idea, we have placed environment as the surround and note that one's environment has a modifying effect on the expression of beliefs, values, capabilities, and behaviors. Changing the environment—say, from one in which mediation is a primary leadership activity to one in which command and control is more common—modifies not only behaviors but also the potential for learning.

This view of ways in which elements of learning most likely interact with one another is less authentic than the earlier image but more like a collage of photographs in figure 2.3. Garmston attempts to mechanically represent the ways in which the different elements interact. Again, identity, shown at the left, is characterized as the primary actor providing direct influence on beliefs and values. This is consistent with the observations by countless Cognitive Coaches to be the case as they assume identities as mediators. When this occurs, beliefs about ways of being helpful to another inevitably shift from advising, recommending, or telling to attentively listening, supporting the other in attaining goal clarity and coaching. Values, capabilities, and behaviors shift.

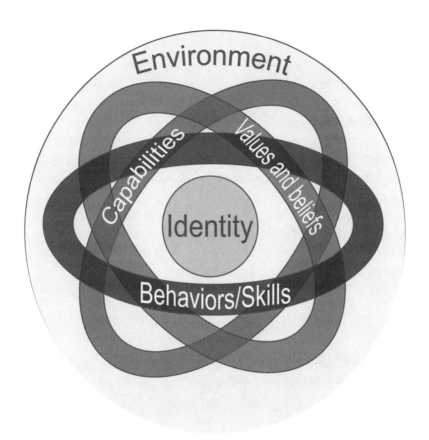

Figure 2.2. Identity Influencing: Dilts's Elements as in a Dynamical System

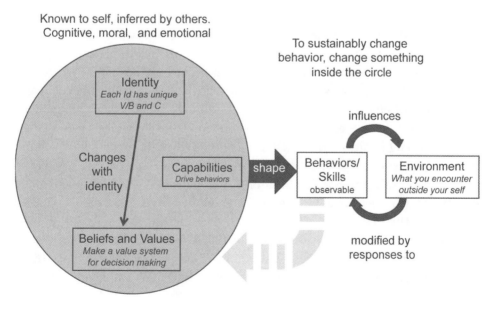

Figure 2.3. Dilts's Dynamical Relationships Shown Linearly

Figures 2.2 and 2.3 display the elements as being relational, existing within a dynamical system in which everything affects everything else. In these views, affecting identity still fosters the greatest change within the individual as it promotes dramatic effects at all the other parts of the system.

Consistent with the Cognitive Coaching goal of being self-directed, one can self-coach oneself into a temporary identity perspective. In one example, an administrator revealed to us that he was distressed at his own committee behaviors as being egocentric, defensive, and argumentative. He resolved to improve this at a subsequent meeting and searched for a strategy to release him from these behaviors. He hit upon the notion of assuming an identity as a student and found he could easily be more responsive to the ideas of others, put his own ideas on the table as "wonderings," and thereby become more interdependent and effective.

IDENTITY CHANGES

Earlier in his career, Bob's son, Kevin, was a teacher coaching student athletics and playing in several adult teams until the birth of his son. Remarkably, at his birth, a transformation occurred. Looking down into his son's face for the first time, Kevin suddenly was a father. In a sense this experience redefined his identity. Changes in values and beliefs consistent with parenthood emerged. He sought and developed the capabilities for behaviors related to fathering a young child. Sports activities dropped away; time with his son became the dominant interest in his life.

Other examples of identity conversions are found in spiritual and addiction literature. A review of spiritual writings report conversion experiences in which a "transformed self is achieved" either suddenly or gradually (Schwartz 2000, 5). In his review, Schwartz notes that the change in self is often "radical" with a new centering of interest, concern, and behaviors. Similarly, in studies of addiction, Sremac and Ganzevoort (2013) draw on 31 autobiographies in which drug addicts develop and sustain a sense of new identity. From this new self they create meaning from conflicting life experiences. Sremak and Ganzevoort report that the subjects they studied literally recreated their identities. They did so with a narrative process in which the construction and continuous reconstruction of a story of the self occurs. AA recoveries, too, report this process. The third of the 12 steps in Alcoholics Anonymous reads, "We made a decision to turn our will and our lives over to the care of God as we understood Him."

These reports are consistent with Damasio's (2010) view of identity being a work in progress. Identity change in these cases comes from personal narrative in which the recovering addict or alcoholic tells stories about self, often employing religious terms. This process runs concurrently with behavior changes, including abstinence and the practice of new socially useful behaviors. The appearance of spirituality in these cases is also consistent with Dilts's perception, that while identity is a prime mover of the other elements of learning, spirituality can inform identity. According to Dilts and McDonald (1997), Spirit is the pattern that connects everything together as a kind of larger mind of which individuals are but a subsystem. Manifestations of this Spirit, they believe, appear as wisdom, vision, mission, and healing. They are the consequence of acknowledging and bringing Spirit more into one's life and actions. Fruits of the Spirit, they say—such as love, compassion, joy, and peace—are the generative and transformative results of being more in harmony with what Einstein called God's thoughts.

At the spiritual level are subjective experiences of being part of a larger system. It is awakening to this greater context that gives our lives meaning and purpose. Anthropologist and systems theorist Gregory Bateson described it as the awareness of the pattern that connects all things together into a larger whole. This awareness reaches beyond ourselves as individuals (Tognetti 2002).

CAN BEHAVIORS CHANGE BELIEFS?

Can behaviors change beliefs? Maybe. Under certain conditions, repeated behaviors may affect beliefs. The broken arrow in figure 2.3 moving from behaviors back to the lower part of the identity circle is meant to indicate that, on occasion, repeated behaviors, over time and with reflection and metacognition, may lead to changes in beliefs and even values. This might be the case, for example, when a teacher gets the same response, over and over again, when pausing after asking students a question. After a number of trials, she/he may begin to associate the wait time with more elaborate student responses, and with this information, form a belief that wait time affects student cognition. A somewhat similar form of this process may be at play in the eight-day Cognitive Coaching foundation seminar.

In these seminars, participants are introduced to information about identity and learn the meaning of vocabulary terms like *holonomy* and *mediation* (concepts do not live without language to describe them). These concepts are fortified by eight days of observations, demonstration, reading, skills practice, personal experiences, and metacognitive reflection. From this, many construct a tentative grasp of themselves with the identity of a mediator. This is fortified in subsequent work in their schools and perhaps by more study.

Previously, using Dilts's hierarchal model, we understood that intervening at a level above the one we wished to affect would accelerate and strengthen learning in the domain below (Garmston 2005). For example, to support a person acquiring new behaviors, one would also teach at the level of capabilities to enhance the metacognitive capacities for decision making about the new behavior. We find this still to be a sound approach. In a dynamical model, we would also say that capabilities and behaviors are inextricably intertwined so that affecting one cannot but help affect the other. While teaching the mental process of behaviors to teach behaviors has merit, we offer three caveats.

1. Designing an intervention at the level of identity, values, or beliefs carries the challenge that these require a more robust and complex learning experience than to change a behavior. To create a values shift, one of the authors conducted the blue eyed–brown eyed activity in his fifth-grade classroom famously invented by a third-grade teacher and used the day after Martin Luther King's assassination (Peters 1987). In the author's class, two days were spent reading and talking about discrimination accompanied by a blue-eyed day in which all blue-eyed children in the class got privileges for a day the brown-eyed students did not. The scenario was reversed the next day. While the blue-eyes children could conceptualize about the effect of discrimination at the end of the first day when they were the privileged group, they were stunned, distraught, and dismayed (hopefully affecting values and beliefs) as they actually experienced the effects of discrimination on the second day. Experiences precede changes in values. Teachers value mediational processes because they have received and reflected about the benefits and processes of Cognitive Coaching, not because they have read about it.
2. Interventions that affect many or all of the domains in a system will be more conclusive than interventions in just one domain. Beliefs and values, as Jim Roussin speculated, are related. Values, beliefs, and behaviors are related to identities, which reinforce and sustain identity adjustments.
3. Finally, in dynamical systems in which everything affects everything else, an intervention in any domain will cause reverberations in others.

AVENUES TO CHANGE

As Damasio (2010) reminds us, identity is always a moving target. While our autobiographical self, or ego, now thought to be physically located deep in the brain and called the default mode network (Pollan 2015), may not notice gradual mutations, environments and experiences affect us. Many routes can lead to identity alterations. Early in his career, one of the authors was fascinated as he learned from notable figures in psychology. These luminaries—Carl Rogers, Fritz Perls, and Wilson Van Dusen, who at the time was chief psychologist at Mendocino State Hospital in California—elicited internal resources for learning. He witnessed similar orientations as he studied with two exceptional educators: Richard Suchman, who originated Inquiry Training, and Caleb Gattegno, who developed the reading program Words in Color and mathematics applications with Cuisenaire Rods. For this author, experience prompted value shifts that inflamed a desire to mediate learning for others in ways these notables had done. Study and practice followed.

Just as the mind is a belief-validating mechanism (Shermer 2011), the brain also seeks recognition and reinforcement of identity. Consider the formation of beliefs. Touching a hot stove is significantly painful and traumatic enough that in a one-experience event the belief is formed that touching stoves can be painful and dangerous. The mind, as a belief-validating mechanism then seeks confirming information from other sources. In another process, children, dependent on others, may accept what adults tell them. In yet another, repeated experiences of observing children from the tenements struggle in learning school subjects, one might form the belief that tenements children cannot learn. Regardless of the source, a one-time event (such as accepting what an authority figure says) or repeated experiences in which one forms generalizations, the next step is that the brain cannot help itself from what Shermer calls *patternicity*. This is the search for validation in patterns of both meaningful and meaningless data. Because the brain must connect the dots, it then infuses the patterns with meaning, intention, and agency.

THE IDENTITY OF A COGNITIVE COACH

The desired identity of a Cognitive Coach is that of a mediator of thinking, one who is devoted to fostering resourcefulness in others. We refer to this as a "default" position. In this state, a Cognitive Coach's ego is set aside when coaching others in planning, reflecting, and problem resolving. Likewise, neuroscientists, as reported by Judson Brewer at Yale (Pollan 2015), found that the default for experienced meditators was the temporary decommissioning of ego during meditation. It is the identity as a mediator of thinking that causes a Cognitive Coach to "default to" or automatically assume the position of supporting the thinking of others. For example, when someone shares a difficult situation, a frequently asked response might be, "What are you going to do?" Having the identity as a mediator of thinking, would cause one, instead, to paraphrase then pose questions: "So what are your goals?" "What are you thinking about this?" Thinking is a core value of a mediator. Thus, mediators evaluate their own performance based on the degree to which they have helped others to become more self-directed.

CONCLUSION

In this chapter, three categories of identity have been explored: role identity, say of boss, parent, sibling, or friend; social identity in which we relate ourselves to relevant groups, say, midwestern white Republican hunter; and personal identity, which is the focus of development in Cognitive Coaching Communicating from any role or personal orientations casts us into a response loop with others. For example, if I sound, feel, and look like a parent to you, you are likely to respond to me as a child (Wheatley and Kellner-Rogers 1999). Assuming the stance of the expert establishes one's responsibility to share one's greater knowledge and experience and to help others develop correct and appropriate performance. The friend values the relationship and will be loath to jeopardize it. The boss, in the traditional sense, wants compliance and feels responsible for the other person's success or failure. The mediator evokes dispositions for reflection and self-directed learning on the part of the coachee.

Identity can be viewed as personal, social, or related to role. In this chapter, we explored the relationships among personal identity, beliefs and values, cognitive and affective abilities, behavior, and environment. Identity is inseparable from our beliefs and values; capabilities are inextricably interrelated with behaviors; behaviors are the instruments with which we live out our identity, beliefs, and values; and the environment provides a context for all of this, often modifying their expressions. One can change the environment to make it more receptive to reflection and self-directedness. Garmston explores this in chapter 15 and in the Adaptive Schools book with Bruce Wellman (Garmston and Wellman 2013). Our identity evolves through the meanings we make from a history of reciprocal interactions and experiences. In a similar manner, our words and labels allow us to see new realities, and new realities affect our words. Our identity defines our reality, and our subjective reality helps to shape our identity.

Cognitive Coaching is not an algorithm to be rigorously applied as if it were a rote process because each person and situation is unique. Instead, Cognitive Coaching incorporates a set of values, beliefs, capabilities, mental maps, skills, and commitments dedicated to the goal of fostering self-directed learning in others. As these are practiced, tested over time, and assimilated into a person's day-to-day interactions, they become part of a Cognitive Coach's identity; ultimately, that identity becomes an outlook on life. A Cognitive Coach's identity drives him/her to be committed to learning, to continually resist complacency, and to share both the humility and the pride of admitting that there is more to learn. Cognitive Coaches dedicate themselves to serving others, and they set aside their egos, devoting their energies to enhancing others' resourcefulness. They commit their time and energies to making a difference by enhancing interdependence, illuminating situations from varied perspectives, and striving to bring consciousness to intentions, thoughts, feelings, and behaviors and their effect on others and the environment.

A new identity may not be something that we build but rather something to which we evolve. In the same manner that Michelangelo carved his sculptures by chipping away the excess stone to reveal the figure that was already inside the marble, our identity may go through a gradual "morphing" process to form new dimensions of itself. Through Cognitive Coaching, we have seen these transformations countless times as people "discover" powers they have through the processes of mediation. Mediating thinking allows what may lie in the subconscious to rise to consciousness. Identity without distinctiveness is imitation.

> It is no less important to recognize and promote individual identity—hence creating and reinforcing what we have labeled *individuation* and *psychological differentiation*. Although the fusion that needs to occur through the mediation of sharing behavior is important, there is also the "companion" need for individuation, to value one's uniqueness. The individual has to be given the opportunity to learn that whatever is done, the way it is done, the way one thinks or expresses oneself, is absolutely acceptable and appropriate (when it is in fact so). This does not offer a major contradiction with what we said above—rather, it is the other side of the same "coin." We want the individual to feel they don't have to mirror the other: "You are you and I am me" (Feuerstein, Falik, and Feuerstein 2014, 91).

The Mediator's Skills

The brain's many regions are connected by some 100,000 miles of fibers called white matter—enough to circle the earth four times.

—Van Weeden as cited in Zimmer 2014, 34

Emerging findings in the neurosciences are providing confirmation about why certain Cognitive Coaching linguistic and nonverbal practices foster cognitive development. A skillful coach uses certain well-crafted verbal and nonverbal mediational tools to facilitate others' cognitive growth. All are used without judgment; all are intended to sustain a colleague's access to the highest possible neocortical intellectual functions and body–mind intelligence. These tools can be categorized into four groups: paralanguage, response behaviors, structuring, and mediative questioning. This chapter examines each of these coaching tools and describes their positive effects on intellectual growth.

Rapport refers to vocal qualities, body gestures, and other verbal and nonverbal behaviors that exist alongside the words we speak. It may modify or nuance meaning of the spoken word and can be expressed consciously or unconsciously. (The prefix *para* means "alongside.") When confronted with conflicting verbal and nonverbal messages, humans inevitably choose the meaning behind the nonverbal behavior. Glaser (2014) notes that conversations trigger physical and emotional changes in our brains and bodies, releasing either oxytocin, which fosters bonding and collaboration, or cortisol, a hormone that evokes stress and fear. Knowing about paralanguage is one important element in establishing rapport with a person being coached, as without a degree of psychological comfort, the coachee is unable to access and express her thinking.

Response behaviors refer to verbal responses to another person's communications. These response behaviors are important because they convey nonjudgmental

coachee is expressing. Five types of verbal responses are help-
ıking:

wait time and listening
both nonverbally and verbally
- rarapnrasing, such as acknowledging, summarizing, organizing, shifting levels,
  and empathizing
- Clarifying, by seeking meanings and specificity
- Providing data and resources, thus facilitating the acquisition of information
  through primary and secondary sources

Structuring establishes the parameters of time, space, and purposes. These structuring decisions allow the coachee to predict consequences and thereby offer him the
elements of safety that are so important to removing perceived threat according to
findings by neuroscientists (Rock 2008; Aguilar 2014).

Mediative questioning helps coaches to construct and pose questions intended to
engage thinking. Mediative questions assume and potentially evoke positivity and a
growth mind-set in the teacher (Dweck 2006).

If the coach is not yet well acquainted with the coachee, it is important to take
time to establish a positive relationship and build a bond of trust between them.
(Establishing trust will be explored in greater detail in chapter 6.) Humble inquiry
(Schein 2013) is the fine art of drawing someone out, of asking questions the answers
to which are not already known by either party, of building a relationship based on
curiosity and interest in the other person. The coach will spend time finding out more
about a colleague's background, philosophy, and interests as well as share personal
information about himself/herself.

Schein asserts that humility refers to granting someone else a higher status than
one claims for oneself. Thus the coach assumes that teachers know more about their
students, more about the content that they teach, and more about their own skills,
styles, preferences, and strengths than does the coach. Thus, the coach, intentionally
employing the processes of listening empathically and questioning instead of telling,
initiates a positive and friendly relationship with the new coachee.

PARALANGUAGE

On average, adults find more meaning in nonverbal cues than in verbal ones. A recent summary of communication literature supports the theory that nearly two-thirds
of meaning in any social situation is derived from nonverbal cues (Burgoon, Buller,
and Woodall 1996). Receptivity and high expectations are communicated with initial

direct eye contact, low breathing, and head nods. Matching voice tone and pace, using the other person's gestures as you speak, matching posture, gesture, breathing energy, and words all contribute to rapport (Zolller and Landry 2010). Silence accompanied with low breathing also fosters rapport (Poyatos 2002).

We use the word *presence* to express a state in which the coach devotes total attention to the other person. Presence is essential but not sufficient for effective conversations. In some remarkable research, it was found that where managers gave unflattering feedback to people while emitting warm feelings to them, the recipients of the feedback still had positive feelings about the interaction (Goleman 2006; Burgoon et al. 1996). In addition to posture, gesture, and use of space, the intonation, rhythms, pacing, and volume of a person's voice all contribute important information about the communication. This is true within all cultures. (Cultural differences will be explored in chapter 10.)

Vocal cues affect attention. In a classic book, top nonverbal-behavior scholars (Burgoon et al. 1996) report studies that demonstrate that "vocal variety, which includes variation in pitch, tempo, intensity and tonal quality, has been shown to result in increased comprehension." In classroom work, Michael Grinder (1991), observing more than 5,000 classrooms, found that teachers tend to use one of two voices that elicit student attention in order to achieve comprehension. One is a credible voice, with which the teacher gains attention and gives direction. This voice is characterized by a limited range of modulation and a tendency to go down in intonation at the end of a sentence. (Imagine a newscaster reporting news.) The second voice, and the one used in coaching, is an approachable voice. This has a wider range of modulation and a tendency, at times, to rise in inflection at the end of a sentence. When delivered in an approachable voice, questions like "How did you decide to . . . ?" or "What are your goals?" signal safety and inquiry. The same questions offered in the credible voice feel like interrogation and lead to shutdowns in thinking.

Mirror Neurons

In 1992, the Italian neuroscientist Giacomo Rizzolatti found neurons in monkeys that fire both when the monkey carries out certain specific hand motions and when it views those specific motions being carried out by someone else. We, too, are built to respond to what others do. "The human brain harbors multiple mirror neuron systems, not just for mimicking actions but also for reading intentions . . . and for reading emotions" (Goleman 2006, 42). Social skills and rapport depend on mirror neurons. Goleman observes that mirror neurons make emotions contagious, letting the feelings we witness flow through us. Amazingly, these same neurons respond to the mere hint of an intention to move one's body. The intentions of the mediator's actions can be conveyed and processed, and the mechanism is the selective activation of the mirror neuron system. What is mirrored is not only the meaning of the actions observed but also the understanding of mediator's intentions (Feuerstein, Feuerstein, and Falik 2010, 137–38).

Frame-by-frame analysis of pairs talking show head and hand movements punctuate their conversational rhythm that coincide with stress points and hesitations in speech (Provine 2000). Microanalysis reveals that mother and baby precisely time the start and end of pauses in baby talk, creating a coupling of rhythm (Glaser 2014, 36). With words as mere sound effects, mothers and babies communicate nonverbally, a smile appearing on mother's face to be reproduced by baby within three seconds.

Humans gesture when they talk. The observant coach will notice that colleagues place people, events, time periods, and concepts in space. "On the one hand . . . and then on the other hand" is a verbal equivalent of this physical talk. Understanding is communicated and rapport is enhanced when the coach points to the space assigned to a concept when verbally responding. A verbal paraphrase is a reflection of what the colleague is thinking and feeling. Repeating a colleague's gesture adds a visual component to the reflection.

Even laughter, as a paralanguage, is valuable in coaching. Researchers at Stanford University and the Loma Linda Medical School found an increase in white blood cell activity while subjects listened to a comedian. This change in blood chemistry may boost the production of the neurotransmitters required for alertness and memory.

During the early 1970s, Richard Bandler and John Grinder conducted a series of investigations to learn why some therapists were almost magically effective, in contrast to others who simply did a good job. They initially studied Fritz Perls, the founder of Gestalt therapy; Virginia Satir, noted for her results in family therapy; and Milton Erickson, generally acknowledged to be the world's leading practitioner of medical hypnosis (Bandler and Grinder 1975).

The researchers discovered that Perls, Satir, and Erickson constantly mirrored their clients. For example, if the client had his legs crossed, the therapist crossed his legs. If the client leaned forward on her elbows, so did the therapist. When the client spoke rapidly, the therapist did too.

Whether they were conscious of it or not, Perls, Satir, and Erickson were modeling the theory of entrainment, which was formulated in 1665 after a Dutch scientist noted that two pendulum clocks mounted side by side on a wall would swing together in a precise rhythm. It was discovered that the clocks were synchronized by a slight impulse through the wall. Figure 3.1 shows manifestations of rapport.

When two people "oscillate" at nearly the same rates, we observe entrainment. Human beings seek this kind of synchronization. Neuroscientists have discovered neural systems (oscillators) that operate as clocks, resetting over and over their rate of firing to coordinate with the periodicity of an incoming signal (Rizzolatti, Fogassi, and Gallese 2006).

Figure 3.1. Rapport

At the Boston University School of Medicine, William Condon studied films of many sets of two people talking. Not only were the bodies of the speakers matched but also another "very startling phenomenon" was observed: entrainment existed between the speaker's words and the listener's movements. As one person would talk, the second person would make tiny corresponding movements. As Condon (1975) expressed it:

Listeners were observed to move in precise shared synchrony with the speaker's speech. This appears to be a form of entrainment since there is no discernible lag even at 1/48 second. It also appears to be a universal characteristic of human communication, and perhaps characterizes much of animal behavior in general. Thus, communication is like a dance, with everyone engaged in intricate and shared movements across many subtle dimensions, yet all strangely oblivious that they are doing so.

Rapport

When entrainment (matching) of several of these processing and communication systems is present, people can be said to be in rapport. Amazingly, neuroscientists have found that conversations trigger physical and emotional changes in the brain, either opening you up for a trusting interchange or closing you down so that you speak from caution and worry (Glaser 2014). In rapport, brains actually synchronize, crossing the skull and skin barriers to beat as one. Brains thus loop outside our awareness (Stern 2004). It is said that in a conversation between Jonas Salk and Gregory Bateson, one said to the other, "Where does thinking occur?" "About here," the other said, pointing to the space between them. However, intentionally mimicking someone to foster rapport can come off as awkward. Since mirror neurons transmit intentions consciously, matching with the intention to be fully present and available to the other person transcends the communication of awkwardness. Surely, this phenomenon is present in "motherese," the universal language of mothers in which high-pitched playful tones are emitted while mother and baby "precisely time the beginning, end, and pauses in their baby talk creating a coupling of rhythm" (Goleman 2006, 36).

An example of entrainment in a school setting might be the following: Lee says, "Fran, I've got some teachers on my staff doing things I am really excited about. I'm trying to figure out how I can get them to share some of their good ideas with other faculty." While Lee is speaking, she gestures broadly with her hands and arms. Then she shrugs her shoulders as she notes her predicament. Now Fran says, "Lee, tell me a little bit about what's so special about these teachers." As Fran speaks, she mirrors Lee's gestures and body movements.

Because Lee's nonverbal behaviors are congruent with her feelings and thoughts, each particular movement and gesture conveys specific meanings. When Fran borrows these same nonverbal cues, Lee subconsciously senses that Fran knows exactly what she's talking about.

At this point you may wish to see these nonverbal responses in action. Please go to www.thinkingcollaborative.com/supporting-audio-videos 3-1.1, 3-1.2, and 3-1.3. There are two segments in which Jane and Monica demonstrate rapport. Observe the matching of their body posture. In the second video, notice the mirror movement of the coach's head in response to Monica's. In the third segment with Carolee as coach, notice the body configurations matching in more casual postures.

RESPONSE BEHAVIORS

The effective Cognitive Coach draws on several response behaviors, depending on the colleague and the setting. These behaviors are silence, acknowledging, paraphrasing, clarifying, and providing data and resources.

Silence

Some coaches may wait only one or two seconds after having asked a question before they ask another question or give an answer to the question themselves. It is easy to feel that unless someone is talking, no one is learning. In actuality, however, silence is an indicator of a productive conversation. After the coach asks a question or after the partner gives an answer, the ensuing silence (1) communicates respect for the other's reflection and processing time, and (2) results in a positive effect on higher-level cognitive processing.

If the coach waits only a short time—one or two seconds—then a brief, one-word type of response will typically result. On the other hand, if the coach waits for a longer period, teachers tend to respond in whole sentences and complete thoughts. There is a perceptible increase in the creativity of the response as shown by greater use of descriptive and modifying words and an increase in speculative thinking (Rowe 1996). There is also an increased feeling of being valued and respected by the coach.

Coaches can communicate expectation through the use of silence. When the coach asks a question and then waits for an answer, it demonstrates that the coach not only expects an answer but also has faith in the other person's ability to perform the complex cognitive task, given enough time. If the coach asks a question, waits only a short time, and then gives the answer or gives hints, it subconsciously communicates that the colleague is inadequate and can't really reason through to an appropriate answer (Good and Brophy 1973).

When the coach pauses after the other person gives an answer, it causes the continuation of thinking about the task or question. Furthermore, when a coach waits after the partner asks a question or gives an answer, it models the same thoughtfulness, reflectiveness, and restraint of impulsivity that are desirable behaviors for others to use (Rowe 1986).

To view an example of wait time, please go to www.thinkingcollaborative.com/supporting-audio -videos 3-2/. The coach, Carolee, uses wait time after having asked Kris a question (Wait time 1). Chris processes internally for over seven seconds before responding the first time. In response to a second question, Chris reflects nearly as long before responding. These responses are consistent with Rowe's research showing that wait time of three to five seconds elicits deeper thinking.

Acknowledging

Acknowledging without making judgments is a response that simply receives and recognizes what the speaker says without making any value judgments. It communicates that the other person's ideas have been heard. The following are examples of this type of response:

- Verbal: "Um-hmm," "That's one possibility," "Could be"
- Nonverbal: Nodding the head

Paraphrasing

Paraphrasing is one of the most valuable and least-used tools in human interaction. A well-crafted paraphrase communicates "I am trying to understand you and therefore I value what you have to say." A paraphrase also establishes a relationship between people and ideas. Paraphrasing aligns the parties and creates a safe environment for thinking.

Mediational paraphrases reflect the speaker's content and emotions about the content. They frame the logical level for holding the content. Questions by themselves, no matter how artfully constructed, put a degree of psychological distance between the asker and the person being asked. Questions, preceded by paraphrases, gain permission to inquire about details and elaboration. Without the paraphrase, probing and questioning may be perceived as interrogation.

To design an effective paraphrase, begin by carefully listening and observing to calibrate the content and emotions of the speaker. Signal your intention to paraphrase by modulating intonation and using an approachable voice. Don't use the pronoun I. (For example, "What I think I hear you saying . . .") The pronoun "I" signals that the speaker's thoughts no longer matter and that the paraphraser is now going to insert his own ideas into the conversation.

Open with a reflective stem. This language structure puts the focus and emphasis on the speaker's ideas, not on the paraphraser's interpretation of those ideas. For example, these stems signal that a paraphrase is coming:

- You're suggesting . . .
- You're proposing . . .
- So you're wondering about . . .
- Your hunch is that . . .

Choose a logical level with which to respond. There are at least three broad categories of logical levels:

1. *Acknowledging* content and emotion. If the paraphrase is not completely accurate, the speaker will offer corrections. For example, "So, you're concerned about the district-adopted standards and how to influence them."

2. *Organizing* by offering themes and categories that relate several extensive topics. For example, "So you are concerned about several issues here. One is the effects of testing on students' higher-level thinking. Another is making inferences about school effectiveness based upon test scores alone. And yet another is how they influence teachers' instructional practices."

3. *Abstracting* by shifting focus to a higher or lower logical level. Paraphrases move to a higher logical level when they name concepts, goals, values, beliefs, identity, and assumptions: "So, a major goal here is to define what constitutes effective learning and to design authentic ways to gather indicators of achievement." Paraphrases move to a lower logical level when abstractions and concepts require operational definitions: "So authentic assessments might include portfolios, performances, and exhibitions."

Please go to http:www.thinkingcollaborative.com/supporting-audio-videos/3-3.1 and 3-3.2. Note the power of an abstracting paraphrase. Carolee and Ochan are engaged in a reflecting conversation. The coach paraphrases, capturing in one word, the essence of what Ochan was describing. Notice Ochan's appreciation of the abstracting paraphrase that led her to transfer the concept of deliberateness to another area of her work. Finally, Carolee summarizes Ochan's learning in this reflecting conversation.

Clarifying

We are prepared to see, and we see easily, things for which our language and culture hand us ready-made labels. When those labels are lacking, even though the phenomena may be all around us, we may quite easily fail to see them at all. The perceptual attractors that we each possess (some coming from without, some coming from within, some on the scale of mere words, some on a much grander scale) are the filters through which we scan and sort reality, and thereby they determine what we perceive on high and low levels.

—D. Hofstadter (2000)

The brain and central nervous system filter an enormous amount of sensory input. A million bits of information per minute come pouring into the brain—and we certainly can't attend to it all. Instead we generalize, fitting information into already stored patterns and categories. Sometimes we delete, or literally stop data from coming in—for example, we see bugs and dirt on a windshield but look through them to focus on the road and other cars. At other times we distort, shaping information to fit our preconceived structures and beliefs. There is a story about a man who thought he was a corpse. After six months, his psychiatrist asked him if corpses could bleed. The patient said, "Of course not!" The doctor pricked the patient's finger and squeezed out some blood. The patient stared at it and exclaimed, "What do you know? Corpses do bleed!"

Because our brains easily filter and distort information, it's important to understand some of the cognitive reasons that clarifying is essential. The skillful coach looks for the speech patterns described in the following.

Vague nouns or pronouns. When the teacher talks about "the textbooks," she has deleted information about which textbooks she's referring to. The coach can ask, "Which textbooks, specifically?" Clarifying vague nouns and pronouns supplies missing data and provides more precise information. When the teacher says "they . . .," the coach can get a better understanding by asking, "Who specifically is 'they'?"

Vague verbs. These refer to unspecified action words, such as *think* and *understand,* and they also signal deleted information. When the coach hears, "I want students to understand," he can ask, "Understand how, specifically?" to clarify the performance. "Which students?" will clarify the audience. "What will you see them doing when they are understanding?" tells the coach exactly what the teacher is looking for.

Comparisons. Distortions or generalities will sometimes be masked by comparisons. Often, the teacher will not elaborate to whom or to what she is comparing students. For example, if the teacher says, "The class is much easier," the coach can clarify, "Easier than what?" Or if the teacher says, "Rebecca's getting along much better," the coach can ask, "Better than what?"

Rule words. Phrases like "I can't," "We shouldn't," "It happens all the time," or "Nobody does that" should alert the coach that the teacher is limiting his thinking and possibly working with distorted perceptions. Skillful coaches challenge this mindset. When a teacher says, "I can't," the coach can respond, "What's stopping you?" When the teacher says, "We shouldn't," one response is "What would happen if we did? Who made up that rule?" "What other options might you have?"

Universal quantifiers. Terms such as *everybody, all the time, nobody, never,* and *always* also limit thinking. Statements employing these terms can be challenged with intonation. If the teacher says, "Nobody around here does that," the coach may simply clarify by saying, "On what data are you basing that?" Most people will recognize that they have overgeneralized. Or the coach can ask for exception: "Nobody around here does that? Can you think of someone who does?" This, too, leads the teacher to see that she has generalized.

Clarifying is an effective skill to call on when the speaker uses a vague concept or a vocabulary that the listener doesn't understand. The purpose of clarifying is to invite the speaker to illuminate, elaborate, and become more precise in his meaning. Clarifying helps both the listener to better understand the speaker and the speaker become clearer about his intentions. Some examples of clarifying include the following:

- Say more about that.
- What will students be doing if they are comprehending the story?
- Which students, specifically?
- When you say this class is better, better than what?
- When you say "the administrators," which administrators do you mean?
- What do you mean by "appreciate"?

PROVIDING DATA AND RESOURCES

One of the main objectives of Cognitive Coaching is to mediate the other person's capacities for processing information by comparing, inferring, or deducing causal relationships. For coaching to be data driven, data must be richly and readily available for the colleague to process. Providing data nonjudgmentally means that the coach makes it possible for the colleague to acquire the data she needs to deduce these relationships. There are several ways to facilitate this process.

Where the Feedback of Data Fits into the Coaching Cycle

The coach meets with a teacher for a planning conversation. The coach invites the teacher to identify what information is desired, determine when it should be collected, and to specify in what format the data will be recorded. As part of the process, coaches and coachees should agree on documentation methods so that the coachee is not only aware of the goals but can also monitor and celebrate his/her successes. As the coachees monitor their goals, they can modify their strategies if they are not producing success. Examples might include the following:

- Tell me what you want me to record about student responses that will be of help to your understanding of their higher-level thinking.
- So, my job is to keep a tally on this seating chart of which students are on task and which students are off task.
- When in the lesson do you want me to record your directions? Just at the beginning when you outline their assignment, or throughout the entire lesson whenever you give a direction?

During the lesson, the coach observes and records data that the teacher requested during the planning conversation. For example, a coach can record data with video, audio, classroom maps, time-on-task charts, and verbatim scripts of teachers' or students' statements. (Note: There are numerous electronic videos as well as audio and written classroom observational tools available on the Web. We recommend those that promote self-directedness, that are nonjudgmental, that make sense to the teacher being observed, that are suitable for their content and grade level, and that are designed to collect the data that is relevant to the teacher.)

Because the ultimate goal of the coach is to mediate another's capacity for self-coaching and self-modification, the coach may also mediate the colleague's gathering of data. For example, during a planning conversation the coach might say the following:

- What information do you have about these students that will guide your lesson design?
- What indicators will you be aware of to let you know you are achieving the meeting's goals?
- How will you monitor your own pacing to accomplish your agenda within the time constraints?

After the lesson, the coach conducts a reflecting conversation. The coach provides the coachee with the desired data nonjudgmentally. (See examples in "Providing Nonjudgmental Data" in the following pages.) The coach poses mediative questions focusing on the information desired by the coachee during the planning conversation and gathered by the coach during the lesson. The coach mediates the acquisition of data by drawing forth and focusing on the information gathered by the coachee during the lesson:

- What indicators were you aware of that let you know the group understood its task?
- What data were you collecting during the lesson that informed your decision to change the plan?
- What indicators were you aware of that let you know the group was engaged?
- Given this data, how do you think the class responded to your lesson?

Efficacious, craftsmanlike teachers remain open to continuous learning. They gather data through conscious observation of feelings attitudes and skills, through observations and interviews with others, and by collecting evidence showing the effects of their efforts in the classroom environment. The data are analyzed, interpreted, and internalized. Based on the analysis, the teachers modify their actions to more closely achieve the goals. Thus individuals become continually self-renewing and self-modifying.

What is known about effective feedback?

- For feedback to be effective, it must be neutral. If it is overly anecdotal, the discussion wanders and becomes personal.
- If it is quantitative based on previously agreed indicators of how well the process works, the chances of hearing it are improved.
- People need feedback so desperately that, in the absence of actual feedback, they will invent it.
- Data energizes learning. Humans learn and grow when they have the opportunity to consider external and internal data. External data include feedback from peers, teachers, coaches, personal experiences, and/or test results.
- Neither external nor internal data are useful for learning unless feedback is frequent and stated in constructive terms. Feedback given too long after an assessment or learning event won't influence a learner in the same way as data offered almost immediately.
- The term "internal data" describes information gained from feedback and personal reflection.

To become self-directed, teachers need opportunities to generate data and to self-evaluate. They must learn how to compare their current performance to previous performances, and they must learn how to analyze their performance in terms of benchmarks for effective performance.

Five Forms of Feedback

Since our goal is to mediate the coachee's cognition and to enhance his capacity for self-directedness, the how and why of providing feedback is critical. We offer five forms of feedback. Their contribution to self-directedness is presented in a descending order. Each has a varying degree of helpfulness in growing self-directedness as follows:

A. Mediative Questions

The intent of mediative questions is to cause the coachee to supply the data from her own internal and external observations. Posing mediating questions has the highest potential for developing self-directedness as their intent is to alert the coachee to the data that will serve to provide self-feedback, to process that feedback, to construct meaning from it, and to set goals for himself to self-modify. For example:

- What indicators were you aware of that let you know the group understood your topic?
- What data were you aware of during the presentation that informed your decision to change your delivery?
- What did you hear your students saying that gave you some feedback about their readiness to move to a higher level of complexity?
- What feelings were you experiencing when students exclaimed their fascination with the topic?
- As you reflect on the purposes of your lesson, what are some indicators that those purposes were met?

B. Providing Nonjudgmental Data

While a Cognitive Coach may invite the coachee to identify, observe for, and collect his/her own data, the coach may also serve as "another set of eyes" by collecting the data that the coachee requested and then, during the reflecting conversation, supporting the coachee by providing the data, nonjudgmentally, that the coachee requested in the planning conversation. The coach provides these data with the intent of inviting the coachee to make meaning—drawing comparisons, inferences, and analyses—and then, based on that interpretation, make application to future lessons, adapting instructional strategies, and so forth. Examples include:

- You asked three questions within the first five minutes of your presentation.
- Of the six students you wanted me to observe, Tyrone spoke four times, Sarah spoke two times, Shaun spoke once, and the remaining three not at all.
- Here is the map of the classroom showing where you stood and moved. Each circle represents students with whom you interacted. During the presentation, you stopped and interacted with Eric five times, Paula three times, and Maryann two times.

C. Inferences, Causality, and Interpretations

The coach may make his/her own interpretations of the lesson or state a causal relationship

- Your explanation helped the students understand their task. (Inference)
- The criteria you developed guided their capacity to self-evaluate. (Causality)
- The concept you were working on became clearer with each example you gave. (Interpretation)

Notice that all the statements above are from the coach's point of view. When the coach makes these statements, it usurps the self-directedness of the coachee and is, therefore, counterproductive and should be avoided. A higher degree of self-directedness is achieved when the coach invites the coachee to make such causal relationships, inferences, and interpretations.

D. Personal Opinions and Preferences

The coach states his/her own opinion or likes/dislikes. Notice again that all the comments are from the coach's perspective and point of view. These types of statements build dependency on the part of the coachee. It signals that success of the event is determined by the effect of the lesson on the likes and dislikes of the coach rather than the achievement of the outcomes of the lesson intended by the coachee. Again, such statements usurp the desired self-directedness of the coachee. For example:

- You really made me laugh with you told that story.
- I like it when you showed the video clip. I was really engaged.
- I think the kids enjoyed the lesson.
- Your story reminded me of when I was in school.

E. Evaluations and Judgments

Although the coach may think that making judgments—either positive or negative—is helpful or reinforcing for the coachee, in actuality, the opposite is true. The coach who makes value-laden comments sends a signal that he/she is judgmental and being evaluative. Such comments shift the focus from coaching to evaluation. They cause mistrust, deplete creativity, and create fear. Such comments halt self-directedness and self-evaluation on the part of the coachee because the coach makes the evaluations rather than inviting the coachee to self-evaluate. Notice that all comments contain value-laden words. For example:

- Your lesson went well.
- You did a great job.
- Your PowerPoint presentation was very effective.
- The way you got kids involved really worked.
- Your questioning strategy could be improved.

Obviously, posing mediating questions and providing nonjudgmental data are the most supportive of self-directedness. Making interpretations, causal relationships, and inferences—along with personal opinions and value judgments—all rob coachees of performing those cognitive processes for themselves and thus are to be minimized and avoided. You are invited to monitor your own language tools of providing feedback. Which are most common with you? How does your choice fit with your intentions?

STRUCTURING

Structuring is defined as the many ways in which a coach clearly communicates expectations about purposes and the use of such resources as time, space, and materials. Where the coach and colleague sit during a coaching conversation can influence a feeling of a co-creating or a supervisory relationship.

Verbal structuring should be clear, conscious, and deliberate. It should be based on a common understanding of the purposes for the coaching, the roles the coach should play, the time allotments, the most desirable location for the conversation, and the placement of the coach during the observation. Examples of verbal structuring behavior include the following:

- Since you want me to observe the group members' participation, I'll need to sit in a place in the room where I can see them all during the meeting. Where would that be?
- It will take us about 15 minutes for a planning conversation. Let's set a time that would be convenient for both of us.
- Here is a copy of the map of the coaching cycle. During the planning conversation, I'll be asking you to share with me your goals and objectives, which instructional strategies you plan to use to achieve those goals, and by what indicators you'll know your students achieved your purposes. I'll also be asking you to give me directions for what to look for during the lesson that will be of assistance to you.

MEDIATIVE QUESTIONING

> It's not the answers that enlighten us, but the questions.
>
> —Decouvertes

Mediative questioning is intentionally designed to engage and transform the other person's thinking and perspective. These kinds of questions meet at least four criteria:

- They are invitational in intonation and form.
- They engage specific complex cognitive operations.
- They are intentional.
- They address content that is either external or internal to the other person.

Invitational

An approachable voice is used with invitational questions. There is a lilt and a melody in the questioner's voice rather than a flat, even tenor. Plurals are used to invite multiple rather than singular concepts (You will also observe this in video 3-4):

- What are your *goals* for this project?
- What *ideas* do you have?
- What *outcomes* do you seek?
- What *alternatives* are you considering?

Words also are selected to express tentativeness:

- What conclusions *might* you draw?
- What *may* indicate his acceptance?
- What *hunches* do you have to explain this situation?

Invitational stems (dependent clauses and prepositional phrases) are used to enable the behavior to be performed:

- As you think about . . .
- As you consider . . .
- As you reflect on . . .
- Given what you know about the children's developmental levels . . .

Both invitational stems and the paraphrase that precedes the question often employ positive presuppositions. Positive presuppositions assume capability and empowerment. A presupposition is something that a native speaker of a language knows is part of the meaning, even if it is not overtly present in the linguistic structure of the communication. We sometimes receive these deeper meanings unconsciously because they are not communicated by the surface structure of the words and syntax.

Subtle, and not-so-subtle, presuppositions embedded in our statements carry meanings that can either hurt or support others. For example, the statement, "Even Bill could pass that class!" conveys the ideas that (1) Bill is not a great student, and

(2) the class is not challenging. Neither of these pieces of information is present in the surface structure of the sentence. Still, the sentence communicates, "Even Bill [who is not a great student] could pass that class [so therefore the class can't be that challenging]!" The two unstated pieces of information are inferred by the listener as presuppositions, or assumptions underlying the sentence (Elgin 1980).

By paying attention to the presuppositions we use and choosing our words with care, we more positively influence the thinking and feelings of others with whom we communicate. For example, consider the positive presuppositions in the following sentences:

- "What are some of the benefits you will derive from this activity?" (Presuppositions: You can anticipate outcomes; you will derive benefits; your thoughts are more important than my own ideas.)
- "As you anticipate your project, what are some indicators that you will be aware of that tell you that they are progressing and succeeding?" (Presuppositions: You are someone who plans and anticipates; you will progress and succeed; your indicators are more important than anything I might suggest; you are alert to cues and are self-monitoring.)
- "Given your experience and knowledge of learning styles, what are some of your ideas about this?" (Presuppositions: You are experienced, your experience has value because you have reflected on and grown from it, you know about learning styles, you can generate ideas; your ideas have great value.) The phrase "your experience and knowledge of learning styles" tends to pull into short-term memory content with which to process the question.

Specific Cognitive Operations

Questions invite different levels of complexity of thinking. Embedded in questions are certain syntactical cues that signal and invite behavior or thinking processes. Skillful coaches deliberately use these linguistic tools to engage and challenge complex thinking. (This will be further elaborated in chapter 8; see figure 8.1).

Table 3.1 illustrates this process. In the first column, Input, are the data-gathering cognitive operations. In the second column, Process, are the cognitive operations by which meaning is made of the data. The third column, Output, invites speculation, elaboration, and application of concepts in new and hypothetical situations.

Table 3.2 shows some examples of questions intended to engage certain cognitive operations.

They Are Intentional

Questions can have a purpose to explore thinking or specify thinking. When listening to the thinking of the coachee, the coach decides which purpose will better serve

Table 3.1. **Cognitive Operations**

Input	Process	Output
Recall	Compare/Contrast	Predict
Define	Infer	Evaluate
Describe	Analyze	Speculate
Identify	Sequence	Imagine
Name	Synthesize	Envision
List	Summarize	Hypothesize

Table 3.2. **Cognitive Operation Questions**

Cognitive Operation	Question
Identify	Who, specifically . . .?
Values/Beliefs	What do you believe about . . . ?
Relevance/Justification	How is this important to . . . ?
Intentionality	For what purposes . . . ? Toward what ends . . . ?
Metacognition	What were you thinking when . . . ?
Behavior	What will you be doing when . . . ?

Temporality:

Simultaneity	While . . . ?
Synchronicity	During . . . ?
Duration	How long . . . ? For what period of time . . . ?
Rhythm	How often? How frequently?
Sequence	What came before? What comes after? What comes first, second, third?

Flexibility:

Perspective	How would you feel if . . . ?
Alternatives	How else might you . . . ?
Evidence	How will you know if . . . ? What evidence supports . . . ?
Predictions	If you were to . . . , what do you predict would happen?
Causality	What did you do to cause . . . ? What produced . . . ?
Data Use	Of what use will you make of these data? What would that information tell you?
Applications	What will you take from this? How will you apply this elsewhere?
Evaluative Criteria	What criteria will you use to . . . ? By what standards will you judge . . . ?

the thinking of the coachee. If the thinking of the coachee is vague and imprecise, a question for specificity might be posed to invite craftsmanship. For example, a person might say, "My goal for this year is to have a more collaborative team." The coach would offer a question inviting clarity about what is meant by collaboration, such as, "What does collaboration mean for your team?" or "What are some things you might focus on in becoming more collaborative?"

In another case, a teacher might know what it means to be collaborative and would be served by a question to broaden thinking. The coach could respond to the same teacher goal related to collaboration by asking, "If you were to become more collaborative as a team, what might be some of the results you would anticipate for your students?"

To view examples of meditational questions, please go to http://www.thinkingcollaborative.com/supporting-audio-videos/3-4. Jane, the coach, asks Shatta, a Planning Central Office director, two meditational questions: "How do you make decisions? What do you look for in the data?" In addition to identifying elements of a meditation question, notice the invitational tone of Jane's voice.

In video 3-4.2, during a calibrating conversation. Sue, the coach, asks Carolyn a mediating question about values. In video 3-4.3., during a reflecting conversation, Carolee, the coach, poses a mediating question for Ochan, exploring Ochan's expression of fear (internal content) of goals (external content).

A coach is more impactful in mediating thinking when there is consciousness about the intention of the question and what kind of thinking is being invited.

In video 3-4.1 the coach's intention is to invite the coachee to think about his decision-making process. Initially the coachee exercises *recall*, then with a follow-up question, he *analyzes* his decision-making process. In video 3-4.2, the coach's question raises the teacher's values to the conscious level and relates those values to his reasons for selecting certain aspects of his teaching to improve. In turn, this permits the teacher to become even more intentional and therefore effective in selecting improvement strategies. In video 3-4.3 the coach helps the teacher access his emotions (fear) and their relationship to personal goals. Fear is located in the amygdala, the emotional center of the brain, and the intention of this question is to shift brain activity to the neocortex, the rational center of the brain.

External or Internal Content

External content is what is going on in the environment around, and thus outside, the person. Internal content is what is going on inside the other person's mind and heart: satisfaction, puzzlement, frustration, thinking processes (metacognition), values, intentions, or decisions. Questions that most effectively mediate thinking link internal content with external content.

Applying Questioning Criteria to Mental Maps

> To learn new habits is everything, for it is to reach the substance of life. Life is but a tissue of habits.

—Henri Fredric Amiel

Drawing on the mental maps of the planning and reflecting conversations (see chapter 11) and on the characteristics of effective questions, appendix C provides some examples of questions that a coach might ask with the intention of eliciting and evoking each of the intentions in the coaching cycle.

CONCLUSION

Mediators are clear about their purposes and intentionally employ certain verbal and nonverbal tools with others to help transform and empower their cognitive functioning. In this chapter we have presented the verbal and nonverbal tools that serve a mediator in achieving the goals of helping others to become increasingly more self-directed. Although these skills are presented in the context of Cognitive Coaching, they are useful in any dialogue between two human beings desiring to grow together, to plan, to reflect, to solve problems together, and to create deeper meaning and understanding.

For some, the behaviors described in this chapter come naturally. For others, it will take time and practice to become proficient and comfortable using these tools. We recommend that coaches isolate certain skills and consciously practice them rather than attempt to learn them all at once.

If we choose to intentionally employ, over time, the Cognitive Coaching values described in part 1, the mental maps described in part 3, and the mediative tools described here, we can become increasingly skillful in facilitating the self-actualization of others. As our colleagues assume greater responsibility for coaching themselves, others will come to identify us as mediators of self-directed learning.

4

Mediating for Self-Directed Learning

The self is not a thing, but a point of view that unifies the flow of experience into a coherent narrative—a narrative striving to connect with other narratives and become richer.

—Jerome Bruner

The ultimate goal of Cognitive Coaching is self-directed learning, which means to self-manage, self-monitor, and self-modify. With Cognitive Coaching, the mediator helps to engage and enhance a colleague's cognitive and emotional capacities to develop self-directed learning. Ultimately, mediators work to modify another person's capacity for continuous learning and the person's propensity to become self-coaching.

THE PROCESSES AND ROLE OF MEDIATION

The word *mediate* is derived from the word *middle*. Therefore, mediators interpose themselves between a person and some event, problem, conflict, challenge, or other perplexing situation. The mediator intervenes in such a way as to enhance another person's self-directed learning.

Human learning is a matter of strengthening internal knowledge structures. Planning for and reflecting on experience activates these knowledge structures. With mediation, existing knowledge structures can be made more complex through more connections. The structures can also be altered to accommodate new understandings, or they can be made obsolete because some new experience has caused the creation of a new knowledge structure. This sifting and winnowing of prior knowledge structures constitutes learning.

Reuven Feuerstein (1999, 279) states the following in "Mediated Learning Experience":

63

Mediated learning is an experience that the learner has that entails not just seeing something, not just doing something, not just understanding something, but also experiencing
that thing at deeper levels of cognitive, emotional, attitudinal, energetic, and affective
impact through the interposition of the mediator between the learner and the experienced object or event (stimuli). In such a context, learning becomes a deeply structured
and often a pervasive and generalizable change.

THE MEDIATOR'S ROLE

Mediators influence the intensity, flow, directionality, importance, excitement, and
impact of information coming to the person being coached. According to Feuerstein,
the mediated learning experience transforms the information that impinges on a
learner and enters his repertoire in a totally different way. Rather than give advice
to or solve problems for another person, a mediator helps the colleague to analyze a
problem and develop his/her own problem-solving strategies.

A mediator helps a colleague to set up strategies for self-monitoring during the
problem-solving process. Acting as a sounding board, a skilled mediator helps another person become more self-directed with learning. A mediator also:

- Is alert to the mediational moment—usually when a colleague is faced with a
  complex task, dilemma, discrepancy, or conflict. Often, the colleague exhibits
  tension and anxiety, the resolution of which is not immediately apparent.
- Facilitates mental processes for others as they solve their own problems, make
  their own decisions, and generate their own creative capacities.
- Invites the colleague to reflect on and learn from the problem-solving process to
  find applications in future problem situations.
- Helps others to become continuous learners and self-mediators
- Maintains faith in the human capacity for continued intellectual, social, and
  emotional growth.
- Possesses a belief in his or her own capacity to serve as an empowering catalyst
  for others' growth.

How Mediation Affects the Brain

> [The] talking cure can physically change the brain and . . . anytime you have a
> change in behavior you have a change in the brain.
>
> —Lewis Baxter, psychiatrist, UCLA

Feuerstein (1999, 276) believes that mediation produces new connections in the
brain. He states:

One of the most interesting and exciting aspects of mediated learning . . . is that the quality of interaction not only changes the structure of behavior of the individual, not only changes the amount and quality of his repertoire, but—according to increasingly powerful sources of evidence from fields of neurophysiology and biochemistry—changes the structure and functioning of the brain itself in very meaningful ways.

This idea finds support from many quarters. Neuroscientist Gerald Edelman proposes that the brain reconstructs itself from experience. One commonly understood example of this is the neural pruning that occurs within the first two years of life (Shonkoff and Phillips 2000), cauterizing neural capacities for distinguishing sounds outside one's own language group. Ornstein (1991) claims, "To make a personal change, we have to be able to observe the automatic workings inside ourselves." This requires the kind of consciousness evoked by mediation. He describes the brain as having a neural selection system that wires up the nervous system differently, depending on the demands on the organism. Managing and developing the mind is to bring automatic processes into consciousness. From another quarter, neuroscientists have found that the flashes of insight that precede "Aha!" moments are accompanied by a large burst of neural activity in the brain's right hemisphere (Pink 2006). Furthermore, Damasio (2010, 271) reports that constant practice produces a skill set that become "second natured" into the cognitive unconscious. Consciously the mediator pays attention to the goal, allowing more automatic skills held at unconscious levels of competence to manage other aspects of the conversation. While this is true about developing as a mediator, it is also true that the person being coached, over repeated events, begins to internalize the coaching maps used by the mediator as his/her own approach to planning and examining his/her teaching.

Three Resources of Mediation

Three resources are constantly monitored and sustained in the mediation process: reciprocity, clear intentions, and vision.

1. *Reciprocity*: Skillful mediators maintain mutually cooperative, trusting relationships. The mediator holds and demonstrates the utmost respect for the individual's feelings, ideas, and perspectives.
2. *Attention to intentions*: When assuming the identity of a mediator, one must also be very clear about intentions. Mediators monitor their own values and realize that although it might be tempting to solve a problem, evaluate, or give advice, they must hold those thoughts in abeyance in order to allow the other person to resolve the situation. Mediators are not "fixers." Instead, they have faith that others can solve their own problems. Trying to "fix" others detracts from a trusting relationship.

3. *Desired state*: The mediator and the individual also need a clear vision of a more desirable state for the individual. For example:

- If the person is experiencing a feeling of helplessness, the vision would be to help him achieve a feeling of efficacy.
- If the person is struggling with vagueness or a lack of clarity, the vision would be for her to become clear about goals, outcomes, vocabulary, strategies, or definitions.
- If the person is unaware of his actions and values and their effects on others, the vision would be to build greater consciousness of those actions, values, and effects.
- If the person is holding on to a rigid stance or narrow perspective, the vision could be for her to think more flexibly, more broadly, and from alternate perspectives.
- If the person is experiencing a feeling of isolation, the vision would be for him/her to develop stronger, interdependent relationships, connectedness, and a sense of affiliation and belonging.

Thus, Cognitive Coaching is conceived as an exercise in strengthening the neural circuitry that enhances a colleague's capacity for cognitive control and self-regulation.

WHY MEDIATION WORKS

Four theorems guide the Cognitive Coach's beliefs and actions and explain why mediation works:

1. The sum of an individual's constructed meanings resides internally at conscious and unconscious levels and serves as the criterion for perceptions, decisions, and behavior.
2. When these meanings are given form in language, they become accessible to both parties in a verbal transaction.
3. Through verbal transaction (mediation), these meanings and the perceptions, decisions, and behaviors related to them can be refined, enriched, and modified.
4. Through Cognitive Coaching, not only are an individual's meanings, decisions, and behaviors refined and modified, with related results in improved performance, but also refined is the individual's capacity to self-mediate and to become more proactive in continuing self-directedness.

Who Can Be a Mediator?

Skilled in the beliefs, maps, principles, and tools of mediation, anyone can assume the role of mediator for another's self-directed learning. Thus, mediation can occur between two students, two teachers, two administrators, a leader and a group member, or a teacher and a student—assuming the skills are present.

When a coaching relationship is established between two professionals with similar roles, or peers, it may be referred to as peer coaching. Cognitive Coaching has a more specific meaning; it refers to the identity that mediators assume, the coaching maps and tools with which they work, the desire for enhancing others' self-directedness that they embrace, and the faith in the human capacity for meaning making that they cherish. Peer coaching describes with whom you coach; Cognitive Coaching defines how you coach.

In an increasing number of educational communities, custodians, school secretaries, bus drivers, parents, students, and cafeteria workers are learning the skills of Cognitive Coaching. In one Michigan school district, the director of maintenance coaches an elementary principal, who in turn coaches a teacher. In a California school district, a superintendent receives monthly coaching from a mentor-teacher. In an international school in Malaysia, the head of schools is coached by a special education teacher.

The Mediational Moment

We all have many opportunities to engage others in problem solving. The challenge, however, is in deciding among these three questions:

- Can I, do I wish to, and should I provide someone with solutions?
- Can I, do I wish to, and should I collaborate and assist others with ideas or information?
- Can I, do I wish to, and should I assist others in learning how to better solve their problems themselves?

For example, it is *not* a mediational moment if a teacher has a problem with a leak in the classroom ceiling; the principal should arrange for the appropriate maintenance personnel to take care of the problem. If, on the other hand, the teacher's challenge has to do with implementing a new teaching strategy, determining a more effective behavior management technique, devising an innovative way to deal with student diversity, or optimizing the use of a new technological tool, mediation would be a better way to help the teacher generate ideas for planning, implementing, and resolving the challenge.

Many people are not accustomed to seeing themselves as mediators. Because they may have been hired because of their history of problem-solving prowess, they may try to solve others' problems instead of facilitating the problem-solving process. They may give advice instead of helping others to self-prescribe. They may evaluate rather than help a person to self-evaluate.

There are advantages and disadvantages to performing as an advisor for others. Some advantages are the following:

- The problem will be solved (or appear to be solved) swiftly and efficiently.
- The solution will be congruent with your beliefs and values.
- You may feel satisfaction for having been of help.
- Others may perceive you as an effective leader.
- You will learn more about the process of problem solving.

There are also some disadvantages to acting as an advisor:

- The problem you have "solved" may be a surface manifestation of deeper issues that have not been resolved.
- The other person may become dependent on you to solve future problems or blame you if a satisfactory resolution is not achieved.
- The other person will have learned little about problem-solving processes.
- The other person may not take (or actually undermine) your suggestions for solving the problem.
- The other person may build resentment because he views himself as inadequate and helpless.
- The other person will be robbed of the joy of experiencing the "Aha!" moment (Treadwell 2014).
- You will miss the opportunity for developing another's capacities for self-directedness.

A Mediator Intervenes

As described earlier in this chapter, Cognitive Coaching is an exercise in developing in others their neural circuitry that enhances their capacity for cognitive control and self-regulation. We defined self-directedness as self-managing, self-monitoring, and self-modifying. Being conscious of their purpose to create self-directedness in others, mediators seize opportunities, both planned and informal, to use their behaviors deliberately to achieve these goals. For example:

Intervening for Self-Managing

When a colleague is planning an event—such as teaching a lesson, conducting a professional development meeting, anticipating a parent-teacher conference, or envisioning a school board meeting—the mediator, understanding the role of language as a meditational tool, carefully selects and composes questions intended to engage the mental processes of self-management. Those questions are invitational, open ended, and cognitively complex. Some examples are shown in table 4.1.

Table 4.1. Mediator's Questions for Self-Management

Self-managing people approach tasks with:	Examples of mediational questions:
Clarity of outcomes	What specifically do you want students to know or be able to do as a result of this lesson?
A strategic plan	How does this lesson today contribute to your long-range goals or outcomes for your students?
Necessary data	Given what you know about your students' various learning styles, what alternative strategies are you considering?
Knowledge drawn from past experiences	What effective strategies have you used before in situations like this?
Awareness of alternatives for accomplishment	What might some alternative instructional strategies you have in mind to accomplish these goals?
Anticipation of success indicators	What might you hear students saying or see them doing that will let you know that you've reached your goals?

Mediator's Questions for Self-Management

Furthermore, the coach poses questions so that during the event, the planner's perceptions will be alerted to cues that would indicate whether the strategic plan is working or not and to assist in the conscious application of metacognitive strategies for deciding to alter the plan and select alternatives (see table 4.2).

Table 4.2. Mediator's Questions for Self-Monitoring

Self-monitoring people:	Examples of mediational questions:
Strive to become spectators of their own thinking and actions.	What will guide your decisions about . . . .? What will you pay attention to in your own teaching behaviors that you are striving to perfect?
Establish metacognitive strategies to alert the perceptions for in-the-moment indicators of whether the strategic plan is working or not and to assist in the decision-making processes of altering the plan.	What are some of your predictions about how this lesson will turn out? What cues will you be aware of as indicators that this lesson is working? What alternatives do you have in mind for this lesson?

Mediator's Questions for Self-Monitoring

After the event, and if conditions are favorable, the mediator will also take advantage of this opportunity to invite reflection and to maximize insightfulness and meaning making from the experience (see table 4.3). Meaning is made by analyzing feelings and data, comparing results with expectations, finding causal factors, and projecting ahead to how these insights may apply to future situations. Self-modifying implies making a commitment to employ those insights and meanings autonomously in future events and situations (Kegan and Lahey 2001; Kegan 1994).

Table 4.3. Mediator's Questions for Self-Modifying

Self-modifying people:	Examples of mediational questions
Reflect on	As you reflect on this lesson, what feelings (or thoughts, or impressions) do you have about it?
Recall	What are you recalling that leads to those impressions? What important decisions did you make during today's lesson? What prompted those decisions?
Evaluate	What is your sense of the lesson's effectiveness?
Analyze	What might be some factors that contributed to the lesson's success? How does what happened in the lesson compare to what you predicted would happen?
Construct meaning from the experience	As a result of this lesson, what insights or new learnings are you forming?
Apply the learning to future activities tasks and challenges	As you anticipate future lessons of this type, what big ideas will you carry forth to those situations?

Mediator's Questions for Self-Modifying

Furthermore, because the mediator has an explicit intention of helping the colleague become self-coaching by internalizing this process, the coach will invite the colleague to reflect on the process that has just been experienced. Questions intended to surface the benefits of the process might be:

"As you reflect on this coaching process, what insights have you gained?"
"What were you aware of in our interaction that produced those insights?"
"What coaching questions might you ask yourself?"
"If, for some reason I was unavailable to you, what parts of this coaching process might you conduct on your own?"

FOUR CAPABILITIES OF A MEDIATOR

The term *capabilities* refer to the mediator's metacognitive processes, particularly the four named later in this section. The settings in which Cognitive Coaching may occur range from informal, spontaneous conversations to more formal, planned conferences. Sometimes opportunities for mediation present themselves spontaneously and informally in day-to-day life, such as during a conversation in the faculty room or in the hall on the way to class. More formal, planned events are also opportunities for mediation, such as a formal classroom observation of teaching and learning or a scheduled department or faculty meeting.

A coach seizes all these opportunities to use specific mediational skills to engage and develop the other person's thinking processes. Mediators need certain skills, attitudes, and capabilities to perform their role well. For example, a mediator uses language with the intent of causing a change in the other person's reality. These linguistic tools include the following: posing questions intended to engage and transform the mind, creating conditions of trust, envisioning a desired state of mind, remaining nonjudgmental, and resisting the tendency to solve the problem for the learner. (A description of these specific tools can be found in chapter 3.) Following are descriptions of four specific capabilities, or metacognitive attributes, of a mediator.

Knowing One's Own Intentions and Choosing Congruent Behaviors

Behaviors can be either reactive or proactive. Proactive behaviors are based on intention or goal awareness. The ultimate goal of mediation is to help an individual become self-mediating. With that end in mind, mediators are clear about their intentions in the moment. Perhaps they want to reflect understanding, clarify a communication, help the colleague to feel comfort, or cause self-examination. Choosing behaviors that support intentions like these requires the following:

- Being conscious of one's intention in the moment and how that intention serves a greater goal
- Being alert to a colleague's verbal and nonverbal cues
- Having a repertoire of mediational tools
- Knowing how a particular tool may serve a specific intention
- Being able to use that tool with a great degree of craftsmanship
- Anticipating and searching for the effects a tool produces in a colleague

Setting Aside Unproductive Patterns of Listening, Responding, and Inquiring

> In trying to really listen, I have often been inspired by the Zen Master Suzuki-roshi, who said: "If your mind is empty, it is always ready for anything; it is open to everything. In the beginner's mind there are many possibilities."
>
> —Sogyal Rinpoche

Mediators monitor and manage their own listening skills by devoting their mental energies to the other person's verbal and nonverbal communications. To listen with such intensity requires holding in abeyance certain normal, tempting, but unproductive behaviors that might interfere with the ability to hear and understand a colleague. These are normal ways that human beings listen, but as a coach, one must discipline oneself to set them aside because they interfere with mediating thinking. For example:

Autobiographical listening occurs when the brain exercises its associative powers and the colleague's story stimulates the coach to think of her own experiences. Derber (1979) refers to this as narcissistic listening. If a colleague mentions recent automobile repairs, and, reminded of your car's most recent breakdown, you talk about that, you are engaging in autobiographical listening. Coaches set this type of listening aside as soon as they become aware that their attention has drifted into their own story. Besides being distracting, autobiographical listening may stimulate judgment in which negative or positive experiences prejudice listening. Autobiographical listening may also stimulate comparison, in which comparing the situations can further distract the coach. Finally, autobiographical listening may spark immersion, in which we are lost in attentiveness to our own story.

Inquisitive listening occurs when we begin to get curious about and delve into portions of the story that are not relevant to the problem at hand. Knowing what information is important is one critical distinction between consulting and coaching. As a consultant, a person needs lots of information in order to "solve the problem." As a coach, a person needs only to understand the colleague's perspective, feelings, and goals and how to pose questions that support self-directed learning. *Scrutinizing* is also a by-product of inquisitive listening. With scrutinizing, curiosity about that which is not relevant to the mediational moment sinks the conversation into a hole of analytical minutiae that may cause a coach and colleague to lose sight of the larger issue.

Mind reading is sometimes a by-product of inquisitive listening. With mind reading, we try to figure out what someone is really thinking and feeling. Mind reading does not allow us to pay sufficient attention to what a partner is saying.

Solution listening is what we have a tendency to do when we serve as problem solvers for another person. Because we may view ourselves as experienced and successful problem solvers, ready with help and eager to give suggestions, we immediately begin searching for the right solution to a problem. When coaching, however, thinking of solution approaches as your colleague speaks interferes with understanding the situation from the colleague's perspective. It also interferes with formulating mediational moves.

Filtering is often a by-product of solution listening in which we listen to some things and not to others, paying attention only to those ideas that support the solution approach we are developing. This is a common problem for physicians, who must work hard to stay open to possibilities during the early stages of diagnostic visits with a patient. Rehearsing, too, can be a by-product of solution listening, as our attention gets focused on preparing and crafting the way that we are going to present a solution.

Adjusting One's Own Style Preferences

Mediators understand a fundamental and common principle: Humans differ. Distinct patterns and cognitive styles of perceiving and processing information are neither good nor bad. They transcend race and culture, characterize males and females equally, and are observable at all age levels (Markus and Conner 2013). A person's style influences not only what information he or she selects to attend to, it also puts it all together into a coherent experience. Style makes your playlists.

Conscious of these differences, Cognitive Coaches strive to be flexible communicators. They recognize their own style preferences and seek to overcome these habits when interacting with someone who operates from a different style.

Setting aside their own style preferences, coaches observe, respect, appreciate, are open and humble, let go of judgment, are authentic, are collaborative, are caring, and are compassionate (Seagal and Horn 1997). To achieve this, coaches constantly sense, search for, and detect linguistic, facial, and bodily cues about the person with whom they are working. Coaches' mirror neurons align themselves naturally with their colleague's. Because of these mirror neurons and understanding the persons they work with, coaches constantly expand their repertoire so they can match their style to a variety of situations and individuals. (Style flexibility is elaborated in chapter 10.)

Navigating through the Coaching Maps and Support Functions

Four basic coaching maps provide the coach with information about the mental territories of planning, reflecting, problem resolving, and calibrating. Similar to a road map, mental maps provide the coach with directions to destinations as well as options for alternate routes. (See part 3.) The coach selects one or more support functions to guide his/her decisions about how best to support self-directedness. The coach's identity as a mediator of thinking creates consciousness about beginning and ending the conversation as a coach.

CONCLUSION

The concepts in this chapter invite a shift from the present paradigms. For many educators, a dominant sense of satisfaction has come from their expertise as problem solvers. The shift to a mediational identity creates a feeling of being rewarded by facilitating others to solve their own problems. The shift is from teaching others to helping others learn from situations; from holding power to empowering others; from telling to inquiring; and from finding strength in holding on to finding strength in letting go. Changing one's identity requires patience, stamina, and courage. Coaches therefore are committed to and believe:

- I have faith that deep within each person lie the solutions to their own problems. I give attention to my intentions as a coach to help my colleagues find the keys to unlocking the doors to where those solutions reside within them.
- My effectiveness as a coach is not measured in how much advice I give and is taken and applied nor is it in providing solutions to others' problems. Rather, my effectiveness is measured by the degree to which others become more autonomous, efficacious problem solvers themselves.
- My coaching toolbox is always with me in my mediative oral and nonverbal language.
- I am constantly alert to opportunities to mediate others.
- As a continuous learner, I, too, seek coaching in an effort to become more efficacious, more flexible, more conscious, more craftsman-like, and more interdependent.
- I am optimistic about the potential for all individuals to achieve higher-order mental processes.

Repertoire of Support Functions

Whatever we can do to facilitate learning on the one hand and loving on the other is important, because those are the most healing forces available to us.

—Na'im Akbar

No single form of support serves all purposes. Elements of Cognitive Coaching have been applied in many different forms of support models. Some models are technical, some are humanistic, and some are developmental or reflective. Since Joyce and Showers's (1980) initial work on "coaching" to support the implementation of new initiatives, coaching models and practices have proliferated. Garmston (1987) described three models of peer coaching—not what you do but who you do it with—as technical, collegial, and challenge. Some terms used today are reading coach, literacy coach, instructional coach, math coach, data coach, and peer coach. Ed Pajak (2000) created a summary of models that distinguishes the unique features of many programs, but all of these approaches have certain tenets in common. Among them is the belief that teaching is "untidy" and uncertain. Structured collegial conversations to help make meaning from complex instructional situations, and reflective conversations help to generate knowledge, expand teaching repertoire, and promote teacher development. Lipton and Wellman (2012, 5) add that:

Skillful teachers manage the social, emotional, and academic needs of increasingly diverse student populations. Total classroom awareness requires attention to these three dimensions while simultaneously tracking the lesson plan, content accuracy, use of examples, clarity of explanations and directions and choice of language to match student readiness. These teachers provide relevant and meaningful tasks, attend to momentum and pacing while purposefully monitoring student understanding, making adjustments as needed. And all of this is orchestrated for individual students, small groups and the full class.

This chapter considers a support provider's need for situational flexibility to achieve the mission of developing high-performing, self-directed individuals. Specifically, we answer a variety of questions about the many different kinds of support provided to teachers. For example, when does one coach, and when does one consult? Can these functions ever be blended? Isn't everything evaluation? Can one person provide support *and* evaluate? If so, how can that be achieved?

SITUATIONAL FLEXIBILITY

In chapter 1, we described four types of support provided to teachers as they mature in the profession: coaching, collaborating, consulting, and evaluating. Table 5.1 summarizes the purposes and intended outcomes of three of these.

As can be seen in tables 1.1 and 5.1, these forms of interaction are guided by consciously chosen purposes and outcomes. These support functions share a set of skills (described in chapter 3) and, for our purposes, exist within a framework of beliefs and concepts that we call Cognitive Coaching. Therefore, each is used as a stance within a support provider's repertoire for its contribution to the development of professional expertise and self-directedness.

Skillful coaches may depart from Cognitive Coaching periodically to conduct these other forms of interactions. Because they continually strive to fulfill their identity as mediators, however, they consciously begin with and return to the beliefs, values, principles, maps, and tools of Cognitive Coaching as their "default position."

AN EXEMPLARY SUPPORT SYSTEM

Sometimes we are asked, "When coaching, is it all right to give a suggestion? What if you sense the teacher's plan will not work? Should you tell him?"

Table 5.1. Purposes of Coaching, Collaborating, and Consulting

Cognitive Coaching—a relationship with a mediator of thinking	Collaborating—a relationship with a colleague	Consulting—a relationship with an expert
• Enhance and habituate self-directed learning: self-managing, self-monitoring and self-modifying.	• Solve instructional problems, apply and test shared ideas, learn together.	• Increase pedagogical and content knowledge and skills. Institutionalize accepted practices and policies.
• Transform effectiveness of instructional decision making to maximize effectiveness of actions toward reaching goals.	• Form ideas, approaches, solutions, and focus for inquiry.	• Inform regarding pedagogy, curriculum, policies, and practices.
• Habituate reflective practice and build capacity for professional effectiveness.	• Gather and interpret data to inform collaborative practice.	• Provide technical assistance and improve teaching.

Coaching, collaborating, and consulting each serves a valued purpose to the teacher, the institution, or both. Ideally, each takes place in transactions devoted to only one of these functions at a time. There are situations, however, that call for an *occasional*, skillful transition to another function. There are no simple rules to guide the coach, but there are some prerequisite conditions.

Of paramount importance is interpersonal trust and trust in the process. Next is a collegial relationship between the coach and the person being coached. Finally, the coach needs exquisite coaching skills; to consult, he/she also needs a rich knowledge base of the topics under consideration.

For one to successfully support teachers across each of these support functions, skill in conducting classroom observations is a necessary component. Being able to recognize, along a developmental continuum, classroom manifestations of whatever set of teaching standards the district has developed is essential. Also needed are skills at using a variety of data-gathering instruments. Because an important goal in working, especially with new teachers, is to develop independence, not dependence, collaboration is used as a vehicle for developing the beginning teacher's professional independence as well as signaling working together as a value in successful schools.

For beginning teachers, teacher coaches are lifelines to survival. Coaches communicate a reaffirming certainty that each beginning teacher is okay, has the capacity to survive and learn, and contributes value to students' education. Such consistently positive presuppositions are especially important during the early months, when beginning teachers encounter feelings of overwhelming self-doubt and inadequacy (Moir, Barron, and Stobbe 2001).

Seamlessness is achieved when many coaching practices are applied in either consulting or collaborating. For example, trust and rapport are fundamental to any helping relationship. Paraphrasing, with its profound influences on the chemistry of resourcefulness, is prominent in each of the support functions of collaborating, consulting, and coaching. In consulting, as in coaching, there is a need for data. Good practices in either function require clarity about what data to gather, how to collect it, and how it will be useful. Data may sometimes be reported within the consulting role without interpretation from the consultant, as is true in coaching (see chapter 3). During consulting, clarity is needed about the teacher's goals. A consultant, like a coach, will use open-ended questions and pause, paraphrase, and pose to elicit thinking. Given these common features, the question "How do you know when to segue from coaching to consulting, and how do you do it?" is sometimes met with puzzlement.

When to Consult

How does a support provider know when to switch from coaching to consulting? One coach describes her thinking process as follows. First, she knows the teacher well enough to detect when the teacher lacks necessary information or ideas. Second, based on her knowledge of what has worked for this teacher in the past (including the teacher's information processing style and the degree of risk to the teacher in acting on possibilities in this situation), she knows the teacher has the capacity to implement ideas she might offer. Third, the teacher may request help.

The following is a description of how coaching, collaborating, and consulting might be woven together over a three-week period with a beginning teacher.

1. Coach: The teacher expresses interest in guided reading. The support person invites reflection about her intention, values, goals, and planning.
2. Consult: The teacher realizes there are gaps in her knowledge about how to conduct guided reading. The support person shares ideas, locates information, and gives the teacher an article about guided reading.
3. Collaborate: The support person discusses the article and engages the teacher in co-planning. She and the teacher share ideas about how to get started.
4. Consult: At the teacher's request, the support person models a guided reading lesson. When asked for strategies, she offers some for consideration.
5. Coach: The teacher asks the support person to observe. The support person, fully into the coaching role, gathers requested data and conducts a reflecting conversation after the lesson.
6. Collaborate: The support person and the beginning teacher examine student data together and determine an area for future instruction. Together they plan the next lesson.

The best coaches see their work as a dance in which they are constantly deciding when to turn, dip, or bop while maintaining a partnership with the teacher.

When coaching, support providers may be especially generous with silence. They also might ask teachers to elaborate on the values and beliefs they hold about learning. Sometimes they allow teachers to "fail forward" in order to develop rich learning from lessons that did not go as the teacher wanted. Efficacious people who "fail forward" learn from their experiences and apply it to new situations. Coaches also ask "take-away" questions for the teacher to ponder after the coaching conversation. Such questions do not require an answer in the moment. Rather, they are intended for later reflection. Occasionally, coaches model their own reflective thought.

Practices unique to collaboration might include physically helping a new teacher to arrange a classroom or supplies, procuring materials, demonstrating, co-planning, or co-teaching.

Even beginning teachers have a wealth of life experiences on which to draw as we engage them in planning and reflecting about instruction. Teachers report far more satisfaction with coaching than with consulting, and even young and inexperienced teachers bring the cognitive capacity for coaching. Some new teachers may require more collaborating and consulting, and the coach does so with a mind-set to move toward gradual release in more coaching and less collaborating and consulting.

We do not regard these functions as points on a continuum; rather, we view them as a mosaic being chosen in the moment, based on perceived appropriateness to meet various intentions and teacher permission. In fact, support providers may want to start interactions with coaching and collaborating and move to consulting only when they see a need. Some supervision models presumed that reflective conversations are only appropriate for teachers with advanced skills and that, prior to this level of development, teachers needed to be "taught" about instruction in "coaching" conversations that are in reality forms of consulting.

Navigating Support Functions during a Conversation

"I don't know. What ideas do you have?" a teacher might ask during a reflecting conversation. In cases like these, the coach is asked to become a consultant. If the teacher is still uncertain, the coach asks permission to switch support functions by saying, "Would you like some options?" The coach would then shift *his body position away from mirroring the teacher*, shift his tone to the credible voice, and say, "Here are three ideas." After describing the ideas, he would switch back to a mirror position and in the approachable voice, he might say, "What thoughts do those ideas trigger for you?" In this case, the coach has remained in the mediative posture, offering not suggestions but options and then inviting the teacher to build on these.

We stress the importance of *modifying the body posture* because this helps the teacher be clear about the support person's intention in the moment, and the changing relationship within the conversation. To not give some kind of signal, the teacher is left in a state of uncertainty, a primary source of anxiety and withdrawal (Rock 2008). While today we know that mirror neurons will help the teacher understand the intentions of the coach, shifting body and voice tone when offering options supports clarity (Feuerstein et al. 2010).

To witness an example of modifying body posture, please access www.thinkingcollaborative .com/supporting-audio-video/5-1/.

Jane, the coach, switches the support function from coaching to consulting about Habits of Mind with a new teacher, Erin. Notice how Jane's body language breaks rapport to signal a change in support function.

Notice how the coach marks her shift in support function by turning and breaking eye contact with the teacher after having received permission to consult. Near the end of the segment the coach returns to the coaching relationship by shifting her body and then asks, "So I'm wondering how something like that might fit."

Isn't Everything Evaluation?

We're often asked, "Isn't everything just evaluation?" This is a complex question to answer. First, it's useful to consider the various meanings of *evaluate*. *Webster's Unabridged Dictionary* (2014) provides this definition: to determine the worth of; to find the amount or value of; to appraise.

The word *evaluate*, used as a verb, is a nominalization. Nominalizations name ideas as events when they are actually processes. Nominalizations are abstractions, separate from the actual doing of the thing. Because abstractions trigger different representations in people's minds, it is useful to ask, "What specifically does one do when one is evaluating?" First, one must be trained to judge the presence, absence, and levels of adequacy of targeted behaviors. The first problem we encounter with evaluation is that in study after study of observers judging teacher classroom performance, little agreement among different observers occurs (Ho and Kane 2014). To get high reliability, observers need to share a common philosophical orientation and leave biases at the classroom door (Gwet 2012).

When we claim Cognitive Coaching should be separated from evaluation, two issues often arise in people's minds. First, to evaluate is to make a judgment, but aren't humans judgment machines? Our very survival depends on rapid assessment of situations and determining the relative safety of our environment. Don't all actions and questions arise from the foundation of judgments we have made, either consciously or unconsciously?

We agree that judgment and action are inextricably intertwined. Yet one can choose to act without communicating judgment about events, behaviors, or choices that another has made if one is free of judgment about the individual. To repeat, when one is free from being judgmental about the individual, judgments about behaviors can be undisclosed. Studies reveal that judgments often depend on whether or not the observer feels in conflict with the person being observed (Massachusetts Institute of Technology 2008). Rebecca Saxe, who joined MIT's faculty in 2006 as an assistant professor of brain and cognitive sciences, specializes in social cognition—how people interpret other people's thoughts. To be nonjudgmental of others is both a perspective and a discipline, developed as one gains in personal maturity. Harvard professor Robert Kegan (2000) identified this stage of adult meaning making as self-transforming mind. In this order of consciousness, which is infrequently reached and rarely reached before the age of 40, individuals see beyond themselves, others, and systems of which they are a part to form an understanding of how all people and systems interconnect. They recognize their interdependence with others. Relationships can be truly intimate in this order, with nurturance and affiliation as the key characteristics. Relationships become a part of one's world rather than the reason for one's existence. Kegan noted that only rarely do work environments provide these conditions and that long-lasting adult love relationships do not necessarily do so either.

Drago-Severson (2004) offers that more typically the level of development known as "self-authoring thought" characterizes our most mature teachers. In this order, self-authorship is the focus. Individuals "have the capacity to take extraordinary action at the individual or collective level, and yet more often we do not" (Kegan and Lahey 2001, 5). Of course, judgment is associated with coaching (and consulting and collaborating). However, the judgment required in these settings is about comparing behaviors against standards, results against goals, not about the worth or motivation of the individual. *Webster's* says that a judge is one who has skills or experience sufficient to decide on the merit, value, or quality of something. Judgments will always be a necessary part of our work. We can make judgments, however, without being judgmental. We can judge without criticizing, censuring, or praising.

Can One Person Evaluate and Support?

Research by Carl Glickman (1985) at the University of Georgia sought an answer to the question "Is it possible for one person to serve as both a supporting supervisor (coach, consultant) and an evaluator of performance?" Glickman found that it can if:

1. Trust exists in the relationship and the process.
2. The teacher is clear about which role the principal is performing in the moment.
3. The principal's behaviors are pure. That is, when evaluating, only evaluating behaviors are used. When consulting, only consulting behaviors are used (providing data, making judgments, interpreting possible relationships, providing options, offering advice, advocating). When coaching, only coaching behaviors are used (paraphrasing, giving data, asking questions, inviting self-assessment, eliciting analysis, inviting synthesis of learning, requesting commitment). Mingling these three classes of behavior sends a mixed message, and the learning potential of the brain shuts down, which is one more indicator of the power of emotions to disrupt thinking. Anxiety signals from the limbic brain can create neural static, sabotaging the ability of the prefrontal lobe to maintain working memory (Goleman 1997).

Glickman's findings become especially important as many school districts encourage teachers to serve in support roles with their colleagues. Among the forms this support takes are beginning teacher induction programs, peer coaching, mentoring, peer assistance, and peer review. Being an outstanding teacher, however, does not automatically translate into being an effective mentor for other teachers. Just one or two teachers unskilled with the functions of mentoring, consulting, coaching, or peer-assisted review can disastrously affect morale and teachers' willingness to be open to collegial support (Costa and Garmston 2000).

On the other hand, many programs have developed thoughtful curricula for preparing teachers to work in support relationships with their peers.

Self-Directed Evaluation

There is little research that indicates traditional evaluation supports growth; there is, however, a substantial body of research indicating Cognitive Coaching supports growth (see chapter 14). Integrating the principle of self-directed learning and the tools and maps of Cognitive Coaching into the evaluation process brings a growth-producing dimension often missing from traditional evaluation processes.

So, while many school districts are adopting and implementing new professional performance review protocols for use in the evaluation process (see chapter 13 and appendix D), the challenge is to successfully implement the evaluation plans in a way that supports growth and learning. The maps and tools of Cognitive Coaching offer the skill set needed for administrators, supervisors, and sometimes peers to conduct self-directed evaluation.

Self-directed evaluators use structures and skills that engage the professional's thinking about his/her performance. In self-directed evaluation, the evaluator intentionally builds the capacity of the professional for self-managing, self-monitoring, and self-modifying behaviors. According to Costa and Kallick (2004, 51–57), self-directed people can:

- Clarify outcomes and gather relevant data
- Think flexibly
- Develop alternative strategies
- Draw on past knowledge
- Think about their thinking
- Persevere in generating alternative action plans
- Know how and where to turn when perplexed
- Reflect on experience and evaluate
- Analyze and construct meaning
- Be open to continuous learning
- Readily admit they have more to learn
- Change themselves

We see these behaviors as the outcomes of self-directed evaluation. When teachers display these behaviors, students achieve at higher levels (Hattie 2003).

The professional performance review protocols offer the what; self-directed evaluation offers the how.

Cognitive Coaching is particularly appropriate for a growth-producing evaluation process for five reasons. Cognitive Coaching is:

1. Procedural and conditional knowledge, in addition to declarative knowledge
2. Focused on the thinking that produces behavior
3. Research based and congruent with current neuroscience
4. A growth mind-set
5. Trust based

The evaluator's primary role, then, is to engage thinking that results in self-modification that will sustain longer-lasting change. Far more important than simply advising people about what to do are developing rapport, listening, and posing questions that support thinking. Cognitive Coaching provides evaluators with the conversation structures and skills to engage teachers' thinking so they can be self-managing, self-monitoring, and self-modifying.

Making In-the-Moment Decisions

During university days for Art and Bob, they were jolted as they realized how valuable it could be to shift supervision styles. They were working with a graduate student administrative intern. Both of them, by nature, tend toward a mediative, inquiry-based style of supporting others. Examination of progress with "Paul," however, revealed little follow-though on his part and insufficient understanding of the administrative position in which he was interning. Recalling Glickman's description of the relationship between levels of abstraction and levels of commitment to supervisor actions, they decided that Paul was relatively low on both continuums and therefore might benefit from a more direct information approach. Paul's performance and understanding immediately improved as he was provided with descriptions of what he should do next. Even though they had switched to a more direct informational style with Paul, they had not lost sight of their mission: extending his capacity for self-directed learning.

Paul's story represents the coach's dilemma of not knowing what kind of support to offer. More often, the issue is not whether to offer information or questions, but *when*. This is a delicate situation, and it is possible that only the most experienced coaches can make sound judgments most of the time. Many factors come into play. Is this a new teacher, a new assignment, or a new relationship? What is the person's general state of efficacy and craftsmanship? What is he asking for? How do you know if what she asks for is what is needed? What degree of stress is present?

Our colleagues Laura Lipton and Bruce Wellman address similar issues related to mentoring. They believe that a mentor's major responsibility is to increase the novice teacher's capacity to generate knowledge. Lipton and Wellman envision three stances that a mentor might take along a learning-focused continuum of interactions: The mentor might consult—that is, inform a teacher regarding processes and protocols, advise based on the mentor's expertise, or advocate for particular choices and actions. Another stance is to collaborate—that is, to participate as equals in planning, reflecting, and problem solving. At other times, the mentor might decide that coaching—that is, the nonjudgmental mediation of thinking and decision making—is the most effective option.

Whether to deviate from coaching is the most critical question for a support provider. Lipton and Wellman suggest that mentors give thought to three related questions: if, when, and how. If their answer to "if I should deviate" is yes, then consideration must be given to when and how.

Decisions about when and how to deviate from coaching are largely driven by the coach's attention to the verbal and nonverbal cues that signal what someone is thinking and feeling. The coach must read the colleague's communication: Is it confidence, confusion, or discomfort? This may move the coach to offer an organizing paraphrase, leave an area of inquiry for another time, or ask a mediative question. Inexperienced coaches sometimes move to consulting because of their own discomfort, not the teacher's. For example, if a coach feels he is not doing his job because the teacher failed to accurately self-evaluate, the novice coach might be tempted to give advice. Whether to move to another support function seems to be the most complex question. In general, one chooses a support function based on a teacher's resourcefulness.

Several factors influence the choice of services to provide. As in the case of Paul, Glickman's concepts about the level of abstraction and commitment may apply. Abstraction refers to the teacher's ability to examine situations from a variety of perspectives, to generate and examine alternative solutions, to test and modify instructional practices, and generally to reflect about her work. Commitment refers to the amount of work the teacher is willing to invest. In brief, the support provider determines how much initiative, thought, and action the teacher expends in his teaching. Glickman (1985) regards these two factors as developmental. He would have support providers increase the ratio of collaborative, or nondirective, work (coaching) as teachers become more highly abstract or committed.

Coaches must be aware of how a foreign-born speaker's language might influence a coaching interaction. Cultural realities must be considered when imagining that a teacher is at a lower level of abstraction when, in fact, the teacher may be translating concepts into his or her first language. The percentage of teachers in the English-speaking countries for whom English is a second language is increasing.

Signaling a Change

When the teacher knows which function is driving an interaction, he/she can respond congruently. The greatest risk of confusion comes when a support provider decides to shift from one function to another. Here are some moves a coach might make. First, seek permission to change functions: "I've been coaching. I'd like to shift roles and offer some ideas to consider. Then you decide for yourself which might be useful. Okay?"

Then the coach physically moves, in essence creating a visual paragraph for a new beginning of the conversation. She/he pauses and silently uses a frozen gesture, which initiates a neutral zone in which the teacher mentally separates from the coaching function. When the coach sees that the teacher's breathing is regular and unlabored, she/he presumes that the teacher has granted a transition into consulting (Grinder 2001).

Underlying the importance of being sensitive to the teacher, Boyatzis and Jack (2010) and Boyatzis, Smith, and Beveridge (2013) compared coaching that focused on a person's goals to models that were critical and discovered how coaching choices affect teacher cognition. When self-directed models were used, fMRI results revealed positive emotions and openness to coaching. Neurotransmitters increased the accuracy of perceptions and cognitive functioning. More critical coaching processes showed evidence of the opposite: defensiveness and decreased capacity for visioning and problem solving. Additionally, when the conversations focused on the person's goals rather than on another's judgments, five to seven days later the parts of the brain that are used for perceptual accuracy and emotional openness were activated, even when discussing difficult topics. Our assumption is that consulting can be perceived as critical if the stage is not set for autonomy and respect when entering the process (Rock 2008).

MENTAL PREREQUISITES FOR FLEXIBILITY

Skillful coaches shuttle among a variety of perceptual orientations. Each provides unique information unavailable in the other two perceptions. These three orientations are as follows:

- Egocentric, the coach's point of view
- Allocentric, the other person's perspective
- Macrocentric, a wide-angle view of the interaction between the coach and other person

The Coach's Perspective

This perceptual frame operates whenever we are intensely aware of our own thoughts, feelings, intentions, place on a coaching map, and physical sensations. To be aware of our own boundaries requires egocentricity. Being egocentrically conscious allows us to monitor our own processes. With consciousness, we can recognize that we are doing autobiographical or solution listening and decide to set it aside to better understand the other person. Without consciousness of our own thinking and feeling processes, we have no other choice but to stay stuck in whatever internal reality is happening at the moment. Although egocentric listening may generate sympathy, it is not helpful in mediating thinking.

The Other Person's View

Shifting focus from myself to the other person characterizes this perspective. It is the mental resource for rapport. With this point of view, we become aware of how a situation looks, sounds, and feels from the other person's experience. To work within this point of view, the coach must be exactingly attentive to the other person. "Listen with your eyes" to the physicality of communication, with your ears to the delivery and tone of the words, and with your feelings to what you sense about the other person's state. Allocentricity is the catalyst for empathy.

Listening with a Wide-Angle View "from the Balcony"

Compassion, or observation without value judgment, is often a by-product of the macrocentric perspective. In the macrocentric mode, one listens from a view outside the perspective of either party—"listening from the balcony." To a degree, you have access to and yet are detached from the feelings of identification you might have been experienced with either the egocentric or the allocentric view.

Coaches gather the most information about an interaction from this macrocentric position. The deeper that coaching maps, tools, and values are internalized, the greater the ease of going to the "balcony." Coaches have some understanding of their own feelings, some understanding of the other person's perspective, and an awareness of the systemic nature of the conversation. To be macrocentric is to observe the interaction from a psychological distance without identifying with either person, imagining you are viewing the interactions from a space above you.

Knowing one's intentions and choosing behaviors from a range of possible options that are congruent with those intentions requires the ability to move in and out of these three positions. This may be the most essential requirement for exercising situational flexibility. It is also the most valuable resource in modifying coaching interactions to match the other person's ways of processing information and making meaning.

CONCLUSION

The mission of Cognitive Coaching is to produce self-directed persons with the cognitive capacities for self-directed learning. Ultimately, the support person with the greatest flexibility in choices and repertoire has the greatest capacity to support growth. It is for this reason that Cognitive Coaches, devoted to mediating thinking, will occasionally draw from the other support services of collaborating and consulting to achieve these ends.

The Cognitive Coach knows that through his support services in collaboration and consulting he can provide data, ideas, and structured experiences that can enhance the teacher's cognitive capacity, teaching alternatives, and instructional skills. The coach also knows that through collaborating and/or consulting, opportunities will arise that can enhance the teacher's development of resourcefulness related to the five states of mind: efficacy, flexibility, consciousness, craftsmanship, and interdependence. (See chapter 7.)

Thus, even though coaches may enter other support functions, their "default position" is always grounded in the values, principles, and purposes of Cognitive Coaching.

Part 2

SOURCES OF EXCELLENCE

This section addresses four major sources of teaching excellence: trust, five states of mind, research on instructional cognition, and the knowledge base of teaching. We define holonomy and describe the five states of mind that drive human resourcefulness. The chapter on teacher cognition includes recent research on complex thinking processes and findings from the neurosciences.

Excellence refers to qualities that are unusually good and exceed ordinary talents. First, trust is a foundation in which Cognitive Coaching thrives. Relational trust is defined by researchers Bryk and Schneider (2002) as fundamental, yet trust in self, in the coaching process, and in the environment are also needed for excellence to develop and flourish.

Five states of mind, the second feature addressed in this section, are drives that propel human performance. Coaches attend to these mental states of efficacy, flexibility, craftsmanship, consciousness, and interdependence to foster the development of self-directed persons who can sustain performance at the level of excellence, both individually and as members of a community. The third aspect, cognition, explores the research on teacher thinking. Since thinking drives teaching decisions and behaviors, the Cognitive Coach dedicates him/herself to the refinement and development of this resource for application in the classroom. This chapter illuminates why having teachers talk about their theories, assumptions, and decisions about teaching improves those same inner qualities. From this emanate refined decisions, more effective teaching, and improved student learning.

Finally, the fourth aspect of excellence relates to continually inquiring into the growing knowledge base of teaching. Although a teacher's knowledge of content is not the major contributor to student learning (Hattie 2012) it is an essential part of teaching excellence. Also critical are pedagogical knowledge, knowledge about student characteristics and developmental stages, and knowledge of self in order to override approaches to teaching borne from personal history rather than informed connections between the students one is teaching and content-specific pedagogy. Finally, this section addresses the importance of continual learning in knowledge of the brain's role in learning and personal capacities for collegial interaction.

6

Developing and Maintaining Trust

In chapter 3, we described a mediator's eight tools, which we clustered in five groups: paralanguage, response behaviors, providing data and resources, structuring, and mediational questioning. In this chapter, we describe how these tools contribute to one of the essential elements of Cognitive Coaching: developing and maintaining trust.

Carol Simoneau, a Cognitive Coach international consultant, told us this story. A person at a school in which Carol served was assigned to coach an experienced teacher. The teacher was not the most effective teacher and had become a real "loner" in the building, isolating herself from teammates. The coach said that it took her three months of daily checking in and helping the teacher with minor tasks (picking up at the end of the day, setting up for lessons) before the teacher felt comfortable enough to confide in her. Over time, the coach gained the teacher's trust and learned that she felt inadequate with all the mandated changes. Carol realized the teacher's cold exterior was self-protective.

As we look at this event, two ideas stand out. First, that trust is not a thing, an event, or something that just happens; rather, it is a process. Next, this coach devoted energies over a long period of time to befriend and build trust with this teacher. It was trust that gained the teacher's confidence in the coach.

CHARACTERISTICS OF TRUST

Research by Bryk and Schneider (2002) both stunned the educational world and confirmed what most believed: that trust was a necessary, but by itself insufficient, resource for improved student learning. Schools whose test scores were increasing over the period of this seven-year study had high levels of relational trust. In contrast, low-performing schools had low levels of trust. The leader, or principal in these cases, was instrumental in fostering climates in which relational trust could flourish. The researchers defined relational trust as the trust that is developed through the interpersonal social exchanges that take place in a school community: principal to teacher, principal to parent, teacher to teacher, teacher to student, and teacher to parent. Relational trust is built on four criteria: respect, competence, personal regard for others, and integrity.

- Respect is valuing the role each person plays in a child's education and being genuinely listened to.
- Competence is how well a person performs a role.
- Personal regard is the perception of how one goes beyond what is required of his/her role in caring for another person.
- Integrity is the consistency between what people say and what they do.

Relational Trust

> The most important thing we can do is to trust and love one another.
>
> —George Land and Beth Jarman (1992)

Common facets of trust across the literature include confidence, vulnerability, benevolence, reliability, competence, honesty, and openness (Tschannen-Moran and Hoy 2001). Relational trust, report Bryk and Schneider (2002), is the "glue" that binds people together when deciding on policies that advance the education and welfare of the students. This finding is echoed elsewhere about dyadic relationships. Rock and Page (2009) report that relationship is responsible for 30 percent of the value in one-on-one relationships. Glaser (2014) reports that trust is essential for transformational conversations.

We've asked thousands of people to describe how they develop trusting relationships. They report behaviors that are strongly consistent with research on the subject. Among the factors they add to the lists above are being visible and accessible, behaving consistently, keeping commitments, sharing personal information about out-of-school activities, keeping confidences, revealing feelings, expressing personal interest in other people, acting nonjudgmentally, listening reflectively, admitting mistakes, and demonstrating professional knowledge and skills.

Trust grows stronger as long as these behaviors continue, but a relationship can be seriously damaged when someone is discourteous or disrespectful, makes value judgments, overreacts, acts arbitrarily, threatens, or is personally insensitive to another person. Restoring trust should be a skill set for each of us. *Senge* is a practice in Japanese Buddhism that involves becoming aware of effects one has created and making amends that restore harmony and develop our own wisdom (Rock and Page 2009). They write that, "Even bad news can be preferable to not knowing and even finding out when we will find out can be calming" (444).

The most effective apologies a) ask forgiveness, b) acknowledge responsibility without "excuses," c) express regard for the other person's hurt, and d) either commit to not repeating the behavior or explore ways to prevent it in the future.

Because Cognitive Coaching relies on trust, any manipulation by the coach is incompatible with the mission of Cognitive Coaching. Should an employer have performance concerns about a staff member, those concerns are best communicated directly outside the coaching process. Coaching should never be about "fixing" another person. Inevitably, that person perceives the motivation to "fix" and becomes defensive. For example, if you believe that your job is to fix Naseem, mirror neurons guarantee she will accurately read your real intention, no matter how many pleasant conversations you share. Naseem clearly detects that you think something is deficient about her or that she is not cared for. To coach without manipulation, you must change the way in which you see a problem employee. This may be difficult, and it is even more necessary when you have been working overtime with someone whose performance is problematic, who seems to be making limited growth, who isn't making any effort toward growth, or who is resisting assistance. In these cases, coaching is even more important because a natural outgrowth of these processes over time is frustration for the coach.

When coaches remind themselves that all behavior is motivated by positive intentions in the other person's point of view, the coach can let go of judgment or negative interpretation. Thus the coach is freed from debilitating emotions and better able to work empathically and rationally with the other person.

Now, setting aside this discussion of maintaining trust in the face of performance concerns, let's focus on what hundreds of educators and the literature have to say about the practical realities of creating trust. Four considerations follow: trust in self, trust between individuals, trust in coaching, and trust in the environment.

What is it like to be a teacher in the East Los Angeles middle school where Frances Gipson is principal? In a recent letter to parents, she wrote, "I believe that parents and community are critical partners in the achievements of our youth." More than words on paper, Gipson communicates this respect with every parent she encounters, regardless of the reason for meeting. She extends the same respect and trust to teachers and students as she does with parents. As a result, Gipson's school has attained statewide awards, measurements of learning have increased in the two years of her tenure, and teachers trust her. Interpersonal trust requires personal contact, and each day, Gipson is in each classroom, walks the campus, and interacts with students. What teachers can trust with Frances is that she is always authentic, respectful, and passionately committed to the best for students. To successfully cognitively coach another, you must be credible in that person's eyes.

We trust those who are believable, of good character, and sensitive to our interests and styles (Tice 1997). Peg, a kindergarten teacher, once told us the story of how she tried to develop trust with another teacher on the staff. "It seems the more I tried to express interest in her, reach out to her, let her know I cared for her, the more she put distance between us." As Peg elaborated on the situation, it became clear that she had an image of how two people behave in a trusting relationship, and *she presumed that the other teacher held the same vision.* As it turned out, Peg's style of trust included far more intimacy than the other teacher expected. Peg's colleague believed that emotional reserve and only a certain amount of self-disclosure characterized a comfortable, trusting relationship.

To see things from another person's point of view requires cognitive flexibility and is essential to any healthy relationship. Seeking to understand is one of the more important ways in which a coach can communicate that another person is valued, even in a setting in which distrust is present. Maureen, a new personnel director in a California district, met with teacher representatives for contract negotiations. The district laid all its fiscal data on the table. Mistrustful, the teacher representatives kept saying, "The district has not done this before. This must be some kind of trick!" Undaunted and without defensiveness, Maureen repeatedly paraphrased these comments. In time, the mood of the entire meeting shifted, and teachers began to believe that the district might be telling the truth, so they started to work collaboratively with district representatives on understanding the figures.

Apart from understanding how the other person processes information, we can attend to four areas.

- What is it that is trustworthy about me? Qualities like dependability, authenticity, honesty, kindness, courtesy, and integrity are found on this list. Keeping people fully informed also generates trust (Freiberg and Freiberg 1996).
- What is important to the person in the long term—what values, goals, ideals, interests, passions, and hobbies?
- What are the information-processing patterns the person displays, such as cognitive style, perceptual filters, and modality preferences? In settings where tension exists because people perceive one another as different, they must find a shared language and common reference points in order to value one another and build a common work culture. (See chapter 10 for more information.)
- What are the person's current reactions, concerns, thoughts, and theories?

An expression of personal regard is also important to interpersonal trust building, especially in a Cognitive Coaching relationship in which praise is withheld because it interferes with thinking (Dweck 2006). While a majority of Americans think it is

---

THE UNEXPECTED EFFECT OF PRAISE

In a series of experiments on 400 New York fifth-graders, Carole Dweck (2006) had researchers administer a one-on-one nonverbal IQ test—puzzles easy enough that all the children would do fairly well. On completion, each child was given his score and a single line of praise. Randomly divided into groups, some were praised for their *intelligence*. They were told, "You must be smart at this." Other students were praised for their *effort*: "You must have worked really hard."

Then the students were given a choice of test for the second round. One choice was a test that would be more difficult than the first, but the researchers told the kids that they'd learn a lot from attempting the puzzles. The other choice, Dweck's team explained, was an easy test, just like the first. Of those praised for their effort, 90 percent chose the *harder* set of puzzles. Of those praised for their intelligence, a majority chose the *easy* test. The "smart" kids took the cop-out.

---

good to tell their kids they are smart, this behavior has counterproductive results and can even promote underperformance.

Praise communicates a value judgment about another person or the person's performance. It infers an unconscious entitlement to evaluate another. At some level, we often feel uncomfortable about receiving praise. Even on occasions when it might feel good to hear "You did a great job," the praise removes any need for one to apply one's own criteria to self-assessment.

When we ask supervisors why they use praise, they often report two reasons. One is to reinforce behavior. As we can see above, this can promote nongrowth. A second is to communicate positive regard. Robert Kegan and Lisa Lahey (2001) implore us to move from the language of prizes and praising to the language of ongoing regard. One attribute of this language is to replace "you" statements with "I" statements. "Abdulla, I appreciate the way you took time to bring me up to date. It made a real difference to me." Kegan and Lahey (2001, 100) say, "Characterizing our own experience, positive or negative, leaves the other informed (not formed) by our words."

Coaches signal personal regard by the following behaviors:

- Spending time with the other person in activities not related to the coaching task.
- Making inquiries or statements related to the other person's personal interests or experiences.
- Practicing all the fundamental behaviors of courtesy and respect revealed in the research on trust building: proximity, touch, courteous language, and personal compliments ("I like that tie"). Sometimes the workplace gets so busy and task oriented that these little personal connections are abandoned, and the day-to-day workplace becomes an emotional wasteland.

Excessive reliance on personal trust without regard to the goals of the organization can lead to paternalism and parochialism. Andy Hargreaves (1994) notes that trust in expertise and in a process that helps to solve problems on a continuing basis maximizes the organization's collective expertise.

Trust in Self

> To do good things in the world, first you must know who you are and what gives meaning to your life.
>
> —Paula Brownlee

The state of mind of consciousness includes being aware of one's actions and the effect they have on others. Self-trust depends on this awareness and is prerequisite to developing trusting relationships with others. The best coaches we've seen are conscious and clear about their identities as mediators and their own values and beliefs in areas such as pedagogy, philosophy, and human development. They manage their behaviors to be congruent with those core values. They experience a well-defined sense of personal identity that comes, in part, from their ability to articulate their beliefs with precision and passion. They function at a high stage of affective development, characterized by a strong sense of values. They maintain the belief that no matter what the situation, they will remain true to themselves. We would say that they have integrity.

Ultimately, these personal core values shape a coach's perceptions about leadership responsibilities, the meaning of learning, the potential for a school or community, and what motivates people. Trusting yourself also means being conscious of the ways in which you process and make meaning of experiences. For example, we relate easily to those with similar cognitive styles, but it requires greater effort to withhold value judgments about other people's attitudes and perceptions when their style differs from ours. Although we use all of our senses all the time, we often pay attention to one sense more than another. One of the capabilities of a skillful coach is being able to adjust one's own style preferences. This requires knowing your own representational system strengths and being conscious of how you understand your experiences through visual, kinesthetic, and auditory channels.

Gender, culture, race, religion, geographical region, childhood experiences, and family history are explored in chapter 10. These factors predispose us to draw certain inferences and to attend to certain stimuli while blocking out others. None of us leaves our emotions on the doorstep when we go to work. At times we function at less than our best for a variety of spiritual or emotional reasons. Flexible coaches know when they're being negatively influenced by their emotions, and they respect the fact that colleagues also experience emotional shifts that sap energy and distract attention. Coaches who trust themselves are more capable of building trust with others.

Trust in the Coaching Relationship

> Be transparent, focus on the other, seek a collaborative partnership, and view the relationship as long term, not a temporary interaction.
>
> —Charles Green

Nearly all relationship difficulties are rooted in conflicting or ambiguous expectations surrounding roles and goals. Paramount are perceptions of unequal status, which can arouse the limbic system and set the neocortex into a defensive mode (Rock 2009). Whether we are assigning tasks at work or choosing a decision-making process for a meeting, we can be certain that unclear expectations will lead to misunderstanding, disappointment, and decreased trust.

As described in the chapter 3, certainty is another social need of the brain (Rock 2009), and coaches provide this by making structuring comments that clarify expectations. For example:

- "I'm not going to give you advice. Let's explore some alternatives together."
- "My job is not to evaluate you but to help you reflect on your teaching."

Clear expectations are most important dealing with the purposes and forms of classroom observations. When principals perform the two supervisory functions of coaching and evaluation, it's important to be clear about the goals of a classroom visit. Confusion, suspicion, and even hostility arise when a teacher is uncertain which support function is occurring, coaching or evaluation.

When mentor teachers are responsible for the dual functions of consulting and coaching, again, clarity about intention is important. Peer assistance and other ongoing staff development efforts send a signal that this is a learning community for all teachers. Within that context, peer assistance may be seen as an expression of continuous learning, with recognition that teachers, too, are individuals. Some may be at different stages in their development and require different types of support.

Strategically, we would recommend that peer assistance programs, in which teachers are mentoring and coaching each other, be part of an ongoing, well-developed professional development system (Woods 1997). This allows time for collective learning about how to give and receive support without the threat of evaluation.

Trust in the Environment

> Just as a picture is drawn by an artist, surroundings are created by the activities of the mind.
>
> —Buddha

Many authors indicate that the culture of the workplace influences staff performance more than teachers' experience, skills, knowledge, training, or staff development (Seashore-Lewis, Marks, and Kruse 1996). Subtle signals from the environment, culture, and climate of the organization influence co-workers' thoughts, behaviors, and perceptions. Employees are quick to perceive the congruity between a learning organization's stated core values and its day-to-day practices. If there is consistency, greater trust will exist; the organization "walks its talk." If there is inconsistency, suspicion and mistrust will arise.

"WALKING THE TALK"

Joe Saban, former superintendent of the Crystal Lake High School District in Illinois, describes how teachers learned the process of using portfolios with students by first keeping their own portfolios. Soon they decided that if portfolios were useful for teachers and students, then they would also be beneficial for school administrators. Thus, the building principals began keeping school portfolios. Soon the practice spread, and then Joe also kept portfolios to demonstrate his accomplishments to the board of education.

The effective coach also works to create, monitor, and maintain a stimulating, mediational, and cooperative environment deliberately designed to sustain and enhance trust. Wheatley (1992) theorizes that the movement toward participation in organizations is rooted, perhaps subconsciously, in our changing perceptions of the universe. Knowing how to network, how to draw on the diverse resources of others, and how to value each person's expertise, diverse views, perceptions, and knowledge base is increasingly essential to survival. We might view this as a new form of interlocking intelligences: collaboratively melding perceptions, modalities, skills, capacities, and expertise into a unified whole that is more efficient than any one of its parts.

Cognitive Coaches, being aware of this, strive to behave consistently in regard to these core values and beliefs. Because life in school is often fragmented, idiosyncratic, fast paced, and unpredictable, coaches' long-term impacts on a school occur through the consistency with which they handle day-to-day interactions.

Researchers at Rutgers University studied teachers' perspectives on what makes principals trustworthy (Kupersmith and Hoy 1989). Three characteristics emerged. First, principals took responsibility for their own behaviors. They admitted mistakes and did not blame others. Second, principals acted as people rather than as "roles." Trusted principals revealed personal information about themselves—their likes and dislikes, their emotions, their history—so others had a sense of who they were away from the job. Third, trusted principals were nonmanipulative. They influenced directly, not covertly, and they had no hidden agendas. This research also found links between interpersonal trust and trust in the environment. Teachers who had faith in the principal often trusted each other and the central office personnel.

We sometimes hear the complaint "But there is no trust here," as if that settled the matter about being able to move forward on any venture. When you are told this, ask what kind of behaviors would people be seeing and hearing if trust were present. With a behavioral description of a desired state, one has targets to work toward rather than wallowing in despair. Acting on those behaviors is the first step in recovering trust.

TRUST IN THE COACHING PROCESS

Trust building begins with the first encounter, when the colleague realizes that the coach is interested, that the coach is an empathic listener, and that the relationship is nonevaluative. Increasingly, as the coach and the teacher work together in a nonthreatening relationship, they both place greater value on the coaching process. They realize that the intent of the process is to grow intellectually, to learn more about learning, and to mutually increase their capacity for self-improvement. They realize that the process is not one that the "superior" does to the "inferior"; rather, they are two dedicated professionals striving to solve problems, improve learning, and make the curriculum more vibrant. Furthermore, the teacher soon realizes that the coach is working at the coaching processes as hard as the teacher is working at the teaching process.

In time, we find that teachers begin to request the coaching process with phrases like the following:

- "That felt good."
- "You really made me think."
- "Could you come back?"
- "Would you teach me how to coach like that?"

Soon the process begins to spread. Word gets around through the school grapevine. Principals call each other, requesting to be Cognitively Coached through an upcoming parent conference. The director of staff development calls a principal, asking to be coached through a forthcoming in-service presentation. Teachers in grade-level or department meetings soon find that they are coaching each other spontaneously, beyond the classroom, situation free.

In chapter 3, we described eight behaviors that contribute to the intellectual growth and the engagement of cognition in others. Many of these same nonjudgmental behaviors contribute to a trusting relationship as well. In fact, we know that the neocortex of the brain shuts down by degrees under stress. The greater the stress, the greater the shutdown. Generally, human males will move toward freeze, fight, flee, or flight; females tend and befriend. Under great stress, we lash out, run away in terror, or freeze up. or we lash out. Physical rapport, subtleties of body language, voice tone, implied value judgments, and embedded presuppositions in our language all have an effect on the comfort and thinking of others. To communicate with skill and grace, we must communicate with the total body-brain system. The systems that process nonverbal signals and feelings are as important to thinking processes as they are to establishing a trusting relationship.

Rapport and Trust

Trust is about the whole of a relationship; rapport is about the moment. Trust is belief in and reliance on another person developed over time. Rapport is comfort with and confidence in someone during a specific interaction. Rapport may be naturally present or you may consciously seek it, even when you are meeting a parent, student, or colleague for the first time. You cannot manipulate someone into a relationship of trust and rapport, but you can draw on specific verbal and nonverbal behaviors to nurture the relationship.

The research reviewed in chapter 3 revealed that the coach's matching of gestures, postures, or voice qualities contributes to rapport. These behaviors have an enormous impact on feelings of connectedness and rapport. Such an alignment permits a non-verbal form of communication that the other person is being "understood" and, in the deepest sense, is "feeling felt" by another person (Damasio 2010; Caine and Caine 2015). Under these conditions, permission is being tacitly given for coaching. The coach recognizes this permission by observing full, rather than shallow, breathing patterns (Zoller and Landry 2010).

Distrust Cannot Be Masked

You can't trick someone into a trusting relationship. Babad, Bernieri, and Rosenthal (1991) conducted a study in which five groups of judges viewed 10-second film clips of teachers talking to or about students. The judges, who ranged from fourth graders to experienced teachers, were asked to rate the students' scholastic excellence and the teacher's love for each student. In some cases, the judges heard teachers talking about students; in other cases they simply watched, with no sound, as the teacher talked to a student (who was not visible). In each case, one was a good student of high potential and one was a weak student of limited potential.

None of the judges had difficulty detecting students' excellence and teachers' love. The negative affect that teachers tried to conceal was detected by observers through "leaks" in communication channels that are not under as much conscious control as what the teachers were saying to the students (Babad, Bernieri, and Rosenthal 1991). The teachers' words gave them away when they talked *about* students; their actions

---

LISTENING FROM YOUR MISSION

There is no effective communication without listening. Listening is the tool that turns words into communication. Physiogically, there is a part of your brain—the thalamus—that decides where to send incoming signals. The thalamus is like a great receptionist in the office of your brain. It looks at the "phone" in your brain, sees three lights on, and says, "No way that brain is going to take another call—I'll just get rid of this guy." And like a good receptionist, the thalamus is highly sensitive to what's going on in the office, sees how tense people are, and it evaluates incoming traffic.

And like great receptionists, you cannot fool your thalamus. If you say, "I am ready to take the call" but mean "I can't believe I've got another call now, this is totally insane, I am still . . . ," your thalamus doesn't let real content into the decision-making parts of your brain because they are already busy.

So, if you actually listen, you have to go beyond the outward steps of "active listening" we all learned as rote procedure for dealing with conflict. You actually have to care. You might not care about the person, you might not care about the conversation or the issue, but you do care that your behavior helps you meet your real goals, your objectives. For many people, that personal mission includes some kind of problem solving, some kind of learning, some kind of personal accountability, some kind of making the world better. Chances are, if you cannot summon one of these commitments—to the person, to your mission, to your organization—than your communication is doomed to shades of mediocrity.

People who are able to bring their hearts online are able to listen to the message beyond the words. They are able to turn conflict into a learning experience. They persevere in spite of the complexity, the messiness, the frustration (Boase 2011).

---

alone gave them away when they talked *to* students. We can consciously take action to nurture trust and rapport, but we can't mask our true feelings.

CONCLUSION

Now that we have described the tools of the mediator as they contribute to both cognitive development and trust, you may wish to watch a video of a planning, reflecting, or calibrating conversation listed in appendix G. In the end, trust is cemented when the coach asks the teacher to evaluate the coaching process and recommend refinements. This specifically signals that the exchange was a coaching situation, not an evaluation, and it shows the teacher that the coach is working to refine her own behaviors, too. This kind of ending question also communicates that the teacher is the authority about which coaching behaviors are most appropriate and useful.

One of the first intentional considerations of a coach, therefore, is to establish and then to maintain trust. Hargreaves states, "Trust in people remains important, but trust in expertise and processes precedes it." Trust is a fragile commodity that must be constantly nurtured and tended. Once lost, trust is even more difficult to reestablish.

States of Mind

As we develop soul in our work we need to recognize our dual identity: we are both individuals and members of a group. Indeed, finding the soul of work involves the balance and integration of apparent opposites, such as head and heart, intellect and intuition, and self and group. This process is not so much based on the "shoulds" but upon "what is." It is my belief that as we attend to the soul of work we will find we feel more complete.

—Daryl Paulson (1995)

In this chapter, we look beyond surface behaviors of thinking and teaching to discover the sources of excellence that drive human performance. Our aim is to focus the coach's capacities on mediating the energy sources that fuel self-directed learning, behavior, and, ultimately, the development of capacities for effective action in holonomous settings. The five states of mind we describe are the wellsprings that nurture all high-performing individuals, groups, and organizations. They are beacons guiding us toward increasingly authentic, congruent, and ethical behavior (Costa and Garmston 2002)

First, we revisit the concept of holonomy (the state of being simultaneously a part and a whole) and the tensions that the paradox of holonomy produces. Next, we explore five states of mind that people draw upon to perform at their best and to resolve such tensions. Finally, we offer ideas for mediating these states of mind.

THE SEARCH FOR WHOLENESS

If we represent knowledge as a tree, we know that things are divided and yet connected. We know that to observe the division and ignore the connections is to destroy the tree.

—Wendell Berry

Every person's story is about moving toward wholeness. Forward-looking professional development programs know and support this. Carl Jung, one of the 20th century's most influential psychologists, describes this as the joining of opposites, which can be depicted by a snake eating its tail or the joining of poles to make a single whole. Others describe the development toward wholeness as an evolving journey of deepening complexity—cognitively, morally, and in ego states (Kegan 1994).

Holonomy, the study of wholeness, was described in chapter 1. Each person strives to resolve a duality—to develop and preserve a unique, separate, and personal identity and, at the same time, to strive for affiliation and reciprocity with others.

Recognizing people as holonomous beings reveals the polarity that drives much of their emotive behavior. Self-assertion finds an outlet in personal ambition, competitiveness, and aggressive or defensive behaviors. On the other hand, the integrative potential finds fulfillment through authenticity, or internal and external congruence, in identification with family, community, or other social groups. This simultaneous drive for self-assertion and affiliation is a continuing source of tension among individuals and within groups (Koestler 1972).

Holonomy: Paradox and Promise

Holonomy is a combination of two Greek words: *holos* meaning whole, and *on* meaning part. Holonomy conveys the notion that an entity is both an autonomous unit and a member of a larger whole simultaneously. Holonomy may be oxymoronic since it implies a combination of opposites—being both a part and a whole—acting autonomously and, at the same time, working interdependently. Because all life forces are simultaneously independent and interdependent, self-assertive and integrative, whole unto themselves yet always a part of systems larger than themselves, holonomy, therefore is paradoxical.

Autonomous human beings are self-referencing, drawing on their own unique systems, experiences, strengths, and origins to continuously grow. This growth includes the capacity to transcend their own original patterns. In autonomous, assertive individuals, growth emanates from within, as do the life forces within a seed that produce growth into a tree.

There is, however, no such thing as a purely autonomous, independent entity. All organisms are at once autonomous and simultaneously in reciprocity with their environment and with other organisms. Even the most effective self-modifying, self-authoring, autonomous individual maintains membership in numerous larger environments and communities: family, a church group, a workplace, a department or staff, a community, or a nation.

The interaction of these two components of holonomy, which produces the paradoxical conflict, is about the human capacity and proclivity to work collaboratively in, and at the same time to learn from, collegial groups. A paradox implies polarity and conflict. It is this duality that gives rise to certain tensions between these opposing internal drives and external forces. Humans strive for self-identity and assertion and simultaneously strive for reciprocity and affiliation with others. The source of human tensions stems from equilibration—the search for balance between such dichotomies as:

- *Ambiguity and certainty.* Humans have a passion for certainty and a simultaneous need for doubt. Our minds tend to distort meanings to fit our own biases, assumptions, and theories. We are self-convincing and believe there is truth in what we know. We are, however, in a constant state of tentativeness, open to the reality that our information may be inadequate, that additional information may dispute our conclusions, that we have not analyzed or defined our conflict thoroughly. Numerous examples can be found in scientific progress as more recent and conclusive evidence disputes obsolescent theories. (The heliocentric universe, global warming, and the definition of photosynthetic life are but a few examples.)
- *Knowledge and action.* Human beings strive for congruence between what we know and believe and how we behave. While we may possess an abundance of knowledge about a subject, we do not always apply that knowledge to our actions. An example might be from education; we know that most boys mature later than girls do, yet we teach a uniform curriculum simultaneously to both.
- *Egocentric and allocentric perspectives.* Humans continually work to ameliorate our own perspectives with others' perspectives. We perceive situations or events from our own point of view but may find we are at odds with the point of view held by others. Any married couple can cite numerous examples of contrasting male and female perspectives; resolving learning/thinking style differences stretches us to consider alternative points of view.
- *Self-assertion and integration.* Humans strive to become autonomous, self-initiating, unique individuals and, at the same time, hold membership and allegiance to the larger community. In meetings, for example, we may ask ourselves, "Is this the time when I need to assert my ideas or should I choose to go along with the will of the group even though the group's decision may be contrary to my own views?"
- *Transparency between inner and outer lives.* What one thinks and feels inside is presented on the outside in behavior. Humans have the capacity to mask emotions. Young children derive great delight from pretending. As we mature, we experience feelings, thoughts, and opinions but choose not to expose them to others. Examples might be when someone tells an offensive joke—we may smile overtly but we feel hurt or put off inside; or while we may be experiencing rage inside, we wish to convey to others the appearance of calmness and rationality.
- *Solitude and interconnectedness.* At times, humans desire to be alone and introspective and yet are driven by the basic need to be interactive and reciprocal

with others. We wish to be alone but are exhilarated by the interpersonal interaction with crowds. There are times when we wish to retreat to our cave and other times when we wish to visit the commons. Those who are adept at presenting to large groups may draw energy from their audience but then find solace in being alone. Most humans dread the thought of being deserted or abandoned, yet we often search for seclusion.

Effectively balancing and resolving tensions and conflicts such as these demand that we draw upon certain internal resources. These internal capacities may be described in terms of five states of mind. Taken together, they are forces that human beings access as they strive for increasingly authentic, congruent, ethical behavior. They are the tools of disciplined choices that guide human actions in resolving the tensions listed above. They are the primary vehicles in the lifelong journey toward integration.

Living holonomously requires a person to continually engage in the process of growing toward wholeness while seeking balance with individuality. A holonomous person is one who becomes increasingly aware of this internal conflict and knows how to manage it with flexibility—being alert to both environmental and internal cues that inform his or her knowing when and how to act autonomously, as well as when and how to act interdependently (Huppke 2014). Table 7.1 describes these tensions.

As stated in chapter 1, the mission of Cognitive Coaching is to produce self-directed persons with the cognitive capacity for excellence, both independently and as members of a community. One purpose of Cognitive Coaching, therefore, is to develop each individual's capability to accept that he or she is simultaneously a whole and a part (whole in terms of self and yet subordinate to a larger system) and then work to realize that capability. Effective teachers, for example, are autonomous individuals: self-asserting, self-motivating, and self-modifying. However, they are

Table 7.1. Tensions and Drives

Self-Assertion and Integration	Our striving to become an autonomous, self-initiating, unique individual and at the same time hold membership in and allegiance to the larger community.
Knowledge and Action	Our striving for congruence between what we know and believe and how we behave.
Egocentricity and Allocentricity	The amelioration of our own perspective with another's perspective.
Ambiguity and Certainty	Our human passion for certainty and the simultaneous need for doubt.
Inner and Outer Lives	What we are feeling inside and what we are presenting as our behavior on the outside.
Solitude and Interconnectedness	Our desire to be alone and introspective and our need to be interactive and in reciprocity with others.

also parts of larger wholes: a department, a school, a district. They are influenced by the norms, attitudes, values, and behaviors of the collective. The school, in turn, is an autonomous unit interacting within the influence of the district and the community. Holonomy is both a goal and an ideal: a vision toward which humans and organizations forever strive.

By resolving these dichotomies, we can develop intellectual, moral, and ego resourcefulness. Russian psycholinguist Lev Vygotsky (1978) suggests that intelligence grows in two ways: through our own experiences and through our interactions with others. Our intelligence actually increases as we justify reasons, resolve differences, listen to another person's point of view, achieve consensus, and receive feedback.

With repeated opportunities to resolve these conflicts, and with purposeful reflection, we develop resourcefulness in coping with future conflicts and tensions. We experience a sense of interconnectedness, unity, and bonding to common goals and shared values. We enlarge their conception from "me" to "us." And we understand that as we transcend the self and become part of the whole, we do not lose our individuality but rather our egocentricity.

To effectively balance and resolve these challenges, we must develop and draw upon certain internal resources. We describe these internal drives as five states of mind.

Five States of Mind

> We all have the extraordinary coded within us, waiting to be released.
>
> —Jean Houston

In his popular book *Drive*, Daniel Pink (2009) describes what motivates us: autonomy, mastery, and purpose. We identify such drives as five states of mind that inform human perception and are the resources that human beings access as they resolve the tensions inherent in holonomous settings. They are the tools of disciplined choice making and the primary vehicles in a lifelong journey toward integration.

These basic human drives influence, motivate, and inspire our intellectual capacities, emotional responsiveness, high performance, and productive human action. We categorize and define them as shown in figure 7.1: efficacy, flexibility, craftsmanship, consciousness, and interdependence.

1. The capacity for *efficacy*: Humans quest for competence, learning, self-empowerment, mastery, and control.
2. The capacity *flexibility:* Humans survive by developing repertoires of response patterns that allow them to see other perspectives and create, adapt, and change.
3. The capacity for *consciousness:* Humans uniquely strive to monitor and reflect on their own and others' thoughts and actions.

4. The capacity for *craftsmanship:* Humans yearn to become clearer, more elegant, precise, congruent, and integrated.
5. The capacity for *interdependence:* Humans grow in relationship to others and are social beings in need of reciprocity and community.

Although some of the states of mind may periodically be more dominant than others, these five distinctly human forces unite the expression of wholeness in an individual. The capacity to access and channel these forces provides the continuing

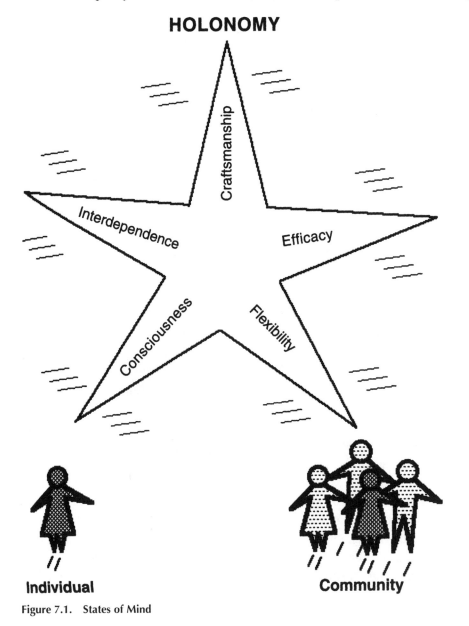

Figure 7.1. States of Mind

sources of excellence in all of life's endeavors, both professional and personal. The states of mind are recognizable in our language and actions, and they are at once dispositional and cognitive. They inform and make possible the application of strategies and the selection of thinking processes appropriate to a situation. They contribute to increased accomplishment and satisfaction.

The five states of mind are innate in our brains and bodies. Being capacities, they are invisible in that they represent a person's perceptions, emotions, and decision-making processes for responding intelligently to the ever-changing, in-the-moment context of life.

> There is a dimension of the universe unavailable to the senses.
>
> —Joseph Campbell

Like gravitational or electromagnetic fields, these states of mind cannot be observed directly; they are known by their effects. The ball falls from our hand; we label gravity as a cause. Likewise, we label invisible powers in professional educators— their states of mind—that generate profound, observable effects in classrooms and throughout the school culture.

The five states of mind are not fixed; rather, there is the potential for being liberated, expanded, or atrophied. Normally, infants are imbued with curiosity—they explore and seek novelty yet, when raised in an impoverished environment, the capacity for normal development is compromised. It is well known that while babies are born with the natural capacity to learn any language, this capacity disappears with disuse. Neuronal connections in their brain are formed through hearing, mimicking, and experimenting with sounds. Through neural pruning, they give up the sounds and patterns they do not hear and strengthen their use of the language they do hear. What they do not hear and use atrophies. Living in a responsive environment and being recognized will reinforce these tendencies.

Similarly, all humans have a natural, innate capacity for efficacy, consciousness, craftsmanship, flexibility, and interdependence. When suppressed, the expression of these factors declines. Even changes in leadership can diminish this capacity from being active. Likewise in adults, these states of mind are subject to many factors: environmental conditions, developmental levels, emotional states, situational conditions, interpersonal relationships, and individual styles. They can be blunted if they are not exercised or practiced or if external (environmental) or internal (personal/physical/emotional) conditions stunt their development. They can become inaccessible or limited.

Efficacy, consciousness, craftsmanship, flexibility, and interdependence, while native to humans, must be refined, nurtured, supported, practiced, and fulfilled on a regular basis if they are to remain constants. When treating humans as if they are intelligent, they actually become more so. When challenging tasks requiring persistence are given, when children and adults experience complex thinking, when they have to solve their own problems and are held responsible for solving them and for experiencing the consequences, these capacities are developed more fully and are more likely to continue developing over time. Cognitive Coaching strengthens and makes these resources more accessible in the moment and over time.

What follows is (1) an exploration of the sources of each of the states of mind as innate, human drives; (2) an explanation of the meaning and manifestations of that state of mind; (3) how Cognitive Coaches mediate by liberating and enhancing that state of mind; and (4) indicators of each state of mind as a source of excellence in educational settings.

Efficacy

> Self-development of personal efficacy requires mastery of knowledge and skills attainable only through long hours of arduous work.
>
> —Albert Bandura (1997)

The Pursuit of Self-Efficacy Is Innate

The main purpose of a brain is survival, and many of the structures in the brain are involved in making certain the individual and the species endure. Although these structures were designed originally to give the person the means to survive attacks from wild beasts or enemies, in contemporary society, efficacious humans quest for competence, continuous lifelong learning, self-empowerment, goal achievement, mastery, and control. For example, as a survival mechanism, endurance running was the main means of stalking prey by outrunning them. Humans today are the only form of life that runs marathons in the heat of the day for which they must endure and persist. We climb to the highest pinnacles, break previous records, and lend ourselves to competitions with ourselves, with each other, and with nature. Each time we experience the satisfaction of goal accomplishment, our brain secretes endorphins—the feel good chemicals (neurotransmitters) that are active in the brain's reward system. In other words, the efficacious brain makes "feel good" chemicals that are released when certain behaviors increase the probability of survival.

Meanings and Manifestations of Efficacy

Particularly catalytic, our sense of efficacy is a determining factor in the resolution of complex problems. Efficacious people are resourceful. They engage in cause-and-effect thinking, devote energy to demanding tasks, set challenging goals, persevere in the face of barriers, and learn from an occasional failure (Dweck 2006). People with efficacy accurately forecast future performances. They are optimistic and confident, and they feel good about themselves. They push beyond self-doubt when it exists. They have sound self-knowledge, and they control performance anxiety. Effort and persistence, in the face of setbacks, are hallmarks of efficacy. The more efficacious we feel, the more flexibly we can engage in critical and creative work. Developing effective thinking, therefore, requires becoming increasingly self-managing, self-monitoring, and self-modifying.

Charles Garfield's study of peak performers found that their primary locus of control is internal, not external. One element that stands out clearly among peak performers is their virtually unassailable belief in the likelihood of their own success. Their track records reinforced their beliefs. Conversely, they also recognized when a task was unattainable, and they moved their energies elsewhere.

Efficacy transcends race, gender, culture, income level, subject matter, and age of students. It is a predictor of group perseverance on special projects. It correlates with higher math achievement, more science learning, improved language skills, and teachers' increased willingness to implement innovation. Efficacy also correlates with positive teacher–administrator relationships, constructive parent–teacher relations, and reductions in teacher stress.

Cognitive Coaching Mediates Efficacy

These descriptors provide a vision of the desired state toward which coaches can facilitate growth. One value of efficacy and the resulting self-confidence is that teachers are free to be more flexible. Flexibility is critically important for any creative work in which one wants to see the overall pattern as well as the details.

Studies of Cognitive Coaching consistently find significant improvements in teacher efficacy. Efficacy may be the most catalytic of the five states of mind because a person's sense of efficacy is a prime factor in determining how complex problems are resolved. If a teacher feels little efficacy, then despair, hopelessness, blame, withdrawal, and rigidity are likely to follow. However, research indicates that teachers with robust efficacy are likely to expend more energy in their work, persevere longer, set more challenging goals, and continue in the face of barriers or failure. In sum, efficacy is characterized by the following:

- Efficacy is related to specific areas, such as teaching math or working with middle school kids.
- Efficacy is related to doing certain types of tasks, such as getting things organized, so a person may feel efficacious in one context and not in another.

- Efficacy changes over time with influences from new information and task experiences.
- Personal efficacy stems from self-assessment of teaching skills and influences the effort that teachers expend working with students.
- Personal efficacy also is achieved by professional development: learning about subject matter, pedagogy, students, and self.
- Outcome efficacy stems from self-assessments of teaching results. When teachers (or students) are involved in developing the criteria for performance, there is a greater likelihood that they will become self-managing and self-monitoring.
- Outcome efficacy influences teachers to modify instruction.
- Outcome efficacy is enhanced through creating learning communities in which teachers collaborate to improve instruction.
- Highly efficacious persons are frequently optimistic, less stressed, able to persevere through difficulties to achieve results, high achievers, self-actualizing, and self-modifying.

Efficacy is a prerequisite for improved student learning. Fullan (1982) regards teacher efficacy as a vital factor in the successful implementation of change. Rosenholtz (1989) found that teachers' efficacy influenced students' basic skills and mastery. The more certain teachers feel about their technical knowledge (personal efficacy), the greater the students' progress in, for example, reading. The more teachers are uncertain, the less students learn.

One study on the relation between efficacy and curriculum implementation showed that teachers' efficacy and interdependence significantly predicted the implementation of new curriculum guides (Poole and Okeafor 1989). Neither efficacy nor teacher interactions (interdependence) alone produced a significant difference in use of the curriculum, but together they brought about change.

In the Rand Corporation's seminal research on school effectiveness, Berman and McLaughlin (1977) found that collective teacher efficacy was the single most consistent variable related to school success. The efficacy identified by the Rand Corporation study is what Fuller, Wood, Rapoport, and Dornbusch (1982) label "organizational efficacy," or the link between people's perceptions of valued goals and their expectations that those goals can be achieved by participating in the organization (outcome efficacy).

Laborde and Saunders (1986) report that people governed by an internal locus of control (see chapter 8) show initiative in controlling their environment. They control their own impulsivity, gather information, are cognitively active, eagerly learn information that will increase their probability for success, and show a sense of humor. Compared to individuals with an external locus of control, they are more trustful, less anxious, less hostile, less angry, less suspicious of others, less prone to suicide, less depressed, and less prone to psychosis.

Efficacy in Educational Settings

To act with confidence, meet challenges, and cope with situations that are new to them, teachers must feel that they are competent to control these situations—to overcome difficulties, become familiar with the new and the unknown, and approach these with the expectation they will master them (Feuerstein, Feuerstein, and Falik 2010, 80). Such teachers:

- Have an internal locus of control
- Operationalize concepts and translate them into deliberate actions
- Pose problems
- Make causal links
- Produce new knowledge
- As continuous learners, establish feedback spirals and continue to learn how to learn seeking to modify themselves through feedback (Bandura 1982)
- Are optimistic and resourceful—self-actualizing and self-modifying
- Implement curriculum in collegial environments and translate concepts into action

Consider the state of mind of efficacy in the following example. Schools A and B serve low-income, minority children. Teachers in each school have similar years of teaching and academic preparation. Both work in old buildings with the same instructional materials. School A's teachers share a growth mind-set, are collaborative, visionary, and upbeat. School B's teachers share a fixed mind-set, believing that intelligence is fixed and cannot be modified. They are frustrated and stressed from the standards and evaluative measures being imposed upon them. They believe that if their textbooks matched their curriculum, if parents would be more supportive, and if students were better motivated, reading achievement would be higher. In School A, most teachers feel confident in their knowledge about reading and believe that if they work hard enough—and smart enough—students will learn. Test scores at each school reflect teachers' beliefs.

School A's teachers have personal efficacy (the belief that they have the knowledge and skills to teach reading) and outcome efficacy (the conviction that when they use their teaching skills, students achieve). Teachers in School B have low efficacy and, as a result, are more prone to blame others for poor student achievement. They also feel more stress and are less positive about teaching (Hoy, Tarter, and Woolfolk Hoy 2006).

Flexibility

> It is change, continuing change, inevitable change, that is the dominant factor in society today. No sensible decision can be made any longer without taking into account not only the world as it is, but the world as it will be.
>
> —Isaac Asimov

Flexibility as an Innate Drive

To live is to change. Humans have survived over eons because of their capacity to adapt to changing environments, needs, and technologies. An amazing discovery about the human brain is its plasticity—its ability to "rewire," change, and even repair itself to become smarter. Researchers have described the brain's role in cognitive flexibility as it adjusts thinking from old situations to the new, overcoming responses that have become habitual (Moore and Malinowski 2009), and its relationship to executive functioning, planning, and working memory (Miyake et al. 2000).

Humans are a unique form of life because they can perceive from multiple perspectives and can deliberately change, adapt, expand, and control their repertoire of response patterns. Located in the prefrontal cortex, response flexibility enables the mind to attend and assess the subtle verbal and nonverbal cues, and then to modify internal external reactions accordingly. Flexibility is the mind's ability to deal with human differences—conflicting ideas, alternative perspectives, divergent points of view, and collective problem solving.

Meanings and Manifestations of Flexibility

Because flexible thinkers are able to see through the diverse perspectives of others, they are considered empathic. They are open and comfortable with ambiguity. They create and seek novel approaches and have a well-developed sense of humor. They envision a range of alternative consequences. They are open-minded, having the capacity to change their minds as they receive additional data. They engage in multiple and simultaneous outcomes and activities. They draw on a repertoire of problem-solving strategies.

Flexible thinkers understand causal relationships, allowing them to work within a rule-bound structure to reengineer rules to help, rather than hinder, their work. They understand not only immediate reactions but are also able to perceive the bigger purposes. Like the queen in the game of chess, the most flexible person is the one with the greatest number of choices and alternatives. Recognizing options and willingness to test them is the hallmark of flexibility.

The peak performers that Garfield (1986) studied had a quality of flexible attention, which he called "micro/macro attention." Microthinking involves logical, analytical computation and seeing cause and effect in methodical steps. This mode is important in the task analysis portion of planning a lesson or a curriculum. It encompasses attention to detail, precision, and orderly progression.

The macro mode is particularly useful for discerning themes and patterns from assortments of information. It is intuitive, holistic, and conceptual. Macrothinking is good for bridging gaps and enables us to perceive a pattern even when some of the pieces are missing. It is useful in searching for patterns in a lesson or in a week of lessons.

As noted in Garfield's (1986) study of peak performers, they practice style flexibility, knowing when it is appropriate to be broad and global in their thinking and when a situation requires detailed precision. Peak performers also trust their intuition, honoring the mind–body relationship. They tolerate confusion and ambiguity up to a point, and they are willing to let go of a problem, trusting that their creative unconscious will work productively (Poole and Okeafor 1989).

Flexible people can live with ambiguity and doubt because they have a great capacity to look upon life as a series of problems to be solved. They enjoy problem solving because it's a challenge. Flexibility, like efficacy, is related to risk taking. David Perkins (1983) describes creative people as "living on the edge." They are not satisfied with living in the middle; they are always pushing the frontier. They generate new knowledge, experiment with new ways, and constantly stretch themselves to grow into new abilities.

Cognitive Coaching Mediates Flexibility

Unless a teacher thinks flexibly, what is the likelihood that, when faced with stubborn problems, she will consider various options and perspectives or even think about the perspective of a struggling parent? Flexibility involves the ability to step beyond and outside oneself and look at a situation from a different perspective. This is what Jean Piaget (1954) called "overcoming egocentrism." Many psychologists believe that this is the highest state of intelligent behavior. As we shall see later, it is also a prerequisite state of mind for functioning interdependently. The mental skills of flexibility include the ability to shift among egocentric, allocentric, and macrocentric views; between a bird's-eye view (detail) and an eagle-eye focus (gestalt); between the immediate present and the long-range future; between logic and intuition; and between an individual and a group.

With clear intentions to enhance flexibility in thinking, coaches deliberately compose such mediative questions as:

- What might be some other options you have?
- As you evaluate these alternatives, which seem most plausible? As you brainstorm all the possibilities, which . . . ? Knowing that John is a kinesthetic learner, what might be some ways to help him with this math concept?
- Because there doesn't seem to be an immediate solution to this problem, what might you be telling yourself, to hold it in abeyance?

Research evidence from Edwards (chapter 14), reporting on the results of teachers being coached by experienced coaches, indicates that teachers increased their creativity and flexibility.

Flexibility in Educational Settings

According to Shulman (1987), expert teachers are more opportunistic and flexible in their teaching. They take advantage of new information, quickly bringing new interpretations and representations of the problem to light. It is this flexibility, and not merely the knowledge or experience of possible scenarios, that makes the difference.

Flexible teachers are aware of and legitimize differences of opinions, tendencies, desires, and styles without necessarily accepting them. Flexible teachers search for and value the differences between individuals and their unique behaviors. They are continually forming a distinct and accepting self-perception in relation to others. Such teachers:

- Are more adept at developing and testing hypotheses about learning difficulties or instructional strategies and look upon each experience as learning opportunities
- Are willing to consider and embrace change (Hattie 2003, 8)
- Adjust to students' diverse styles, preferences, cultures, and developmental levels
- Tolerate ambiguity, seeking further information rather than being satisfied with available data
- Are more adept at anticipating problems and then improvising, seeking, and generating alternatives
- See through multiple perspectives
- Make distinctions between surface and deep learning; surface learning is more about content and deep learning more about understanding and imposing personal meaning (Hattie 2003, 9)

Consciousness

The White people think the whole body is controlled by the brain. We have a word, *umbelini* (the whole intestines): that is what controls the body. My *umbelini* tells me what is going to happen: have you never experienced it?

—Mongezi Tiso, Xhosa tribesman, South Africa

Consciousness as an Innate Drive

The neurobiology of consciousness is organized around what Damasio (2010) calls the triad of wakefulness, mind, and self. The brain stem, the thalamus, and the cerebral cortex are the major players. Although consciousness may not be unique to *Homo sapiens*, cognitive ethnologists claim that some animals also have intentions, beliefs, and self-awareness and that they consciously think about alternative courses of action and make plans (Shettleworth 1998). The great advantage that humans have over animals, however, is that we can strengthen and direct our capacity for consciousness because of the neural anatomy of our brain, which permits development of an autobiographical self. Human consciousness exceeds that of other life forms in that the human conscious mind encompasses actual and imaginary contents effortlessly, thus allowing planning and reflection. One way of describing human consciousness is that we are aware that certain events are occurring, and we are able to direct their course (metacognition). As humans, we can direct or control whatever we are conscious of within ourselves.

The function of consciousness in humans is to represent information about what is happening outside and inside the body in such a way that it can be evaluated and acted upon by the body. To be conscious is to be aware of one's thoughts, feelings, viewpoints, and behaviors and the effect they have on oneself, on others, and on the environment. A unique function of human consciousness is to reflect on what is past, imagine what could be, and plan for its achievement.

Over centuries of evolution, the human nervous system has become so complex that it is able to affect its own states. We can make ourselves sad or happy, regardless of what is happening outside. Yet the human nervous system is still limited to managing very limited bits of information at any one time, such as differentiated sounds, visual stimuli, or nuances of emotion or thought (Rock 2009). If conscious capacity is limited, however, how do teachers, parents, coaches, or students focus their consciousness to serve them in their goals?

Damasio (2010) suggests that consciousness begins when brains acquire the power

> of telling a story without words, the story that life is ticking away in an organism, and the states of the living organism, within body bounds, are continuously being altered by encounters with objects or events in its environment, or, for that matter, by thoughts and by internal adjustments of the life process. Consciousness begins when this primordial story—the story of an object causally changing the state of the body—can be told using the universal nonverbal vocabulary of body signals. The apparent self emerges as the feeling of a feeling. (30–31)

Meaning and Manifestations of Consciousness

The mark of people who are in control of consciousness is their ability to focus attention at will, to pay attention to their intentions, to deflect distractions, to concentrate for as long as it takes to achieve a goal. Expanding consciousness informs improvement and helps to expose blind spots and bring up new ideas. Developing effective thinking therefore requires the development of consciousness, a priceless resource (Csikszentmihalyi 1990).

Those who exercise consciousness monitor their own values, intentions, thoughts, and behaviors and their effects on others and the environment. "Moral behaviors are a skill set, acquired over repeated practice sessions and over a long time, informed by consciously articulated principles and reasons but otherwise 'second natured' into the cognitive unconscious" (Damasio 2010, 271). This is an apt description on the evolution of a mediator. They have well-defined value systems that they can articulate. They generate, hold, and apply internal criteria for decisions they make, and they can articulate the rationale behind their actions and thoughts. They practice mental rehearsal and the editing of mental pictures in the process of seeking improved strategies.

Consciousness means knowing what and how we are thinking about our work in the moment and being aware of our actions and their effects on others and on the environment. Consciousness is the central clearinghouse for executive decision making. This state of mind is a prerequisite for self-control and self-direction. Through consciousness we are metacognitively aware that certain events are occurring, and we are able to try and direct their course.

Consciousness is the medium in which all states of mind are mediated. To make personal change, one must be conscious of one's own inner workings (Vaida 2014). To mediate consciousness in others, we must be attentive to both verbal and nonverbal communications in order to help others locate the words that represent their experiences.

Recently Bob was on safari in the Ngorongoro Crater in Tanzania. Late one morning, he stepped from his Range Rover to observe some plant life more closely. Suddenly, a glowering hippopotamus—weighing 5,000 to 6,000 pounds and with lower canine teeth 18 inches long—emerged from a waterhole. Grunting and snorting, glistening and frightening, the hippo charged. Terrified, Bob dashed for the Range Rover and safety.

What happened in those few seconds? Bob's brain detected threat and, with lightning speed, conjured up a few response options, selected one, and acted. Neural and chemical aspects of the brain's response caused profound bodily changes—surges in metabolic rate and energy. The overall biochemical profile of his body fluctuated rapidly, evoking changes in both brain and body, skeletal muscles contracting. His frightened eyes absorbed the scene connecting the retina and the brain's visual cortices to reactions in his viscera. In time, this memory will become a neural record of many of the changes that occurred, some in the brain itself and some in the body.

Although teaching is rarely this exciting, the same neurochemical processes ignite when a threat or any intense experience is perceived. Cognitive Coaches must realize that "cognitive" always has an "emotional" component. Emotions and the viscera often perform an important role in perception and decision making. This is true in situations of clear and immediate danger, like a charging hippopotamus. It is also true in less obvious dangers, like perceived ego risk in a conversation.

The following are some characteristics of consciousness:

- The capacity for consciousness lies in our neurobiology.
- The development of consciousness is influenced by mediational relationships, personal intention, and practice.
- Consciousness is involved when deliberate, rather than automatic, control or intervention is needed.
- We are consciously aware of only a small part of what our minds are taking in at any given time.
- Conscious processes occur one at a time, take effort, and are inefficient.
- Consciousness is not devalued by the presence of nonconscious processes; that is, both are important sources guiding human experience.
- Conscious processes are necessary transit zones in which to unlearn old patterns and learn new ones at an unconscious level.
- Consciousness provides us with a unique, consistent interpretation while unconscious processes produce contradictory interpretations of experience.
- We can be aware of something without really being aware of it. As an example, you are probably not aware of the position of your left ear until we've called attention to it, yet it is never out of your unconscious awareness.
- Consciousness can be strengthened through self-observation.

Cognitive Coaching Mediates Consciousness

Coaches strive to help colleagues bring to awareness how they feel about a situation. Met in awareness, these feelings sometimes evoke gut reactions that open or close some options to deliberations. Once these feelings are brought to consciousness, colleagues are in a better position to apply their intuitions within the more cognitive processes of decision making. Feelings are translated through images—held in the brain as pictures, smells, feelings, or sounds related to the body's perceptual systems—and as abstract patterns as well. When feelings are brought to the surface, they can also be examined or challenged to determine if they are interpretations of the real event being worked with or a memory of another event that therefore eliminates viable options from subsequent consideration.

Coaches often shine a spotlight on data previously not noticed by the teacher. Three stimuli bring data to consciousness: a discrepant event, a flashlight shined by another person, or intention. Imagine you are in traffic, hardly noticing the shapes or occupants of the cars around you, when a car with flashing lights (a discrepant event) suddenly breaks into your consciousness. Or imagine that a friend, serving as a spotlight, asks a question that causes you to examine a feeling or theory or action of yours. Finally, notice that whenever you are clear about your intentions, your attention is riveted in that direction.

People engaging in an enriched state of consciousness often think about their own thinking and feeling. They monitor their values, thoughts, behaviors, and progress toward goals. They articulate well-defined value systems, and they generate and apply internal criteria for decisions. They practice mental rehearsal and edit mental pictures as they seek to improve performance.

Consciousness in Educational Settings

The complexity of classroom life, the conditions in which teachers are called on to react, makes the ability to regulate behavior in socially and culturally appropriate ways critical for teachers. The regulation of behavior is a product of an individual's ability to impose thinking on actions—to examine oneself, to assess the situation, and to decide how and when to react. Such teachers:

- Are aware that certain events are occurring and are able to direct their course
- Monitor their own values, intentions, thoughts, and behaviors and their effects on others and the environment
- Have well-defined value systems that they can articulate and generate
- Hold and apply internal criteria for decisions they make
- Seek improved strategies through practicing mental rehearsal and editing of mental pictures

Christina Linder, a middle-school English teacher, reports: "In learning to more clearly project outcomes for my lessons, I found that I was planning more precisely, and that smaller, more tangible goals were appearing in my lesson plans. I had often suspected that at times specific student needs had been sacrificed in attainment of the broader goal."

At the same school, Jan Whitaker wrote the following: "After a few Cognitive Coaching sessions, I realized it wasn't the material that was important but how it caused thinking in my students. Missing in my very detailed and organized lessons was the opportunity for students to develop creativity, craftsmanship, application, analysis, and higher levels of thinking."

Both teachers commented on how this awareness led to changes in their teaching. Consciousness is the monitoring arena for each of the other states of mind. If one's personal efficacy becomes depleted, if one is rigidly seeing from only a single perspective, or if one is acting impulsively at the moment of consciousness of these responses, one has choice among alternate responses.

Craftsmanship

> Learn to do uncommon things in an uncommon manner. Learn to do a thing so thoroughly that no one can improve upon what has been done.
>
> —Booker T. Washington

Craftsmanship as an Innate Drive

When the cave people had their bellies full and their fires lit, they turned their attention to decorating the cave walls. Craftsmanship drove them to embellish, hone, refine, and constantly work to improve themselves and their environment. Craftsmanship is the healthy dissatisfaction that humans feel with their own accomplishments. Humans yearn to become clearer, more linguistically precise, congruent in beliefs and actions, and integrated with an aligned work life.

Meanings and Manifestations of Craftsmanship

Excellence in performance is the soul of craftsmanship. People working toward craftsmanship strive for precision. They seek perfection, elegance, refinement, and specificity. They generate and hold clear visions and goals. They monitor progress toward those goals. Achievement of these goals generates an increase in dopamine, the brain's reward system. The holonomous person who is both flexible and craftsman-like attends to the big picture and to details. In one setting, a holonomous teacher may be excruciatingly detailed. In another setting, that teacher might be artfully vague. For example, when giving classroom instructions, the teacher may give detailed instructions ("Fold your papers in half and then half again . . .") whereas when seeking cooperation from a parent, the teacher may be deliberately vague: "Because we both want Rafael to succeed . . . ."

Craftsman-like people are characterized by the following:

- They assess their own performance and results.
- They value seeking data about their work in order to study and improve it.
- They envision and set high standards for themselves.
- They may appear to perform flawlessly and easily, but underneath that appearance is a great understanding of the complexity of their work and practice.
- They calibrate and monitor progress toward goals.
- They strive for continuous improvement.
- They persevere to close the gap between existing and desired states.
- They monitor and manage refinements in thought and language.
- They predict, monitor, and manage time.
- They distinguish between perfection and excellence.

Cognitive Coaching Mediates Craftsmanship

Language and thinking are inseparable. Language plays a critical role in enhancing a person's cognitive maps and has a great impact on an individual's ability to think critically, create a knowledge base for action, and feel efficacious. The skillful coach, recognizing that language and thinking are closely entwined, consistently strives to enhance the clarity and specificity of the teacher's language. Coaches enhance craftsmanship by posing such questions as:

- So how do your actions compare with your expectations of yourself?
- So you're gratified with your progress so far. What will be your next task to master?
- As you consider your desire to meet students' individual needs, what might that look like in your classroom?
- When you say that your students "understand," what do you see them doing or hear them saying that may indicate they understand?
- Based on the results of this experiment, what refinements might you make in your approach?
- So as you learn more, you are finding that there is still more to learn. What more do you want to learn about this topic?

Craftsmanship in Educational Settings

To further appreciate craftsmanship, consider studies from the League of Professional Schools (Lunsford 1995). They found that in schools where teachers are the most successful, these same teachers have the highest dissatisfaction with their work. The drive for elaboration, clarity, refinement, and precision—all attributes of craftsmanship—serve as energy sources from which faculties improve instruction and student learning.

Teachers are required to cope with complex, often unexpected tasks. Meeting a challenge requires readiness not only to excel in a familiar area but also to confront unfamiliar and complex problems. Meeting a challenge relates to something that does not already exist; it entails anticipating potential outcomes and strategies. Teachers who exhibit craftsmanship can meet challenges. Such teachers:

- Strive to continually perfect their craft
- Set and work to attain personal high standards
- Pursue ongoing learning
- Seek precision, mastery, refinement, and pride in their artistry
- Generate and hold clear visions and goals

- Strive for exactness of critical thought processes and communication
- Test and revise, constantly honing strategies to reach goals
- Attend to what they know and what they still need to learn

The craftsman-like teacher is also precise in managing temporal dimensions. Teachers orchestrate across six time dimensions: sequence (in what order), duration (for how long), rhythm (at what tempo, pattern, and speed), simultaneity (along with what else), synchronization (with whom and what), and short- and long-term perspectives. A craftsman-like teacher strives for excellence in each of these dimensions.

Interdependence

> We've each been invited to this present moment by design. Our lives are joined together like the tiles of a mosaic; none of us contributes the whole of the picture, but each of us is necessary for its completion.
>
> —Karen Casey and Martha Vanceburg

Interdependence as an Innate Drive

The brain's function is to regulate life. Somewhere in human evolution this emerged as managing well-being in relationship to others, in social systems (Damasio 2010). Humans desire reciprocity, belonging, and connectedness. We are inclined to become one with the larger system and community of which we are a part.

In human evolution, successful hunters and gatherers had a better chance of survival if they worked together with others. Humans are relatively weak, with no fur or natural armor with which to protect themselves. Being a weak, lone individual competing for scarce resources would certainly be a disadvantage. Establishing cooperative agreements with others would make survival more likely when resources were limited. Eventually our ancestors' brains began to change from purely survival brains into social brains. Most scientists now agree that our social brains have been shaped by natural selection because being social enhances survival (Cozolino 2006).

Further evidence to support the concept of the social brain comes from studies of very young children. Tomasello (2009) makes a strong case for an innate capacity that children have for empathy and cooperation. At an astonishingly early age, children begin to help one another and to share information. It appears that we are born ready to cooperate. (See also Costa and O'Leary 2013; Lieberman 2013; Goleman 2006.)

Meanings and Manifestations of Interdependence

Interdependent people have a sense of community: "we-ness" as much as "me-ness" (Sergiovanni 1994). They are altruistic. They value consensus, being willing to influence and be influenced by the group in service of group goals. They contribute to a common good, seek collegiality, and draw on the resources of others. They regard conflict as valuable, trusting their ability to manage differences in a group in productive ways. The more highly evolved, the more humility they possess, and the more they view conflict as opportunities to learn about and modify themselves (Kegan and Lahey 2009), the more they exhibit interdependence. They continue to learn from feedback from others and from their consciousness of their own actions and effect on others. They seek collaborative engagement, knowing that we are more effective together than individually.

Interdependence means knowing that we will benefit from participating in and contributing to work. Learning unfolds within social contexts, and hence interdependence increases intelligence. Every function in cultural development appears twice: first, on the social level, and later on the individual level; first between people (interpsychological), and then inside (intrapsychological). This applies equally to voluntary attention, to logical memory, and to the formation of concepts. All the higher functions originate as actual relationships between individuals (Vygotsky 1978). Interdependent people envision the expanding capacities of the group and its members, and they value and draw on the resources of others to enhance their own personal competencies.

Margaret Mead's statement "Never doubt that a small group of thoughtful, committed citizens can change the world; indeed, it's the only thing that ever has" captures this thought.

Cognitive Coaching Mediates Interdependence

Learning to think interdependently and work collaboratively requires coaching (Knowles, Holton, and Swanson 2005, 3). Coaches help others to grow and develop as collaborative thinkers. Developing interdependent thinkers requires the initiated and sustained facilitation of adaptive learning. According to Knowles, adults need:

- To know the "why, what, and how" of things
- To be self-directed
- The resource of prior experiences as influences on their mental models
- A readiness to learn when information is connected to their life and the learning is developmental
- An orientation to learning that is problem centered and contextual
- A motivation to learn for both intrinsic values and personal payoff

Coaches, therefore, intentionally direct their mediative questions to cause others to think interdependently. For example:

- What do you see as some benefits of allowing the group to come to consensus about the agenda?
- What might you have to give up in order to have the group unified on this approach?
- What might be some of the common values that you and Maryann share?
- What might it take to build group commitment to this plan?
- How might you lend your knowledge and resources to the group's accomplishment of its goals?
- With whom might you team to work on this task?
- As you consider John's strengths, how might he make his best contribution to this task?
- While you and Maryann disagree about how best to work with this student, how might you transcend your differences and learn from each other?
- Because you want as many interested parties involved as possible, who else might you involve in solving this problem?

Interdependence in Educational Settings

Interdependence is an essential state of mind for modern schools. For example, Rosenholtz (1989), in her study of elementary schools in Tennessee, found that the single most important characteristic of successful schools was goal consensus. Saphier, Haley-Spica, and Gower (2008) of Research for Better Teaching report that the issue of good schools is so simple that it's embarrassing: In good schools, people have common goals, and they work together toward those goals.

Interdependence modifies working relationships. Sarason (1991) suggests that among the most complex tasks humans undertake is recognizing and trying to change power relationships. Sarason believes this is especially true in complicated traditional institutions. The school-improvement agenda of the new millennium is moving all educators into more interactive and professional roles. So from the perspective of what's good for schools as well as the perspective of what's good for individuals within the schools, we find that interdependence is a rich and essential resource.

There is great educational value in sharing our emotional and cognitive/mental treasury of behaviors. In some large schools and institutions, however, several hundred, or even thousand, adults and children can spend time in the same building in the closest of physical proximity and not know one another or greet one another when they meet by chance. Interdependent teachers

- Know that they will benefit from working collaboratively
- Are altruistic and willing to change relationships to benefit the larger good

- Value consensus, while being able to hold their own values and actions in abeyance
- Lend their energies and resources to the achievement of group goals
- Contribute to a common good
- Seek collegiality
- Draw on the resources of others
- Regard conflict as valuable and manage group differences in productive ways
- Seek collaborative engagement, knowing that we are more effective together than individually

MEDIATING RESOURCES FOR HOLONOMY

We either make ourselves miserable, or we make ourselves strong. The amount of work is the same.

—Carlos Castaneda

Holonomy describes a characteristic of dynamic or nonlinear systems in which everything influences everything else. Tiny events can cause major disturbances in the system, and you do not need to touch everyone to make a difference. Each autonomous unit in a holonomous system influences the system in addition to being influenced by it (Garmston and Wellman 2013).

Of course, the goals of holonomy are never fully achieved. In addition, there is no such thing as perfect attainment of the states of efficacy, flexibility, craftsmanship, consciousness, and interdependence. These are utopian energies toward which we constantly aspire. The journey toward holonomy and the five states of mind is the destination.

Performing intelligently in holonomous systems means living in tensions: whole and part, event and context, individual and group. It is a life work, described by Jung as individuation—an integrated development of the physical, emotional, spiritual, intellectual, and social being. It is a human journey everywhere, across all continents, ages, and cultures. It is most complex in the 21st century in postindustrial societies such as ours, in which Robert Kegan (1994) estimates that perhaps as much as 50 percent of Americans have not achieved the cognitive complexity needed to cope with daily life. Our dream is that all educators will one day possess the complexity to provide the very best instruction for students. As described earlier, our mission is to foster autonomous, self-directed learners, capable of intelligent, values-based living within holonomous systems. Although there are many important aspects related to mediation, the most critical is to remember one's intention in a mediational transaction.

There is no doubt that school staffs that exercise intelligent membership in holonomous systems also work collaboratively and produce better results with students than do faculties whose members work in isolation from one another. Such workplaces are characterized by clear values and norms, reflective dialogue, deprivatized teaching, and an atmosphere in which the primary focus of energy, attention, and conversation is on student learning. Richard Elmore and Elizabeth City (2007, 3) note that "The discipline of school improvement lies in developing strong internal processes for self-monitoring and reflection." This requires self-directedness—an enduring goal in the work of Cognitive Coaching.

Mediating holonomy pursues three outcomes: developing capacity among the parts, strengthening the whole, and creating the consciousness and skills with which part and whole inform and support the continuing development of each another. To achieve the first outcome is to support people in becoming autonomous and self-actualizing. Efficacy and craftsmanship especially contribute to this; so do supporting teachers in a continual inquiry into the knowledge base about teaching and developing commitment to setting and working toward high standards.

The second outcome calls for members of the school community to function interdependently, recognizing their capacity to both self-regulate and be regulated by the norms, values, and goals of the larger system. Consciousness, craftsmanship, and interdependence are primary energy sources. To attain this requires clear standards for collaborative work and for teaching, if necessary, the structural, communication, and metacognitive skills required for productive collaboration.

The third outcome is achieved by developing each part and whole to value the other component. This outcome also is achieved by devising and using strategies that draw on the advantages of each while avoiding the shortcomings. The entire set of states of mind contributes to each of these outcomes and, perhaps more than anywhere else, is true for this third outcome.

CONCLUSION

The five states of mind can serve as diagnostic constructs through which we can assess the resourcefulness of others and plan interventions. However, assisting others toward refinement and expression starts first with your own states of mind. From there, you can work with others, the system of which you are a part, and even students.

Keep in mind the following three attributes of the states of mind:

- They are transitory; they come and go. A person's state of mind varies depending on a number of factors, including familiarity, experience, knowledge, fatigue, and emotion.

- They are transformational. Dramatic increases in performance accompany heightened states of mind. Just as one's current state of confidence will affect in-the-moment competence, the states of mind access the personal resources required for peak performance.
- They are transformable, either by oneself or by another. A colleague can cheer you up or encourage you, changing in that moment your state of mind and your capacity. Similarly, your own conscious awareness of your states of mind allows you to choose and to change them.

Holonomy captures two universal human tendencies: the self-assertive desire to preserve individual autonomy and the integrative drive to part something larger than the self. When the five states of mind are thought of as a whole, they create an inner tension that asks one to consider individual needs, the drive for autonomy, in the context of community. Together they form a thoughtful scaffold for thinking that is both interwoven and interdependent. Each of these concepts supports and contributes to the success of the other, and hence we resist the urge to separate them and attempt to observe them as independent capacities.

8

Teacher Cognition and Instruction

It is obvious that what teachers do is directed in no small measure by what they think. . . . To the extent that observed or intended teaching behavior is "thoughtless," it makes no use of the human teacher's most unique attributes. In so doing, it becomes mechanical and might well be done by a machine. If, however, teaching is done and, in all likelihood, will continue to be done by human teachers, the question of relationships between thought and action becomes crucial.

—National Institute of Education (1975)

Let's take an imaginary tour of a elementary classroom to observe cognition at work during teaching. As you walk in, you are immediately struck by all the activity around you. A group of students works on a papier-mâché relief map. Other students work at the computer. A few students are helping each other with their homework. Still others are working at the science table. The teacher sits at a round table working with a small reading group. She listens to Rebecca read; next she listens to and corrects Diane. She diagnoses Tyrell, then Leroy. At that moment, an office monitor enters and delivers a note. The teacher, still listening to Rebecca, reads the note, commends the monitor for coming in so quietly, composes a response, sends him out, and never misses a beat with Rebecca.

Or consider this high school classroom, where George arrives late, noisy, and looking very dejected. Glancing at George, the teacher instantly makes an assessment of his emotional state, what's happening with the class discussion, and George's potential for disruption. She tosses George a humorous greeting: "Hi, George! Good to see you. Make friends with that chair over there and turn to the chart on page 70." She then locates in her mind some appropriate open-ended questions for students to work on. She instructs the students to think-pair-share about these questions for two minutes. In this way, she has achieved two goals simultaneously: providing George a chance to settle in and offering other students the opportunity to check their under-standing of the concept just taught.

Danielson (2007, 2) notes that "More recent research has confirmed that teaching is also cognitively demanding; a teacher makes hundreds of nontrivial decision daily." One of the great cognitive capacities of teachers is this ability to simultaneously man-age a multitude of activities in the classroom. Professionals have an extensive and con-tinually expanding body of knowledge and skills, and they make decisions about when and how to apply which portions of that knowledge in diverse situations. This is what makes teaching among the most intellectually complex and demanding professions.

RESEARCH ON TEACHERS' THINKING

Teacher cognition research is concerned with understanding what teachers think, know, and believe as they engage in the acts of planning and implementing instruc-tion, and reflecting on their teaching. Its primary concern, therefore, lies with the unobservable qualities of teaching—teachers' mental processes. During the mid-1970s, great advances were made in understanding the ways that cognitive systems perform in learning (Borg 2006). According to Borg (2009), however, little research has focused on teachers' cognitive processes since then. Thus, many of the references cited in this chapter date from that era.

In the 1960s, research on teaching focused on the search for effective teaching behaviors as correlated with student achievement tests scores. The purpose was to identify certain effective behaviors in the belief that teachers could then apply them universally. In the 1970s, however, this view of teaching became questionable. Cog-nitive psychologists found complex relationships between what teachers do and what they know, feel, and believe. We became more aware of the fact that teachers' cog-nition influenced their instructional decisions. Teachers were not pawns that simply implemented curricula designed by others in an unthinking manner. Rather, teachers exerted agency in their work (Bandura 1982). They made decisions, both before and while teaching a lesson. They reflected upon and modified their decisions (Schon 1983). Such cognitive processes became a new focus for educational researchers. The questions they asked were not simply "What do teachers do," but also "What do teachers think?" "What decisions do they make?" and "What prompted those decisions?" The notion of universally applicable teaching behaviors was viewed with increasing skepticism as the uniqueness of classroom and school cultures, the uniqueness of teachers and of learners became acknowledged (Borg 2009; Costa, Garmston, and Zimmerman 2014).

Teachers who possess cognitive systems with highly developed levels of percep-tion, abstraction, complexity, and decision making consistently have students who perform well on both lower and higher cognitive tasks (Clark and Peterson 1986). In this chapter, we describe the act of teaching as a highly intellectual process that involves continual decision making before, during, and after instruction. We also examine how linguistic, nonlinguistic, and affective systems affect cognition and teacher thinking, including the implications for Cognitive Coaching.

FINDINGS ABOUT TEACHER COGNITION

Many researchers have contributed to an understanding of teacher thought and the stages of cognitive complexity that adults can attain. Cognitive Coaching is a process of engaging, enhancing, mediating, and the intellectual functions of teaching, or, as Marzano (2011, 90–91) would describe it, a cognitive system informing teaching. As we shall see, however, the intellectual interrelates with nonlinguistic ways of knowing, including the emotions, biases, perceptions, and reflexes (Bargh 2014). What teachers think affects how they behave, and their behavior directly affects student achievement. Coaches cannot achieve their goals without some understanding of the processes of teacher thinking: how teachers reason as they plan for, execute, and evaluate instruction.

Through research on teacher cognition, we have gained valuable insights into what teachers think about as they teach. Research on human information processing, emotional intelligence, effective problem solving, human intelligence, brain capacity, and brain function has revealed many major findings.

1. *All behavior is based on rather simple cognitive maps of reality.* Teachers make decisions based on these maps. Swedish researchers Dahllof and Lundgren (1970) discovered that teachers used a subset of the class, ranging from the 10th to the 25th percentile, as an informal reference group for decisions about pacing a lesson or unit. These mental constructs guided the teacher's behavior toward the entire class.

2. *Invisible cognitive skills and unconscious reflexive reactions also inform teaching performance.* These skills and emotions influence the behavior of teachers and students, student achievement, and, reciprocally, teachers' thought processes.

3. *Talking aloud about their thinking and their decisions about teaching causes teachers to refine their cognitive maps, perceptions, and cognitive skills.* Since existing maps hold implicit theories about teaching and learning, refining these causes changes in theories and in instructionsl behaviors. All cognitive maps are incomplete and teachers' maps are sufficiently pragmatic to have gotten them where they are (Clark and Peterson 1986). Talking aloud about teaching decisions not only causes examination, refinement, and development of new theories and practices, but also engages teachers emotionally. Many report stasifaction and gratitude for the conversations.

4. *Invisible cognitive skills of teaching can be categorized in four domains. Preactive* thought occurs as the teacher plans before teaching. *Interactive* thought occurs during teaching. *Reflective* thought occurs when teachers recall and analyze a lesson. *Projective* thought is used to synthesize and apply learnings and plan next steps.

5. *"Thinking" is influenced at the intersection of three ways of knowing.* These ways of knowing are linguistic, nonlinguistic, and affective (Marzano 1998; Berry et al. 2003). All information that is perceived by the senses passes through these three processors and is encoded in linguistic representations, sensory images, or affective representations. For example, while teachers' thought processes are influenced by deeply held theories of learning, beliefs about their own efficacy, perceptions of the system in which they work, and beliefs about education or student conduct they are represented thrice.

- In the linguistic mode, these are stored as propositions in the form of abstract concepts (e.g., "A quiet student is a good student."). To describe the concept, the teacher must use token words to express herself.
- The same concept is also stored in permanent memory in a nonlinguistic mode: in images, sounds, smells, or kinesthetic sensations.
- Such propositions about students are also stored in an affective mode. *Affect* includes the unconscious mind—feelings, emotions, and mood. Feelings are purely physiological, the result of chemical reactions in the body. *Rational thinking is impossible without the involvement of feelings and emotions.* Emotion occurs in relationship to what our thoughts have to say about the feelings.

Coaching, then, must take into account and utilize these ways of knowing. Because Cognitive Coaching uses processes that attend to a teacher's language, sensory memories, and feelings, it can raise these deep-structure forms of knowing to a more conscious level so that the teacher can elaborate, clarify, evaluate, and alter them. Ornstein (1991) reminds us, "To make a personal change, we have to be able to observe the automatic workings inside ourselves."

An unexpected by-product occurs when teachers talk about their reasons for doing things and respond to questions about their perceptions and teaching decisions. They often experience a sense of professional excitement, and they also speak of renewed joy and energy related to their work (Edwards and Newton 1995).

BASIC MODEL OF INTELLECTUAL FUNCTIONING

If Cognitive Coaches intend to develop strategies, learning activities, and assessment indicators that focus on the intellect rather than on superficial behaviors, it is helpful to have a basic model of human intellectual functioning in mind to guide them. We

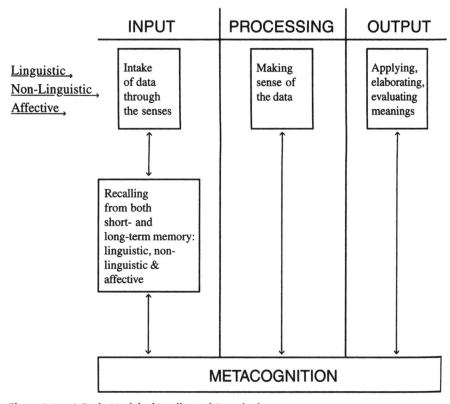

Figure 8.1.   A Basic Model of Intellectual Functioning

draw upon Feuerstein's and other authors' models of human intellectual functioning that distinguish four basic thought clusters (Feuerstein, Falik, and Feuerstein 2015, 125; Lomofsky 2014; Costa 1991; Taba 1962).

- Input of data through the senses and from memory (the more senses that are engaged, the richer the input)
- Processing those data into meaningful relationships
- Output, or application, of those relationships in new or novel situations
- Metacognition, or self-monitoring, regulating, and directing one's own thoughts, intentions, actions, beliefs, and emotions

Every event a person experiences causes the brain to call up meaningful, related information from storage, whether the event is commonplace or rare. The more meaningful, relevant, and complex the experience is, the more actively the brain attempts to integrate and assimilate it into its existing storehouse of programs and structures. The most complex thinking occurs when an external stimulus or problem challenges the brain to:

- Draw on the greatest amount of information, conceptual systems, schema, and structured meanings already in storage
- Expand already existing conceptual systems
- Develop new structures

A *problem* may be defined as any stimulus or challenge that has no readily apparent resolution. If there is a ready match between what is perceived by the senses and what is already in storage, no problem exists. For example, there is not much need to process information when you are asked your name because the response to the challenge is readily available. Jean Piaget (1954) called this *assimilation.*

If, however, the challenge or problem cannot be explained or resolved with existing knowledge in short or long-term memory, the information must be processed. Some action must be taken to gather more information to resolve the discrepancy, and the resolution must be evaluated for its "fit" with reality. Piaget called this *accommodation*, and he saw it as the process by which new knowledge is constructed. Coaching in the area of cognition mediates accommodation.

CONSCIOUSNESS AND COGNITION

Until recently the quality of consciousness, the awareness of processing information, was not studied. However, new studies reveal several striking phenomena. People apparently process and retain information without awareness. The essence of these studies shows "that people may deny seeing or remembering something—that is, be unaware of it—while at the same time showing through their behavior that they do see or remember the thing in question" (Shettleworth 1998).

As stated earlier, current views of information processing hold that memories are stored within the cognitive system in three forms: linguistic, nonlinguistic, and affective. Information is acquired, stored, retrieved, and acted upon according to the types of memories held in these systems.

TEACHER KNOWLEDGE

Marzano (1998) explains that knowledge is composed of information, mental processes, and psychomotor experience. A Cognitive Coach elicits a variety of information in all these areas: subject matter, pedagogy, students, or the teacher herself. The teacher then represents the information linguistically as declarative propositional networks. These facts, time sequences, episodes, vocabulary, generalizations, principles, and concepts usually also have strong nonlinguistic representations. Rarely, however, will craft information have strong affective representations. Through me-diational conversations, such information can be integrated and refined.

Teachers' Mental Processes

Mental processes are primarily procedural, and they are normally held in linguistic and nonlinguistic modes. Some mental processes of teaching have a wide diversity of uses and products, and they involve the execution of many interrelated subprocesses. For example, teaching a concept attainment lesson to students requires getting students' attention, focusing attention on a conceptual challenge, giving clear directions, classifying student responses as either positive or negative, and performing within acceptable boundaries of teacher responses related to the outcome of the lesson. Other mental processes of teaching are less complex and are made up of a general set of principles that can be applied in several settings. In getting student attention, for example, the teacher can position himself in the room, assume a still and grounded body posture, call for attention in a credible voice, and hold his body and gesture still until complete attention is achieved.

Experienced teachers will have additional models and mental processes for fo-cusing student attention. Some simple mental processes have narrow uses and are learned to the point of automaticity, such as distributing materials for student work and are, therefore, performed with less effort. Cognitive skills of teaching and learn-ing become automatic with continued practice. The more automatic some mental processes become, the more "cognitive space" is available for complex decision making during the lesson (Hattie 2003). Coaching helps teachers streamline existing mental models and form new ones.

Teachers' Psychomotor Knowledge

Psychomotor knowledge includes the physical skills and abilities a teacher uses to negotiate daily life in the classroom. Most psychomotor skills are stored in memory in a fashion identical to mental processes, and they are stored in an "if–then" syntax. As such, they have a strong linguistic base: For example, *if* the students are mis-behaving, *then* the teacher silently moves to a space reserved for group discipline, stands with authority and a serious expression, and directs students in a credible voice about proper behavior. *If* she believes that students have listened and under-stood, *then* she pauses, walks silently back to the instructional position, changes to a more melodic teaching voice, and resumes teaching (Grinder 1993).

Because psychomotor skills are usually learned through observation, instruction, and behavioral feedback, they have strong nonlinguistic representations. The af-fective system is usually not engaged unless a teacher had some strong negative or positive emotional experience connected to the skills. Psychomotor learning is not the domain of Cognitive Coaching; rather, it is the function of a consulting relationship. (See chapter 5 for the distinctions between consulting and Cognitive Coaching.)

Our brains constantly strive to satisfy and resolve discrepancies perceived in the environment. Information that the brain has not processed remains in memory for very short periods of time. Merely experiencing something or memorizing without acting on that information commits it only to short-term memory. This fact may explain the unusual power of Cognitive Coaching.

Most data perceived by the brain are either not attended to (and therefore never enter short-term memory) or are not processed and are forgotten within 18 seconds (Wolfe 1987). Much of the daily experience of teaching, therefore, is forgotten and not explained, its potential influence for teacher learning lost. Processing the instructional experience allows the teacher to facilitate construction of new meanings and insights.

The coach's role is to engage and mediate the teacher's cognitive systems. When Cognitive Coaching is used, teachers may process the same teaching event as least six times, providing an effective form of mental rehearsal. Mental rehearsal techniques can help improve performance even in areas as complex as teaching (Williams 2012).

1. Before the planning conversation, teachers formulate their objectives and plans.
2. During the planning conversation, they engage in deep, proactive mental rehearsal during which the coach questions, paraphrases, and clarifies in ways that help the teacher to be more precise with lesson strategies.
3. Instruction then occurs, with the teacher's conscious awareness about key elements greatly heightened.
4. The teacher recalls the instructional event in preparation for the reflecting conversation.
5. The teacher recalls the instructional event during the reflecting conversation.
6. Finally, the teacher continues to reflect and refine after the reflecting conversation, especially when closure was not reached on a subject or issue.

The long-range outcome of Cognitive Coaching is to habituate the intellectual formation of effective thinking: acquiring the habits and dispositions of self-directed learning. The teacher must aim to continue growing intellectually by internalizing these intellectual processes and experiencing them when the coach is not present: to be self-managing, self-monitoring, and self-modifying. In this way, the intellectual functions and mental processes of effective teaching become automatic.

STATES OF MIND AS QUALITIES OF TEACHER EXCELLENCE

In chapter 7, we introduced and explored five states of mind—efficacy, flexibility, consciousness, craftsmanship, and interdependence—as sources of excellence that drive human performance. Our aim is to focus the coach's capacities on mediating the energy sources that fuel self-directed learning, behavior, and, ultimately, the development of capacities for effective action in holonomous settings. The five states of mind we described are also the wellsprings that nurture teachers, professional learning communities, and schools.

Cognitive Coaches, being intensely concerned with teacher quality, realize the manifestations of those states of mind in the complexity of everyday classroom situations:

- What is the likelihood teachers will persevere through disappointment and challenge in their teaching, spending additional time, consulting others, and reviewing and revising their decisions when hoped for results have not occurred? (Efficacy)
- What is the likelihood that faced with a stubborn student problem the teacher will look at it from various perspectives and can see the viewpoints of a student, a disappointed parent, or a teacher from years past? (Flexibility)
- What is the likelihood that in the process of teaching and concentrating on the responses of an individual student, the teacher can also cast a sharp eye over the classroom to monitor the full class? (Consciousness)
- What is the likelihood that the teacher will insist on quality performance from both themselves and from students—and accept nothing less? (Craftsmanship)
- What is the likelihood that the teacher will regard teaching as a collaborative pursuit, engaging with others in applying systemic solutions to persistent problems? (Interdependence) (Costa, Garmston, and Zimmerman 2014)

This vision of quality teaching requires that Cognitive Coaches focus more deeply than those observable teacher behaviors by understanding that quality has more to do with the invisible mental attitudes, thinking dispositions, and innate drives than with discrete moments of success. When teachers and Cognitive Coaches focus collectively on the five states of mind, they become more positive, productive, and powerful.

This same belief is true for teachers' work in classrooms with their students as well. For example: Flexible teachers can make a distinction between and hold both surface and deep learning, immediate and long-range outcomes in mind when planning, conducting, and reflecting upon a lesson. Surface learning is more about the content: Do students know the content, the ideas for this lesson, and doing what is needed to demonstrate mastery? Deep learning, however, is more about understanding, relating, and extending ideas, and an intention to understand and transfer meaning for oneself and to other situations in school, life, and work.

FOUR STAGES OF INSTRUCTIONAL THOUGHT

When we compare the research on human information processing with research on teacher cognition, we are struck by the natural similarities. *Planning* (preactive) consists of all the intellectual functions performed before instruction. *Teaching* (interactive) includes the multiple decisions made during teaching. *Analyzing and evaluating* (reflective) consists of all those mental processes used to reflect, reconstruct, explain, interpret, find patterns, and judge instruction. Finally, *applying* (projective) abstracts from the experience synthesizes new generalizations, and carries them to future situations (Clark and Peterson 1986).

Planning—The Preactive Phase

Planning may well include the most important decisions teachers make because it is the phase upon which all other decisions rest. Planning basically involves four components (Shavelson 1976):

1. Anticipating, predicting, and developing precise descriptions of students' learnings that are to result from instruction. Ironically, this can be a low priority for many teachers (Zahorick 1975).
2. Identifying students' present capabilities or entry knowledge. This information is drawn from previous teaching and learning experience, data from school records, test scores, and clues from previous teachers, parents, and counselors (Shavelson 1977).
3. Envisioning precisely the characteristics of an instructional sequence or strategy that will most likely move students from their present capabilities toward immediate and long-range instructional outcomes. This sequence is derived from whatever theories, beliefs, or models of teaching logic, learning, or motivation that the teacher has adopted. The sequential structure of a lesson is deeply embedded in teachers' plans for allocating their most precious and limited resource—time. When a person thinks about an action, the brain sends electrical impulses to the nerves and muscles in the corresponding locations of the mind and body associated with the action. This is called the "carpenter effect" (Ulich 1967), and it is the scientific basis for the practice of mental

rehearsal. (See the sidebar "More about Mental Rehearsal.") Mental rehearsal is also beneficial for overcoming the psychological and emotional problems associated with performance (Whetten and Cameron 2010.).

4. Anticipating a method of assessing outcomes. The outcomes of this assessment provide a basis for evaluating and making decisions about the design of the next cycle of instruction.

There is reasonable agreement among researchers that teachers do not adapt a linear form of planning (e.g., specifying objectives first, then selecting learning activities). Instead, teachers seem to enter the planning process from multiple and various entry points. In fact, several studies find that many teachers think first about content, materials, resources, students' interests, and abilities. Aims and purposes are considered last (Zahorick 1975).

In their study comparing expert and novice teachers, Borko and Livingston (1989) found that, although none of the expert teachers had written lesson plans, all could easily describe mental plans for their lessons. These mental plans typically included a general sequence of lesson components and content, although they did not include details such as timing or pacing the exact number of examples and problems. These aspects of instruction were determined during the class session on the basis of student questions and responses. When asked what would be happening in class each day, the experts described plans that explicitly anticipated contingencies that were dependent on student performance. They were skillful in keeping the lesson on track and accomplishing their objectives, while also allowing students' questions and comments as springboards for discussions.

Even underneath this structure, what is going on in teachers' minds as they plan? Psychologists know that the human intellect has a limited capacity for handling variables. Miller (1963) describes this as "M-space," or "memory space." He found that humans have a capacity for handling and coordinating seven different variables, decisions, or disparate pieces of information (plus or minus two) at any one time. This assumes that the person has attained Piaget's stage of formal operations, which not all adults achieve. Other researchers have concluded that most adults can operate on about four disparate variables simultaneously (Rock 2009).

When humans approach the outer limits of their capacity, a state of stress begins to set in and they feel a loss of control. Most intellectual energy appears to be invested in techniques and systems to simplify, reduce, and select the number of variables. For teachers, certain planning strategies help to reduce this stress.

During planning, a teacher envisions cues, or definitions of acceptable forms of student performance for learning. This simplifies judgments about appropriate and inappropriate student behaviors. The teacher also selects potential solutions, backup procedures, and alternative strategies for times when the activity has to be redirected, changed, or terminated. This kind of planning causes thought experiments during which a teacher can mentally rehearse activities to anticipate possible events and consequences. This improves the coordination and efficiency of subsequent performance because systematic mental rehearsal prepares the mind and body for the activity. Rehearsal also is the main mechanism for focusing attention on critical factors relevant to the task.

MORE ABOUT MENTAL REHEARSAL

Much of the literature and research on mental rehearsal is drawn from the field of athletics. Olympic pentathlete Marilyn King reports using a combination of three kinds of visualization: visual imagery (being able to project a visual image of herself), kinetic imagery (feeling or physically sensing), and auditory imagery. These three forms of visualization are used in three different areas: long-term goal orientation, envisioning a goal step by step, and centering (or concentrating) to eliminate outside distractions and focus inward. The famous French alpine skier Jean-Claude Killy mentally rehearsed ski slopes with a stopwatch. He concentrated on every turn of the slope, timing himself as he envisioned every inch of his performance from start to finish. Killy claimed that the recorded time for his mental performance closely paralleled the time he actually achieved in the competition that followed. With mental rehearsal, a pre-performance routine often is created. This routine may consist of a smorgasbord of psychological skills that the athlete feels most comfortable using. The skills are put together in a set order and used routinely before each and every performance.

For example, while on the diving platform, Olympic gold medalist Greg Louganis mentally rehearsed each dive using a self-talk, pre-performance routine, which served as an attentional cue or trigger right before his forward three-and-a-half somersault dive. He used the following cue words: "Relax, see the platform, spot the water, spot the water, spot the water, kick out, spot the water again." Note that he is doing this before the dive and not during the dive. The activity prepared his mind for the coming dive and his body for the series of movements. Rehearsal also enabled him to forget everything else in his surroundings.

One common problem as we work is failing to maintain effective concentration. Through mental rehearsal, we learn to direct attention to cues that are most important for performance, and at the same time to close down the perceptions of distracting external stimuli (Jansson 1983). Much research concerns performance during short physical activities, but we also know that mental rehearsal of complex, interactive activities of longer duration also improves performance. This means that it is helpful to rehearse whole races, entire games, and, of course, entire lesson sequences (Landers, Maxwell, Butler, and Fagen 2001).

Mental rehearsals of conversations have been found to improve performance. (See "Inner Coaching" in chapter 11.) Williams (2012) reports an example in which a manager mentally rehearsed possible exchanges for a performance review with an employee. She visualized the employee raising objections about his performance review, saw herself paraphrasing his concerns and tactfully explaining that she had a different perception of his performance. She imagined that her efforts to listen to her employee, even though she couldn't always agree with him, helped to keep him from becoming argumentative and baiting her into a quarrel. Finally, she imagined herself standing up, smiling, and very professionally shaking the employee's hand as their meeting came to a close. Her meeting, in fact, went just like that.

Planning also demands that the teacher exercise perceptual flexibility by engaging multiple perspectives, multiple and simultaneous outcomes, and multiple pathways. However, viewing learning from multiple perspectives requires a certain degree of disassociation from the instruction in order to assume alternative perspectives. Highly flexible teachers have the capacity to view a lesson in both the immediate and the long range. Not only are they analytical about the details of this lesson (the micro mode), but they can also see connections between this lesson and other related learnings. They know where this lesson is leading and how it is connected to broader curriculum goals (the macro mode). Less flexible, episodic teachers may view today's activity as a separate and discrete episode, unrelated to other learning events.

Accomplished teachers teach toward multiple outcomes simultaneously. They have immediate goals for this lesson, but they also know that this lesson leads to a long-term, more pervasive outcome as well. They keep in mind the school's or district's standards of learning. For example, teachers may wish to have students understand the causes of World War II. Simultaneously, they want students to learn interdependence by working in groups, to experience long-range planning for their group project, and to practice the habits of mind of persisting, striving for accuracy, and developing a questioning attitude (Costa and Garmston 1998).

Experienced, effective, flexible teachers draw on multiple strategies to achieve their lesson goals. Based on their knowledge of the content to be taught, the learners in their charge, and the available resources, they can design multiple alternative instructional strategies for achieving their goals. (This becomes even more apparent during teaching. the interactive phase, described below.)

Planning a teaching strategy also requires the cognitive function of analysis, both structural and operational. Structural analysis is the process of breaking down the learning of the content into its component parts. Operational analysis involves putting a series of events into a logical, sequential order. To handle this information overload, teachers often synthesize much of this information into hypotheses, or best guesses about student readiness for learning. They estimate the probability of successful student behavior as a result of instruction.

Although many studies report on teacher practices regarding planning, Clark and Peterson (1986) point out that research is silent on what planning processes are most effective. From our experience, we agree with Shavelson (1976) that of the four stages of teachers' thought, planning is the most important because it sets the standard for the remaining three phases. We also place particular emphasis on the value of specifying clear learning objectives. We find that the more clearly the teacher envisions and mentally rehearses the plan, the greater the chances that the lesson will achieve its purposes. Chances are also greater that the teacher will self-monitor during the lesson and will more critically analyze the lesson during the reflective phase, thus assuming a greater internal locus of control.

Hattie (2003) reports that the number of requests for information made by expert and experienced teachers during the time they were planning instruction was about the same, but experts needed to know about the ability, experience, and background of the students they were to teach, and they needed to know about the facility in which they would be teaching.

Expert teachers anticipate and prevent disturbances from occurring whereas non-experts tend to correct already existing disturbances. This is because expert teachers have a wider scope of anticipation and more selective information gathering (Hattie 2003). Because of their responsiveness to students, experts can detect when students lose interest and do not understand.

Expert teachers are more likely to set challenging rather than "do your best" goals. They set challenging and not merely time consuming activities; they invite students to engage rather than copy; and they aim to encourage students to share commitment to these challenging goals. Eighty percent of most class time is spent with teachers talking and students listening, whereas expert teachers have students engage in challenging tasks to a greater extent of the time (Hattie 2003, 9).

Teaching—The Interactive Phase

Teachers are constantly interacting with students in an environment of uncertainty (Harvey 1966). Teachers are constantly making decisions that may be subconscious, spontaneous, planned, or some mixture of these. Teachers' in-the-moment decisions are probably modifications of decisions made during the planning phase, but changes made during the fast-paced interaction of the classroom are probably not as well defined or as thoroughly considered as the decisions made during the calmer stage of planning. In-the-moment decisions are spontaneously influenced by teacher information, mental processes, student cues, and automated psychomotor processes. Once a lesson begins, teachers have little time to consider alternative teaching strategies and the consequences of each.

Six kinds of temporal dimensions interact constantly with teachers' other thoughts and values to influence their daily decisions. Two of these time dimensions are sequence and simultaneity. *Sequence* refers to ordering instructional events within a lesson. *Simultaneity* refers to the capacity to operate under multiple classification systems at the same time. This means that an instructor can teach toward multiple objectives, coordinate numerous and varied classroom activities at the same time, plan a lesson incorporating several learning modalities, and think about multiple time frames. In addition, teachers plan within a variety of time horizons—daily, weekly, long range, yearly, and the school term. Effective teachers relate information from all those time frames as they prepare daily lessons (Yinger 1977

Keeping the standard of learning and the planned strategy in mind while teaching provides the teacher with a backdrop against which to make new decisions. During the beginning of a lesson, for example, the teacher may emphasize structuring the task and motivating students to become curious, involved, and focused. Later in the sequence, the teacher may use recall types of questions to cue students' review of previously learned information and to gather data to be considered later. Further into the lesson, the teacher may invite students' higher-level thinking and, finally, provide tasks for transference and application.

Clark and Peterson (1986) describe teacher thinking related to changing planned content during teaching. They examined influences on those decisions; cues that teachers read in order to make decisions; and the relationships between and among teachers' interactive decisions, teachers' behaviors, and, ultimately, student outcomes. A relatively small portion of teachers' interactive thought deals with instructional objectives. A greater percentage of teachers' interactive thought deals with the content or subject matter. A still greater percentage of interactive thought deals with the instructional process. The largest percentage of teachers' interactive thought concerns learning and the learner.

One of the great mental skills of teaching is simply the teacher's ability to remember the lesson plan during the pressure of interaction. Teachers often suffer cognitive overload—too many things going on all at the same time. Yet skillful teachers respond immediately, intuitively, and spontaneously to how the lesson plan is playing out. Highly conscious teachers are alert to what is going on in the classroom; less conscious teachers continue their lessons regardless of what occurs among students.

For example, alert teachers search for clues that students are prepared. Have the students acted on the information, digested it, and made meaning out of it or used it? Are students staring vacantly, or do body language and facial cues indicate attention? The alert teacher constantly observes, questions, clarifies, and interprets students' behaviors to make decisions about moving ahead in the sequence or remaining at the present step longer.

Metacognition refers to teachers' critically important capacities to consciously "stand outside themselves" and reflect on themselves as they manage instruction—to be spectators of their own thinking actions and beliefs; to examine their own reasoning. to monitor their own assumptions and mental models. (See also chapter 9, Domains of Inquiry.) During a lesson, teachers may ask themselves such managerial questions as: "Are my directions clear? Can students see the smartboard? Am I using precise words to make sure that the students understand? Should I speed up?" (What Factors Maximize Learning?). They also are aware of their dispositional concerns: What am I learning about myself? (Self-knowledge) Where is this lesson leading? How are my students gaining a better appreciation of beauty, fairness, justice and kindness? (What's worth learning?) Such internal dialogue means the teacher is constantly monitoring his/her own and students' learning behaviors during instruction.

Metacognition is also the ability to know what we know and what we don't know. It is the ability to plan a strategy for producing needed information, to be conscious of steps and strategies, and to reflect on and evaluate the productivity of thinking. David Perkins (1983) has elaborated four increasingly complex levels of metacognition:

- The *tacit* level, being unaware of our metacognitive knowledge
- The *awareness* level, knowing about some of the kinds of thinking we do (generating ideas, finding evidence) but not being strategic
- The *strategic* level, organizing our thinking by using problem solving, decision making, evidence seeking, and other techniques
- The *reflective* level, not only being strategic but reflecting on our thinking in progress, pondering strategies, and revising them accordingly

The metacognitive skills necessary to successful teaching, which a Coach may want to be alert for, include the following:

- Keeping place in a long sequence of operations
- Knowing that a subgoal has been attained
- Detecting errors and recovering from them by making a quick fix or retreating to the last known correct operation

This kind of monitoring involves both looking ahead and looking back. Looking ahead includes the following:

- Learning the structure of a sequence of operations and identifying areas where errors are likely
- Choosing a strategy that will reduce the possibility of error and will provide easy recovery
- Identifying the kinds of feedback that will be available at various points and evaluating the usefulness of that feedback

Looking back includes the following:

- Detecting errors previously made
- Keeping a history of what has been done until the present and thereby determining what should come next
- Assessing the reasonableness of the present and the immediate outcome of task performance

Teachers monitor the classroom for conscious and subconscious cues, and sometimes the cues build up so much they disrupt conscious information processing.

Flexible teachers manage their impulsivity by avoiding strong emotional reactions to classroom events. This is an efficient strategy to reserve the limited capacity for conscious processing of immediate classroom decisions: being inclined to find humor, for example, in their errors rather than becoming defensive.

Flexible teachers have a vast repertoire of instructional strategies and techniques, and they call forth alternative strategies as needed. They are aware that in a math lesson, for example, Sam obviously isn't getting the concept of sequence. Sensing that Sam is a kinesthetic learner, they give Sam a set of blocks and ask him to put them in order.

Many classes are filled with students with a heterogeneous array of languages, cultures, interests, and learning styles. Teachers may have one or more students, or even whole groups, who are Hmong, Vietnamese, Arab, East Indian, and Latino, all in the same classroom. Each must be dealt with employing different strategies, cultural experiences, vocabulary, examples, and techniques. Efficacious and flexible teachers continually cast about in their vast repertoire for, or invent, strategies that may prove effective.

Routines are also helpful in dealing with the multidimensionality of information-processing demands in the classroom. Routines reduce the need to attend to the abundance of simultaneous cues from the environment. Efficacious teachers develop a repertoire of routine systems for dealing with classroom management functions (taking roll, distributing papers and books). They also have systematic lesson designs (e.g., spelling and math drills) and teaching strategies (e.g., questioning sequences and structuring). Efficacious teachers are more effective scanners of classroom behavior, make greater references to the language of instruction and learning of students, whereas less efficacious teachers concentrate more on what the teacher is doing and saying to the class and concentrate more on student behavior.

Analyzing and Evaluating: The Reflective Phase

> Memories are not stored by facts but by the emotions attached to the facts. When accessing the corresponding emotion, the brain reconstructs the memory that is stored as an image from its dispositional representations. Therefore, remembering is first and foremost a matter of accessing the correct emotion.
>
> —Ronald Kotulak (1997)

After teaching the lesson, the teacher now has two sources of information: the lesson that was envisioned during planning and the actual lesson as performed during teaching. Analyzing involves collecting and using understandings derived from the comparison between actual and intended outcomes. If the teacher finds a great similarity between the two, there is a match. A discrepancy exists when there is a mismatch between what was observed and what was planned. Teachers generate reasons to explain the discrepancy or cause-and-effect relationships between instructional situations and behavioral outcomes.

Teachers can either assume responsibility for their own actions or they can place the blame on external forces. Teachers with an external locus of control tend to misplace responsibility or situations or persons beyond their control. For example: "How can I teach these kids anything? Look at the home backgrounds they come from. Their parents just don't prepare them to learn."

Efficacious teachers have an internal locus of control. They assume responsibility for their own successes or failures: "Of course the students were confused. Did you hear my directions? They were all garbled. I've got to give more precise directions." Seligman (1990) suggests that persons develop "explanatory styles" with which they subconsciously form internal hypotheses or rationalizations to explain to themselves good and bad events in their environment. Hartoonian and Yarger (1981) found that some teachers might dismiss or distort information that indicates that students did not learn as a result of their teaching strategy. Less efficacious teachers may give themselves credit for student improvement but misplace blame when performance is inadequate.

Even with this analysis, the cycle of instructional decision making is not yet complete. The learnings must be constructed, synthesized, and applied or transferred to other learning contexts, content areas, or life situations. Commitments must be made to take action applying the learnings to future situations.

Applying: The Projective Phase

> We do not learn from experience . . . we learn from reflecting on experience.
>
> —John Dewey, 1938

At the projective stage, the teacher takes the new knowledge constructed through analysis and applies that knowledge to future instructional situations or content. Experience can bring change, but experience alone is not enough. Experience is actually constructed: compared, differentiated, categorized, and labeled. This allows the teacher to recognize and interpret classroom events, departures from routines, and novel occurrences. Thus, the teacher can predict the consequences of possible alternatives and activities. Without this conceptual system, the teacher's perception of the classroom remains chaotic.

As continuous learners, self-directed teachers consciously reflect upon, conceptualize, and apply understandings from one classroom experience to the next. As a result of this analysis and reflection, the teachers synthesize new knowledge about teaching and learning. As experiences with teaching and learning accumulate, concepts are derived and constructed. Teachers' practice thus becomes more routinized, particularized, and refined. They are capable of predicting consequences of their decisions, and they experiment more often and take more risks. They expand their repertoire of techniques and strategies to be used in different settings with varying content and unique situations. Richard Shavelson (1976) states, "Any teaching act is the result of a decision, whether conscious or unconscious, that the teacher makes after the complex cognitive processing of available information. This reasoning leads us to the hypothesis that *the basic teaching skill is decision making*."

CONCLUSION

In this chapter we have elaborated our beliefs that the craft of teaching is cognitively complex. All the behaviors we see in the classroom are artifacts of these internal mental processes. The myriad decisions are driven by even more deeply embedded conscious or subconscious beliefs, styles, metaphors, perceptions, and habits.

The research on teacher cognition presented here supports our attempt to refocus the definition of teaching away from an archaic, behavioristic model to a more modern and viable cognitive model and thus direct the coach's focus away from overt behaviors of teaching to concentrate on influencing those invisible, inner thought processes of teaching.

These cognitive processes were presented in four clusters of thought: planning, the preactive phase; teaching, the interactive phase; analyzing, the reflective phase; and applying, the projective phase. Based on this conception of teaching, the model of Cognitive Coaching, described in chapter 11, directly parallels and is intended to enhance teachers' growth toward even more thoughtful teaching. If teachers do not possess these mental capacities, no amount of experience alone will create it. It is through mediated processing and reflecting upon experience that these capacities will be developed (Feuerstein and Feuerstein 1991). As Cognitive Coaches, therefore, we are interested in operating on the inner thought processes—in "coaching cognition." Teachers possess wide and expanding bodies of information and skills, and they make decisions about when to use what from the extensive range of their repertoire. Cognitive Coaching assists teachers in becoming more conscious, efficacious, precise, flexible, informed, and skillful decision makers. Together, teachers and coaches create greater student learning.

The ultimate purpose, however, is not only to enhance teaching and the resulting student learning but also to capacitate the teacher for self-directed learning and to cause the teacher to grow toward higher states of mind. In addition, our intent is to support the development of holonomous school cultures that, because they value and have skills in the dual goals of autonomy and interdependence, promote the professional development of teachers, reinvent instruction as appropriate, and increase student learning. Such is the vision of agile organizations. Thus the coach continually focuses on these long-range goals in the use of Cognitive Coaching skills.

9

The Knowledge Base of Teaching and Learning

Teaching is one of the most complex human endeavors imaginable.

—Saphier, Haley-Spica, and Gower (2008, 1)

The process of teaching is like the flow of busy traffic at an unmarked intersection in Istanbul: Only resident drivers know the unstated rules of the road. Similarly, only accomplished teachers know the unspoken knowledge and perceptions that guide their work in the classroom. Cognitive Coaching draws upon that knowledge and *mediates it* to inform teacher practice. Knowing the knowledge base of teaching is a fundamental mission of each school and is achieved through multiple approaches. Continual inquiry into the knowledge base of teaching and learning is important to persons providing support to teachers. In all forms of professional development and in performing the functions of collaborating or consulting (see chapter 5), such knowledge is essential to sound guidance. It is also important in coaching, for through these interactions teachers clarify theories of learning and integrate their knowledge of content, pedagogy, and information about students and themselves into a coherent whole. This depth of learning is available nowhere else—not in universities, in books, or from co-workers.

We are often asked if coaches need to know the subject matter that the coachee teaches—that is, for example, if the coach knows the content of physics to better coach a physics teacher. We find that it is both a help and a hindrance. It may be helpful in that the coach and the teacher "share a common conceptual language." On the other hand, not knowing the content allows both the teacher and the coach to focus on the process of teaching rather than on the content. It promotes the coach as a co-learner along with the teacher in that it provides an opportunity for the coach to learn more about physics, for example. Furthermore, when the coach is very familiar with the content, there is a tendency to presume one has common understanding of discipline-related terms when that may not be the case, or to be evaluative—"That's not the way I'd teach it."

In schools that value transdisciplinary learning, when a teacher from one department or grade level coaches another teacher from a different department or grade level, there is a great possibility that they both find a common ground beyond the content. For example, the physics teacher has a larger goal of not only teaching physics principles but also teaching students to "think like a scientist." The coach, a "health and well-being" teacher from a lower grade level, finds that "thinking like a scientist" is her main goal as well. They decide to share some resources and have the upper-grade physics students visit the lower-grade health class to explore their understanding of what it means to think like a scientist.

In the following section, we define teaching as a complex, nonlinear form of inquiry drawn upon a rich knowledge base about instruction and learning. We think of it as a form of inquiry rather than a body of knowledge. "Inquiry" suggests a constant, dynamical process of ongoing, creative problem solving, questioning, and experimentation leading to continuous spirals of learning, whereas a "body of knowledge" suggests fixed content that experienced educators will know and draw upon. Cognitive Coaching mediates, integrates, and extends this knowledge base by focusing on teacher decisions, knowledge, perceptions, beliefs, and values. In this chapter, we specifically consider six essential questions (figure 9.1) that provide the basis for continuous learning, research, and problem solving in education. Those domains for inquiry include:

1. What's worth learning? (Content knowledge)
2. What works in teaching? (Pedagogy)
3. What factors influence student learning? (Knowledge of students and how they learn)
4. Who am I and who am I becoming? (Self-knowledge)
5. How might I stretch to higher levels of cognitive complexity? (Knowledge of cognitive processes of instruction)
6. How might collegial interactions be continually strengthened and enhanced? (Knowledge of collegial interactions)

WHAT IS TEACHING?

Teaching is knowing about, inquiring into, and experimenting with content and curriculum, learners and learning, fellow teachers and teaching, and educational philosophies and goals. Teaching is also about clear values and working within the contexts of various classrooms, schools, communities, and cultures. Teaching is about applying all of this integrated knowledge in a unit of instruction, a lesson, or an in-the-moment interaction.

Teachers enter the profession as novices, and a small percentage of these develop into experts. Sternberg and Horvath (1995) and Hattie (2003) characterize the expert stage as the point at which teachers have greater knowledge, are more efficient, and are more insightful than teachers at any earlier stage of development. Teachers maintain a capacity for learning throughout their careers. However, experience alone is not enough; growth is not automatic. Growth occurs by reflecting on experience and effectively using higher-order thinking processes to plan, monitor, evaluate, and modify educational tasks. *Self-directed teachers practice careful reflection. They learn lessons each time they teach, evaluating what they do and using these self-critical evaluations to adjust what they do next time* (Office for Standards in Education 2004, 10; Tokuhama-Espinosa 2014). Teachers are more likely to increase their capacity for learning when their thinking is mediated in a learning-centered work culture.

Expert teachers continue to develop and grow as they construct meaning, reinvest their cognitive resources, and apply new learning. As discussed in chapter 8, higher-concept teachers produce students who achieve at higher academic levels. Cognitive Coaching mediates this type of career-long development.

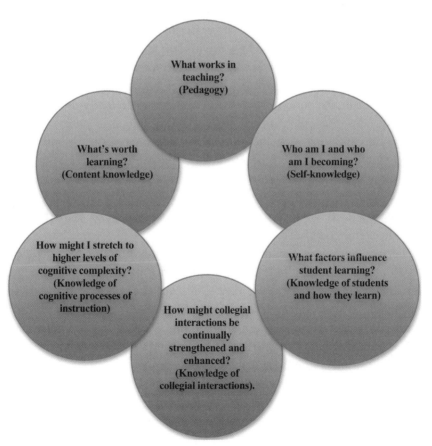

Figure 9.1. Domains of Inquiry

SIX DOMAINS OF PROFESSIONAL INQUIRY

The true delight is in the finding out rather than in the knowing.

—Isaac Asimov, Russian-born American author and biochemist

Although the accumulated knowledge in each domain is vast, there is still more to learn—it is never "mastered." In this chapter, we organize and address the interacting fields of teaching knowledge under six domains of continuing inquiry. The effective coach is familiar with each of these domains and how each one influences a teacher's work in and out of the classroom. Coaches, therefore, encourage teachers to continuously experiment, innovate, learn more about, and synthesize the results into new patterns and practices.

1. What's Worth Learning? (Content Knowledge)

Expert teachers have a deep reservoir of knowledge about, and continue their learning of, the subject matter they teach. They know the structure of their discipline—its organizing principles and concepts. The greater their subject-matter knowledge, the more flexible and student-oriented they are in their teaching.

For example, when a biology teacher presents a lesson about which she has deep personal knowledge, she places fewer boundaries on class discussion and how lab work proceeds. She encourages students to explore more freely. She knows which significant questions to ask, and she knows which problems to pose so that students will inquire into and synthesize the most salient concepts of the discipline.

Teachers have always faced the problem of having more content to teach than they have time. Teachers who know the organizing principles of a discipline make better decisions about what to teach and what to omit. They also make better decisions about how to develop instruction, which effectively serves both short- and long-term learning goals.

For example, some organizing principles of literature are characterization, plot, setting, rising action, and falling action. Many forms of literature can be examined for these dimensions, such as plays, novels, or short stories. An example of an organizing principle in science instruction is the interaction of cycles. Rather than teach about butterflies in one unit, frogs in another, and weather in yet another, a teacher can use her understanding of cycles as an organizing theme to invite learning about comparisons, causal relations, and interactions among the cycles of many living organisms and environmental conditions. This provides the greater possibility of connecting to student interests, past knowledge, and learning styles.

Skillful teachers see their content not as an end but as a vehicle for fostering life skills. They use, for example, the teaching of calculus (or social studies or art) as a means of helping students learn to solve problems, to strive for accuracy, as a creative tool, or as a way to create intrigue, wonderment, and awe (Costa and Kallick 2014). One professor of mathematics considers his purpose is to have students "fall in love" with math (Suri 2013).

Skillful coaches use the following guidelines to invite teacher reflection about content:

- Organizing and adapting the curriculum to facilitate an in-depth understanding of broad themes, recurring concepts, universal patterns, and thinking dispositions in each content area, and demonstrating relationships across the curriculum
- Connecting key concepts and themes within and across the curriculum, and building a wide range of diverse perspectives to extend learning for all students
- Devising and articulating short- and long-term goals with high expectations for learning
- Sequencing instruction to help students synthesize and apply new knowledge and to see relationships and connections across subject matter areas

For example, with regard to finding the recurring dispositions of scholars in various disciplines: How do scientists, mathematicians, historians, or economists test their hypotheses and determine their accuracy? What are some examples of the results of persisting in physics, athletics, the arts, and literature? What can be learned from these examples?

2. What Works in Teaching? (Pedagogy)

The queen on a chessboard is a useful metaphor for the knowledge domain of pedagogy. The queen is the piece with the greatest flexibility and range of movement. So, too, an experienced, effective teacher flexibly draws on a variety of moves during classroom work. This is the knowledge area of pedagogy.

Saphier et al. (2008), for example, identify 53 ways that a teacher can gain students' attention and keep them on task. Although an individual teacher might not know all 53 of these moves, experienced teachers draw from a large personal repertoire to select the best match for each situation. Teachers also manage momentum, space, time, routines, and discipline. They apply principles of learning, hold high expectations, and design and guide learning experiences that enable students to reach objectives (Marzano, Pickering, and Pollock 2001).

Teacher clarity about pedagogy correlates highly with student learning. Clarity is required when teachers make learning objectives explicit, present new material, explain concepts, give directions, elaborate on directions, activate prior knowledge, re-explain old material, address confusion, and make instructional connections.

Again, expert teachers develop a repertoire in this area and make best-fit matches for individual students and situations. Following are some pedagogical guidelines:

- Consistently drawing on an extensive repertoire of appropriate strategies and resources, and adapting and refining strategies while teaching in response to students' needs
- Using questions to revise activities and extend students' abilities to synthesize what they know to achieve learning goals
- Facilitating challenging learning experiences that promote collaboration, independent learning, thinking dispositions, and choice for all students
- Using a variety of formative, ongoing self-assessments that are embedded in instruction to guide short- and long-term plans and to support learning for all students

---

EXPANDING REPERTOIRE

Master teachers, like concert artists, consciously expand their performance repertoires. They develop and assemble microroutines that can be combined and reconstituted to fit a wide variety of settings. These experts also habitualize many routines to free cognitive space for more complex perceiving and more sophisticated instructional problem solving. This unconscious competence is the hallmark of the expert in the classroom (Garmston and Wellman 2013).

Stanford professor Lee Shulman (1987) defined a special form of craft knowledge as "pedagogical-content knowledge." This knowledge is formed from the intersection of knowledge about content and pedagogy. Grimmet and MacKinnon (1992) claim that pedagogical-content knowledge is a different kind of knowing. They agree with Schulman's assertion that it is derived from a considered response to experience in the practice setting, and though it is related to knowledge that can be taught in the lecture hall, it is formed over time in the minds of teachers through reflection.

3. What Factors Maximize Learning? (Knowledge of Students and How They Learn)

Since the onset of neuroscientific research and its implications for learning and teaching, there has been increasing support for constructivism as an underlying principle (Tokuhama-Espinosa 2014; Caine and Caine 1991; Jensen 1998; Wolfe and Brandt 1998; Sigel 1984; Kuhn 1962; Dewey 1938; Brooks and Brooks 2001; Treadwell 2014).

Although constructivism is described here as it relates to classroom teachers, these principles of constructivism may also serve as a basis for understanding why Cog-

> Meaning making is not a spectator sport. Knowledge is a constructive process rather than a finding; it is not the content that gets stored in memory but the activity of constructing it that gets stored. Humans don't *get* ideas; they *make* ideas.

nitive Coaches do what they do. They provide the theoretical underpinnings of the model of Cognitive Coaching, and they can be employed as a template to determine the degree to which schools embrace a constructivist philosophy.

Constructivism is a theory of learning that is based in a belief that human beings have an innate quest to make meaning from their experiences. When humans are perplexed by anomalies, discrepancies, or new information, they have a natural inclination to make sense of it. This process of making sense is enhanced in instructional environments when certain behaviors are practiced:

- Seeking and valuing others' points of view
- Challenging other's suppositions
- Posing problems of emerging relevance
- Focusing on "big" ideas and long-range outcomes
- Creating conditions for self-assessment
- Providing opportunities for reflection on experience
- Focusing on the mental processes of problem solving and decision making

The following principles of constructivism guide teachers as mediators of a learning-centered environment:

1. Encouraging Self-Directedness

Autonomy empowers the pursuit of connection making among ideas and concepts. When participants raise issues and frame questions and then analyze and answer them themselves, they take responsibility for their own learning, become problem solvers and problem finders. In the pursuit of new understandings, they are led by their own ideas and informed by the ideas of others.

As such, constructivist environments must be safe, providing freedom to experiment with ideas, explore issues, and encounter new information. When learners anticipate that their thoughts will be judged, their thinking mode shifts from open to closed. Nonjudgmental feedback encourages others to pursue their ideas more deeply and less defensively, and develops the capacity for evaluating the worth of their own and others' ideas. Listening to, paraphrasing, and clarifying ideas indicates that their brains have the power to produce meaning.

2. Mediating Students' Meaning Making

Learning is viewed as a continual process of engaging and transforming the mind. Constructivist teachers cast students in the role of producers of knowledge rather than consumers of knowledge. They conceive their role as a facilitator of meaning making. They intervene between the learner and the learning so as to cause students to approach activities in a strategic way, to help them monitor their own progress in the learning, to construct meaning from the learning and from the process of learning, and then to apply the learnings to other contexts and settings (Feuerstein, Falik, and Feuerstein 2015).

Constructivist teachers' meditational questioning and problem posing stimulate the brain to engage in higher-order and creative cognitive functions. They raise and illuminate perplexing situations, problems to solve, discrepancies, and intriguing phenomena—the answers to which are not readily apparent. Day-to-day, real-life problems are the best way to practice problem solving.

Constructivist teachers seek elaboration of initial responses. Initial responses are just that—*initial*. People's first thoughts about issues are not necessarily their final thoughts or their best thoughts. Inviting elaboration causes others to reconceptualize and reassess their own language, concepts, and strategies.

In dialogue, constructivist teachers use cognitive terminology. The words we hear and use in dialogue affect our way of thinking and ultimately our actions. Constructivist teachers deliberately choose words intended to activate and engage mental processes. Finding relationships, predicting outcomes, analyzing, and synthesizing are mental processes that require others to draw forth their knowledge, make connections, and create new understandings (Costa and Kallick 2008, 124–25).

3. Constructivist Approaches Begin with Raw Data and Direct Experiences from Which to Make Abstractions

Concepts, theorems, laws, criteria, and guidelines are abstractions that the human mind generates through interaction with previous information, new ideas, experiences, and data. The constructivist approach draws on primary sources, information from memory and past experiences, manipulatives, and real-life situations, and then helps learners generate abstractions from them. Learners are presented with data from a variety of sources and are then asked to compare, analyze, synthesize, and evaluate. Learning becomes the result of research related to real problems.

Providing information is also valid for constructivists. However, the provision of information is appropriate when:

- The learner feels a need for it, is interested in it, or requests it.
- The information is linked (scaffolded) to previous learnings.

- It is delivered in a manner that matches the learner's learning style.
- It is relevant to the learner's developmental levels.
- It is drawn from multiple sensory sources, such as visual, auditory, kinesthetic, and tactile.
- It is acted upon and processed by making inferences, forming concepts, synthesizing theorems, inducing algorithms, generating laws, and creating guidelines.
- Such generalizations are applied to contexts beyond the one in which it was learned.

4. Constructivism Suggests a Sequence of Learning

Constructivists such as Taba (1965), Bruner (1960), Piaget (1950) and Feuerstein, Falik, and Feuerstein (2015) believe that there is a sequence to concept formation in which learning begins with gathering sensory data, then making sense of the data (elaborating) by finding patterns and relationships, and then applying those concepts to new problems.

The first stage is the *input* or "discovery" stage in which data are taken in through the senses and drawn from memory. It is an open-ended opportunity to "mess around" and become familiar with materials and to generate questions and hypotheses.

Next is the *processing* stage in which patterns are found, meaningful relationships are elaborated, and concepts are constructed from those data. New labels for concepts are formed, hypotheses are drawn, and inferences are made.

The third step, the *output* or "concept application" stage, is when further iterations of the discovery are made, generalizations and principles are formed, and these are then used as models to explain problems and predict consequences in settings beyond the one in which it was formed (see chapter 8).

5. Constructivist Teachers Challenge Others' Mental Models

Humans often develop and refine ideas about phenomena and then tenaciously hold onto these ideas as eternal truths. Much of what we think happens simply by virtue of our agreement that it should, not because our close examination of our bounded assumptions, limited history, and existing mental models. Constructivists realize that cognitive growth occurs when individuals revisit and reformulate a current perspective. Therefore, constructivist teachers might provide data, present realities, and pose questions for the purpose of engendering contradictions to others' initial hypotheses, challenging present conceptions, illuminating another perspective, and breaching crystallized thinking. Constructivists must listen to and understand the other person's present conceptions or points of view to help them understand which notions might be accepted or rejected as contradictory.

6. *Constructivists Support On-Demand Learning*

In the meaning-making process, humans direct the need for information in the moment. Because they feel a need for information or skills to fill the gaps and inadequacies in their process of meaning making, there is a greater openness to learning. (Computers assist in this process as they allow the pursuit of interests through browsing, tapping varied databases in a random and nonlinear way.) Constructivists, therefore, allow others' responses to drive the interaction, shift the focus, and alter the content.

Being alert to such verbal and nonverbal signals—such as eye movements, facial expressions, postural shifts, gestures, breathing rates, and changes in pitch, speed, and volume of speech, as well as overt expressions of interest and enthusiasms—the teacher may shift the topic or approach. (We all know of the "teachable moment" when our planned lesson is derailed by an event of greater interest.) Learners' in-the-moment insights, experiences, motivations, and interests may intersect around an urgent theme. When magnetic events exert an irresistible pull on learners' minds, the teacher may hold his or her planned strategy in abeyance and "go with the flow."

7. *Constructivism Involves Communities of Learners*

Constructivist learning is a reciprocal process of the individual's influence on the group and the group's influence on the individual. Meaning making is not only an individual operation. The individual also constructs meaning interactively with others to construct shared knowledge. Meaning making may be an individual experience—one's unique way of constructing knowledge—and it is a sociocultural phenomenon. Social processes of interaction and participation enhance, refine, and amplify meanings. An empowering way to change conceptions is to present one's own ideas to others, as well as being permitted to hear and reflect on the ideas of others. The benefit of discourse with others, particularly with peers, activates the meaning-making process (Willis 2013). The constructivist environment encourages questioning of each other, rich dialogue within and among groups and teams of learners, resolving differences in perceptions and styles, and synthesizing alternative viewpoints (Vygotsky 1978).

8. *Process Is the Content*

Constructivist curriculum challenges the basic educational views of "knowledge" and of "learning" with which most schools are comfortable. It causes us to expand our focus from educational outcomes that are primarily collections of subskills to include successful processes of participation in socially organized activities and the development of students' identities as conscious, flexible, efficacious, interdependent, and continual learners. A constructivist curriculum lets go of having learners acquire predetermined meaning and has faith in the processes of individuals' construction of shared meanings through social interaction. Constructivists, therefore, inquire about their colleagues' understandings of concepts before sharing their own understandings of those concepts. When constructivists share their ideas and theories before others have an opportunity to develop their own, questioning of their own theories is essentially eliminated. Others assume that the teacher knows more than they do; consequently, most stop thinking about a concept or theory once they hear "the correct answer."

9. Constructing Meaning Takes Time

Constructivists invite colleagues to seek relationships and create metaphors. When sufficient time is given to activities, learners go beyond initial relationships to create novel relationships, find patterns, and generate theories for themselves. This also means that teachers allow wait time after posing questions.

Failures are never dismissed as mistakes. Rather, time is taken to reflect on their learning, to compare intended with actual outcomes, to analyze and draw causal relationships, to synthesize meanings, and to apply their learnings to new and novel situations.

Unlike many other "quick-fix" educational innovations and experiments, constructivist educators remain focused on the longer view. They realize that assisting others to habituate self-directed learning takes years, well-defined instruction with qualified teachers, and carefully constructed curriculum before a significant and enduring change is observed.

We know that the amount of time on task affects learning. As self-directed learning becomes a goal of education, greater value is placed on allocating time for learning activities intended to stimulate thinking and to practice the construction or personal meaning.

10. Constructivists Invite Metacognition

Constructivist teachers invite students to think about their thinking. Time is taken to monitor and discuss thinking skills and problem-solving processes. They are invited to share their metacognition—to reveal their intentions, strategies, and plans for solving a problem; to describe their mental maps for monitoring their strategy during the problem-solving; and to reflect on the strategy to determine its adequacy (Costa and Kallick 2008, 23–25).

11. Assessment Is Viewed as Another Opportunity to Accelerate Self-Directed Learning

In the constructivist environment, what matters most is whether the inhabitants are learning to become increasingly more self-managing, self-monitoring, and self-modifying. Constructivists help others design diverse ways of gathering, organizing, and reporting evidence of continual learning and meaning making.

In the current politics of education, the key to school success is higher test scores. Most states in the United States have developed standards for learning. Many have implemented new assessments tied to these standards and linked high school graduation to passage of these new assessments; others are beginning to link teacher evaluation and even merit pay to increases in test scores. These educational practices shift the focus toward the transmission of test-related information and judging teachers on how well their students score. In such a climate, it is difficult to embrace and sustain constructive practices designed for individual meaning making.

The intent of constructivist assessment is to support others in becoming self-managing, self-monitoring, and self-modifying. Self-knowledge is the first step in self-assessment. Much of the work of self-evaluation is developed through the metacognitive process of reflection. This process of self-assessment provides internal and external data that promotes one's own learning and growth (Costa and Kallick 2014, 100–18).

12. Modeling

Imitation and emulation are the most basic forms of constructivist learning. Teachers, coaches, parents, and administrators realize the importance of their own display of desirable learning behaviors in the presence of others with whom they work. For example, if coaches want their colleagues to value inquiry, then they must also value it; if teachers pose questions with the orientation that there is only one correct response, then how can students be expected to develop either the interest in or the analytic skills necessary for more diverse modes of inquiry?

In the day-to-day events and when problems arise in schools, classrooms, and homes, the same types of constructivist learning listed above are employed. Without this consistency, there is likely to be a credibility gap (Costa and Kallick 2008, 283–84).

4. Who Am I and Who Am I Becoming? (Self-Knowledge)

Self-knowledge includes the areas of personal values, standards, and beliefs. A teacher with self-knowledge is able to overcome egocentric patterns and teach in the ways that students learn best. A teacher with self-knowledge does not teach only in the way that he or she learns.

With awareness of their own cognitive styles, many teachers discover that the students who distress them the most are the ones with cognitive styles different from their own. For example, students who are abstract and random may drive a teacher whose own learning patterns are concrete and sequential to distraction. This discovery provides another intersection in the domains of knowledge of the discipline being taught, self-knowledge, and pedagogy. Awareness permits the teacher to set aside emotional responses to student style to provide the best learning experiences possible.

Self-knowledge includes information about one's own preferred learning styles, preferred representational systems, moral and ethical considerations in teaching, assumptions and theories about learning, and professional mission and values. Deciding what and how to teach is informed by one's beliefs and values. A calculus teacher, very clear about his own values, once said to us, "I do not teach calculus. I teach life through calculus." The following descriptors related to self-knowledge can support teacher reflection by asking themselves, How am I:

- Using understanding of self and individual students' development to meet learning goals?
- Assessing teaching practice and extending professional development through professional dialogue within the professional community?
- Engaging all students in practicing self- and peer-assessment, identifying their own learning goals, and self-monitoring progress over time?

Furthermore, teachers are influenced by their own plans for future growth and development: retirement, advanced degrees, career ladder mobility, leisure-time hobbies, and so on. Teachers, too, value, exhibit, and therefore teach such dispositions as continual learning, persistence, curiosity, craftsmanship, wonderment, passion, and so forth.

5. How Might I Stretch to Higher Levels of Cognitive Complexity? (Knowledge of Cognitive Processes of Instruction)

Teachers' developmental levels have a direct correlation to their performance in the classroom. Teachers who function at higher conceptual levels are capable of greater degrees of complexity in the classroom and are more effective with students. Teachers with more advanced conceptual levels are more flexible, tolerant of stress, and adaptive in their teaching styles. Thus, they are able to assume multiple perspectives.

Conceptually advanced teachers are responsive to a wider range of learning styles and culturally diverse classes. Teachers at higher levels of conceptual development employ multiple perspectives, including the students' perspectives (Kegan 1994). They also apply the learners' frames of reference within their own frame of reference for instructional planning, teaching, and evaluation. Furthermore, expert teachers not only seem to perform better than novices do, they also seem to do so with less effort. Numerous researchers (Shulman 1987; Berliner 1994; Sternberg and Horvath 1995; Tsui 2003; Marzano 2007; Hattie 2012; Danielson 2013) outline ways that expert teachers differ from nonexpert teachers in knowledge, efficiency, and insight.

Though teachers' cognitive skills differ from one another on entering the profession, all teachers can develop greater cognitive capacity in the four phases of instructional thought described in chapter 8. One major intention of Cognitive Coaching is to develop these invisible skills of teaching through reflection that utilizes and extends the domains of craft knowledge described so far.

6. How Might Collegial Interactions Be Continually Strengthened and Enhanced? (Knowledge of Collegial Interactions)

In high-performing schools, faculty collaboration is the norm (Garmston and Wellman 2013). Many studies demonstrate the unusual potency of teachers' collaborative action on student learning. Seashore-Lewis, Marks, and Kruse (1996) found that collective responsibility for student learning emerged and schoolwide achievement gains were made when five conditions were present and interacting with one another: (1) a shared sense of purpose, (2) a collective focus on student learning, (3) collaborative activities, (4) "deprivatized practice," and (5) reflective dialogue.

Many school renewal efforts today regard trust, collaboration, networking, and disaggregated democratic governance of instructional issues as an important factor in their success. They are committed to shared leadership in determining what students should be able to know and do, how students will demonstrate those competencies, and to continuous action research to test and refine teaching practices. Such agile schools show sustained improvements in student learning over a period of years.

Teaching is one of the most private professions. For most of its history, teaching has been something that a person does in isolation from other teachers. As long as teachers work in isolation, they do not need a high degree of skill in collective problem solving, consensus seeking, or the technical and interpersonal aspects for collectively analyzing student work. This reality is changing. The work of teacher is now being conceived as networking and collaborating with other teachers, support staff, and community members for the benefit of students (Garmston and Wellman 2013).

TEACHERS AS CONTINUING LEARNERS

The pursuit of perfection in a craft is as old as the human species. Bill Powell, formerly superintendent of the International School of Kuala Lumpur, reminds us that our myths are replete with stories that praise the skill and wisdom of master craftsmen. Whether in art, music, sports, or business, the master craftsman in his or her pursuit of refinement and improvement continues to be featured among our most popular heroes. Powell suggests that craftsmanship is both an attribute and an energy source. The constant striving for improvement is certainly a quality of the master teacher, but it is more than a static characteristic. It is also a dynamic motivating force that compels the teacher toward greater refinement, greater specificity, and greater precision (Powell 2000).

The master teacher has always constructed clear visions and specific goals. In the mid-1980s, education began an unprecedented journey toward setting standards for students, teachers, and administrators. In schools that already had a norm of self-assessment, this provided new resources with which teachers could inform their goals for teaching.

Several characteristics can usually be found in these standards for teaching. First, standards reflect a holistic conception of career development, describing teacher competencies in ways that ensure consistency with emerging visions of teaching development. Second, the core standards describe essential efforts in teaching, regardless of the grade level or students being taught. Third, the standards are performance based, describing what teachers should know or be able to do. Finally, the standards are linked directly to current views of what students should know and be able to do to learn challenging subject matter.

Because teaching standards most generally describe generic teaching competencies unrelated to subject matter and age of student—and because our knowledge of teaching, learning, and content continues to evolve—such standards are useful mainly as instruments for coaching and self-appraisal. As instruments of evaluation, they fall far short of reflecting the complexity of decision making in which teachers engage as they integrate and apply information from multiple domains of knowledge (Popham 2014, 47–52).

CONCLUSION

Teaching is one of the most cognitively complex professions. Although the knowledge base of teaching and learning is vast and continually expanding, there is still uncertainty as to what works in various schools in diverse communities with each unique group of students. Intersecting with teachers' cognitive and professional development are stages of adult growth and teaching standards. We believe, therefore, that what makes teaching a profession is the continual inquiry, expansion of repertoire, and accumulation of knowledge through practice.

The professional educator continually experiments, inquires, tests, gathers data, revises, and modifies thought and practice. Such accumulated knowledge influences teaching decisions in at least the six domains described here. Continuing inquiry into these areas constitutes career-long agendas for professional growth. Cognitive Coaching helps teachers to integrate, extend, and apply this information in the crucible of classroom work.

Part 3

ENGAGING IN COACHING

In part 3, we examine four maps of Cognitive Coaching: planning, reflecting, problem resolving, and calibrating. This will be the first time the calibrating map appears in print. Coaches use these coaching maps to guide conversations, recognizing that planned conversations are significantly more effective than those without prethought structure. The coach considers all that has gone before in this book that is congruent with findings about human cognition and learning.

We also examine human diversity as it relates to coaching. Humans differ in many ways, and the accomplished coach is sensitive to these differences as she/he interacts with others. In this section, we describe seven of these distinctions: (1) representational systems, (2) cognitive styles, (3) educational belief systems, (4) gender, (5) race, ethnicity, nationality, and culture, (6) adult stages of development, and (7) generational differences. Generalizations must inform this reporting because endless variations in some of these categories—race, ethnicity, nationality, and culture, for example—are so numerous as to defy description in so short a treatment.

Perception is reality, and accomplished coaches make efforts to understand the world their coachees see in order to increase their coaching effectiveness. They adapt their communications to persons with whom they work, drawing upon their own state of mind of flexibility to do this. Language choices often offer cues as to the ways the coachee is processing information and the alert coach is sensitive to these nuances.

Coaches use the four Cognitive Coaching maps as guides to conversations, recognizing that planned conversations are significantly more effective than those without deliberate structure. Within each of the functions, the maps support the coach and teacher in meaningful conversation. The structure and coaching processes used with each map align with what the research community is discovering in brain science. As one example, Boyastzis and Jack (2010) used fMRI to illuminate reactions in the brain. When self-directed models of coaching were used, the researchers observed positive emotions and openness to coaching. Neurotransmitters increased the accuracy of perceptions and cognitive functioning. More-critical coaching processes showed evidence of the opposite: defensiveness and decreased capacity for visioning and problem solving. Additionally, they found that thirty-minute conversations that focused on a person's goals activated, for as long as five to seven days, the parts of the brain associated with cognitive, perceptual, and emotional openness and better functioning.

Each of the four maps—whether devoted to planning, reflecting, calibrating performance against a standard, or resolving problems—remains dedicated to fostering self-directed learning.

10

Human Diversity in Meaning Making

Where Cognitive Coaching provides a cognitive framework for our work as educators, the tools of Cultural Proficiency provide a moral framework. The culturally proficient educator recognizes that the disparities that exist among demographic groups are human-made and human-maintained. Recognizing these human-made disparities, the culturally proficient educator sees it as her moral obligation to create and influence change both within herself and her school.

—Lindsey, Martinez, and Lindsey (2007, 34)

When you pass strangers on the street, the unfamiliar faces blur. When you let your lives touch and make the effort of asking questions and listening to the stories they tell, you discover the intricate patterns of their differences and, at the same time, the underlying themes that all members of our species have in common.

—Mary Catherine Bateson

Humans differ in culture, temperament, tempo, signature, thumbprint, and patterns of veins in the hand. We hold different values, like different books, people, music, work, and leisure. Some of us are casual; some of us are fastidious. Many of us are serene; others among us are stressed. We are light skinned, dark skinned, slim and stout, young and old. Genetic, family, cultural, and environmental histories combine to make each of us truly unique (Markus and Conner 2013).

In this chapter, we present information about seven of the human variables that influence the human quest for meaning with a caution to the reader. It seems almost all comments related to this territory, unless exceptions are elaborately specified, are generalizations. Such is the nature of being human; in all dimensions, no single statement can represent everyone in any category being addressed. Readers are invited to keep this in mind as you read the discussions about (1) representational systems, (2) cognitive styles, and (3) educational belief systems. Naturally, generalizations also occur in the accounts of (4) gender, (5) race, ethnicity, nationality, and culture, (6) adult stages of development, and (7) career stages. These final four set the broader context within which humans interact. Skilled coaches are sensitive to differences in each of these areas, and they adapt the coaching interactions for each person with whom they work. No person can attend to all these variations at once nor possibly know each of them in a detailed way. Curiosity, openness, and a willingness to learn from others is a start.

Coaching effectiveness increases with each new perceptual filter held by the coach. We've found that having multiple filters to consider, we can try on various perspectives as we work with others, testing which might apply.

Skillful coaches draw upon the states of mind of flexibility and consciousness to enhance coaching effectiveness by attending and adapting to differences in the ways that humans process and make meaning. For the first three filters, we describe ways coaches expand their repertoire and match style to a variety of situations and individuals.

We start by exploring the vagaries of perception and describe three perceptual filters common to all educators: representational systems (the way we represent thoughts in our minds), cognitive style (a biologically informed way of taking in information), and five educational belief systems representing differing orientations toward learning that inform teacher practices and that, for decades, have been the background for each phase of "educational reform" and federally funded programs in education. When coaches are attentive to these three variables—their own and those of the person they are coaching—they gain influence by attending to preferences within the systems of the person they are coaching.

UNDERSTANDING PERCEPTION

Although we tend to believe what we perceive, our perceptions are not trustworthy. Perception is the *interpretation* of sensory information. The same tableau, seen through other eyes, reveals different findings. Perception is more than the passive receipt of information we receive through our senses. Instead, the meanings of signals from the nervous system are shaped by memory, expectation, learning, and attention.

Perception, in large part, is the result of culture, genetics, and neurology. East Asians view the world differently than do Westerners, each group moving their eyes in patterns different from the other when looking at pictures (Azar 2010). Chinese see context while Americans see details. Neurological differences account for differences in perception between men and women. Women, according to studies, distinguish colors better than men, and men detect movement better than women. Abramov, Gordon, Feldman, and Chavarga (2012) explain that these elements of vision are linked to specific sets of thalamic neurons in the brain's primary visual cortex. They further note that there are also marked gender differences in hearing and the olfactory system.

A surprising number of animals see in color, but the colors they see are often different from the ones we see. Prairie dogs are colorblind to red and green; ants can't see red. Deer and rabbits see humans in shades of gray and by smelling. This is why human processing is often discussed in terms of "representational systems"—that is, what we hear is a representation in our mind of the original sound, and what we see is a representation of a picture. The reality is created deep inside our brain (O'Connor and Seymour 1990). In short, what each species perceives, including human groups and individual humans, does not exist "out there" but rather is constructed in ways unique to the physiology of each.

Neuroscientist Antonio Damasio explains that "Representations are constructed based on the momentary selection of neurons and circuits engaged by the interaction. In other words the building blocks exist in the brain, available to be picked up and assembled" (1999). The part that remains in memory is built according to the same principles. So, when we think, we reexperience (reconstruct) the information in the sensory form in which we first experienced it: pictures, sounds, feelings, tastes, or smells. "Images," says Damasio, "are the main currency of our minds" (2010, 160). Damasio uses the term *images* to include sights, sounds, feelings, and smells.

For Cognitive Coaches, understanding that each person creates perceptions unique to that person allows them to accept others' points of view as simply different, not wrong. They come to understand that they should be curious, not judgmental, about other peoples' impressions and understandings. The more we all understand how someone else processes information, the better we can communicate with him or her.

REPRESENTATIONAL SYSTEMS

In Western culture, the primary nonlinguistic systems in which we think are the visual, auditory, and kinesthetic. All persons think using these three primary representational systems, and each of us favors one or two, even though we are often unaware of our thoughts in more than one system. In addition to having a highly preferred representational system, each person has a lead system. Reviews of a number of studies and conversations with observers in this field lead us to believe that roughly 40 percent of the U.S. population uses the visual system as the preferred mode. Another 40 percent is primarily kinesthetic, and 20 percent is auditory. Figure 10.1 shows these systems.

Stop right now and think back to your last good meal. If you first saw pictures, this is your lead system. If you are right handed, your eyes probably went up to the left accessing an image. Whatever you accessed first—pictures, sounds, or feelings—this is your lead system, the jump-start to your thinking process, and it is likely that your

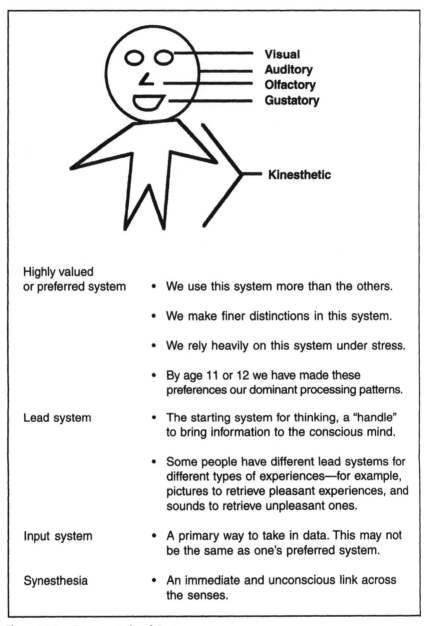

Figure 10.1. Representational Systems

eyes signaled this to an attentive other, eyes down for feelings, eyes aimed at ears for sounds. Perhaps you saw a picture of a friend's face, followed by feelings of relaxation. In that case, your lead system would be visual and your preferred system kinesthetic. As shown in figure 10.1,

- *Each person has a preferred system.* We use this system more than the others, we make finer distinctions in this system, and we rely heavily on this system under stress. By age 11 or 12, we have made these preferences our dominant processing patterns.
- *Each person also has a lead system.* This is the starting system for thinking, a "handle" to bring information to the conscious mind. Some people have different lead systems for different types of experiences—for example, pictures to retrieve pleasant experiences, and sounds to retrieve unpleasant ones.
- *Each person also has an input system.* This is a primary way to take in data. This may not be the same as one's preferred system. Synesthesia is an immediate and unconscious link across the senses.

The person with the most flexibility has the greatest influence in any interaction, because having greater flexibility opens more choices, and more choices allow greater chance of attaining outcomes. Cognitive Coaches watch for cues that can guide their choices. Eye movements can alert a coach to be silent, as mentioned above, or inflectional stress on a word allows the coach to know an idea has special importance to the speaker, inviting a coach's follow-up. In a research review of leaders' influences on student learning, authors reported that successful leaders modify their behaviors to match circumstances and that "We need to be developing leaders with large repertoires of practices and the capacity to choose from that repertoire as needed (Leithwood, Seashore-Louis, Anderson, and Wahlstrom 2004,10). The reason behind the eye movements is unknown. Ehrlichman and Micic (2012) suggested they might be related to how our higher mental functions, such as searching our memories for information, might have been involved. Eye movements are useful cues guiding coaches' use of pausing. When another person's eyes move up to the left or the right, for example, the coach infers the person is processing internally, most likely in pictures, and anything the coach says at that moment will interrupt that person's thinking.

The most effective coaches are conscious of their own and others' personal styles, patterns, and predispositions. They flexibly work in different ways with each individual according to the ways that person is unique. This trait is important for the development of trust because we all tend to be more comfortable with, and therefore trust more easily, people with similar styles. A teacher whose primary strength is visual may best be served by coaching practices that employ visual terms—show, see, picture, enlarge, and so forth—while the same coach might work with an auditorily oriented teacher using more hearing words—tone, rhythm, or resonate, for example.

In most cases, flexible coaches personalize each moment of communication to the individual, responding to the representational systems the other person is using. When a coach uses the other person's representational system, trust is enhanced, rapport in the moment is strengthened, and the communications require no translation.

Exceptions to this principal exist. For example, sometimes a coach wants to stretch a person's thinking into a system not being used. In recalling what occurred in a lesson, a teacher's recollection might be limited to what she/he saw. Noticing this, a coach might ask, "And what were you hearing that indicated students were succeeding?" More data now becomes available to the teacher.

Two types of cues indicate which representation system a person is using in the moment: language and eye movement.

One of the authors, as a teacher, had an administrator who was biased against the author. The author was working on gaining skill in detecting representational system language. By writing down words participants used in staff meetings, the author discovered that the administrator primarily used auditory phrases. (Other cues were there but were unnoticed until then, such as dimmed lights in the administrator's office and classical music in the background.)

The author was highly visual but knew the research on representational system matching and trust. Wanting a better working relationship with the administrator, the author began to use auditory terms when he was around the administrator. In a remarkably short amount of time, the administrator's attitude toward this author became positive. After that, the author no longer had to match language systems with his boss all the time; the connection had been made.

Language Cues

Vocally affiliated pairs and groups may converge their speech patterns. By matching a partner's speech style, a person may emphasize similarities and enhance attraction (Burgoon, Buller, and Woodall 1996). Specific behaviors on which people match speech include accent, rate, volume, pause length, vocalization length, and language choice. When a parent says, "I see what you are saying," we might infer that the parent is making images in her mind. When a student says, "I don't see what you are saying," we might infer the student is having difficulty making pictures out of the teacher's words. Until the student can make a picture, understanding remains elusive (Dilts and Epstein 1995). In this sense, we take people literally and adjust our communication styles to theirs.

Because words can be clues to thinking processes, Cognitive Coaches become familiar with word patterns from the sensory-based representational systems listed in figure 10.2. These words are an indicator of how the person is representing an experience. Words can also be cues for metaphoric matching. We refer to this as language congruence. For example, when someone says, "They play hardball around here," you might respond by saying, "Yeah, it's really tough. Bases are loaded, you go up to bat, and the team is counting on you."

Many methodological errors can be found in the research on language congruence, unfortunately, but several solid studies—and our own experience—have affirmed the relationship between language and representational systems (Einspruch and Forman 1985).

As you listen to a person talk, you may discover there are times when a majority of his or her predicates (descriptive words and phrases—primarily verbs, adverbs and adjectives) are from one of the modality or representational systems listed below. This person is choosing, usually at an unconscious level, to isolate one system from his or her ongoing stream of representational system experiences. This is an indicator for you of how this person is best understanding his or her experiences and how you can best communicate.

Visual	Auditory	Kinesthetic
see	hear	feel
look	listen	grasp
observe	speak	handle
watch	tell myself	energetic
clear	verbalize	in touch
viewpoint	told	gut feeling
perspective	talk	firm
point of view	say	foundation
visualize	clear as a bell	on the level
eyeball	tune in	relaxed
hazy	resonate	tense
fuzzy	tone	weighty
murky	harmonious	heavy
vivid	volume	come to grips
light	loud	lightweight
transparent	dissonant	raise an issue
lighten up	pitch	grasp the situation
look something up	high-pitched	let go
picture	low-key	sleep on it
reflect	squeaky	hurt
acuity	singsong	touchy
see the light	ring my chimes	irrational
focus	unheard of	pushy
image	well said	pain in the neck
mirror	answer	itchy
insight	so to speak	foot the bill
foreshadow	drum it in	shoulder the blame
red	mellifluous	soft touch
purple		

	Gustatory	**Nonspecific**
Olfactory	taste	think
smell	tasteless	experience
odor	tasteful	know
scent	salivate	intellectualize
aroma	mouthwatering	understand
fragrant	tip of my tongue	perceive
rotten	delicious	respond
fresh	lip smacker	accurate
	sweet	solution
	spicy	resolve
	bitter pill to swallow	strategy
	bit off more than she or	logical
	he could chew	

Figure 10.2.   Representational Systems: Language Indicators of Modality Preferences

---

HOW WE SEE

Damasio observes, "[The] images you and I see in our minds are not facsimiles of the particular object, but rather images of the interactions between each of us and an object which engaged our organisms, constructed in neural pattern form according to the organism's design" (1999, 321). The object is real, and the images are as real as anything can be. Yet the structure and properties in the image we end up seeing are brain constructions prompted by an object.

As you listen to a person talk, you may discover there are times when a majority of his or her predicates (descriptive words and phrases—primarily verbs, adverbs, and adjectives) are from one of the modalities or representational systems listed in figure 10.2. This person is choosing, usually at an unconscious level, to isolate one system from his or her ongoing stream of representational system experiences. This is an indicator for you of how this person best understands his or her experiences and how you can best communicate.

Eye Movement

With the advent of research in brain specialization, scientists have investigated the relationship between particular types of eye-movement patterns and variations in brain functioning (Garmston, Lipton, and Kaiser 1998). Although controlled laboratory testing has yielded inconsistent results, the relationship between eye movements and cognitive functioning has been well documented (Jensen 1996; Ornstein 1991, 39; Kinsbourne 1972, 1973; Kocel, Galin, Ornstein, and Merrin 1972). Coaches use this information to guide interactions, as we explore in this chapter.

In 1977 Robert Dilts conducted a study at the Langley Porter Neuropsychiatric Institute in San Francisco that correlated eye movements to specific cognitive processes. Electrodes tracked both the eye movements and brain-wave characteristics of subjects who were asked questions related to using the various senses of sight, hearing, and feeling for tasks involving both memory and mental construction. Dilts's findings tended to confirm other tests that showed that lateralization of eye movements accompanied brain activity during different cognitive tasks. This pattern also seemed to hold for tasks requiring different senses.

Cognitive activity in one hemisphere triggers eye movements in the opposite hemisphere. The following eye-movement patterns provide a description of internal sensory activity for 90 percent of right-handed persons (figure 10.3). This information is reversed for many (but not all) left-handed persons.

- Looking up and to one's left allows us to access stored pictures (visual recall). This is seen often in reflecting conversations (figure 10.4).

Figure 10.3.   Eye Movements and Representational Systems

Figure 10.4.   Accessing Her Visual Memory

Figure 10.5. Constructing Visual Images

- Looking up and to one's right indicates the creation of new images (visual construct). This is seen often in planning conversations (figure 10.5).
- Eyes go horizontally to one's left when we are accessing stored sounds (auditory recall), as when we are recalling in a reflecting conversation.
- Eyes go horizontally to one's right when we are creating new sounds, as in mentally rehearsing before we speak (auditory construct).
- Eyes down and to one's left is common when we are talking to ourselves (internal dialogue). This position and the next must be differentiated from another cause of looking down: a learned behavior that indicates respect, in some cultures.
- Looking down and to one's right is common when we are experiencing feelings (kinesthetic).
- Looking straight ahead when answering a question may mean accessing memorized information, such as "What is your name?"

The coach's question invites the teacher to locate internal resources to employ with her challenge. The coach waits while the teacher's eyes indicate for the coach that an internal search is occurring. She searches at the 10 and 2 o'clock positions, pauses at 2:00, returns to 10:00, then drops her eyes to what appears to be 7:00. Since this teacher is a normally organized right-hander, the coach can infer that she has been accessing visual memory and visual construction with a final check to emotions. Readers wishing to gain skills in inferring cognitive processes from observing eye movements might observe the entire reflecting conversation with this teacher in chapter 11.

Coaches use eye-movement information to select language congruent with the representational system the teacher is using. More important, because eye movements reveal the duration of cognitive processing, the coach knows when to use wait time and how long to wait. Eye movements signal that someone is focusing attention internally, when additional comments or questions can't be processed.

Auditory processing is quick. Persons who process information visually take slightly longer, and processing kinesthetically takes the longest. Science educator Mary Budd Rowe found that teachers who extended their wait time after asking students questions (from the normal one-half second to three to five seconds) produced impressive gains in students' higher-order thinking (Rowe 1996). Later neurological research finds that it takes that long to receive information, process it, and formulate a response.

An example of eye movements may be seen in video 10.1, which can be found at www.thinkingcollaborative.com/supporting-audio-videos/10-1/. Carolee asks Ochan a question, which initiates Ochan's internal search. Notice her eye movements. From what you have learned about eye movements and representational systems, what are some inferences you might make about Ochan's internal search?

COGNITIVE STYLES

Two persons looking at the same invitation to a dinner party may literally see something different. Huong will notice the time, date, and location of the affair as well as the recommended dress. John will register with pleasure that a friend has invited him to a party. He notices that the event is planned for Thursday, an evening he's usually free. It wouldn't be unheard of for him to arrive at the party on the wrong Thursday or get lost on the way because he didn't pay attention to details. Huong's and John's readings of the party invitation illustrate differences in cognitive styles.

Cognitive style is the cognitive functioning related to acquiring and processing information. Research on cognitive style was launched with the pioneering work of Herman Witkin's examination of intellectual styles. Research surged and peaked during the period 1940 to 1970. Today numerous models and theories explain stylistic differences (Lozano 2001), yet we have found that Herman Witkin et al.'s field dependence–independence theory is a simple yet effective construct for learning how teachers' perceptions, intellectual processing, and instructional behaviors differ (Guild and Garger 1985). Currently researchers are examining the neural activities related to cognitive styles (Kozhevnikov 2007). Most of us lean toward one of two poles: exacting attention to detail where perception is not influenced by the background (field *independence*) or a strong influence by the context of information (field *dependence*). Investigators have explored the relationship of cognitive style to personality (Zhang 2000) and intelligence but found limited correlations to the latter (Kozhevnikov 2007).

Over time, Witkin and his associates became convinced that field dependence–independence influences not only perceptual and intellectual functioning but also personality traits such as social behavior, career choice, body concept, and defense.

Field Independence

In general, the field-independent person is task oriented and competitive and may be drawn to occupations in mathematics, engineering, or science. The term *field independence* is used because such a person doesn't need to view the whole system to make sense of things. She/he can isolate and work effectively with the details. This person often likes to work alone and emphasizes getting the job done. She/he works part to whole, perceiving analytically. The field-independent teacher is logical, rational, and likes to figure things out for herself. She learns through books, computers, and audiotapes. She/he is good at sequence and details, and she likes theoretical and abstract ideas.

The field-independent person who's placed on a committee might say, "Just give me a job, and I'll go off and do it. Why are we talking so much? We don't need to get to know each other. We need to complete the task."

The field-independent person wants the coach to focus on tasks. She/he wants independence and flexibility to make decisions based on data and analysis. She looks to the coach to be knowledgeable about curriculum and instruction, to maintain a professional distance, and to give messages directly and articulately. Although these traits seem to be inborn, many people learn to be flexible for different tasks—filling out income tax forms, for example, or hiking in the wilderness.

Jan, a middle-school field-independent teacher, looked back on four months of Cognitive Coaching and reflected that previously she had been extremely curriculum focused in her teaching, devoting her time and attention to the details of each activity. She had been "administering knowledge" and successfully keeping students busy, but at the expense of some aspects of the students' cognitive development and personal growth. Through coaching sessions, Jan developed more balance in her teaching. The coach matched her interest in details and her language of specificity, which gave the coach and the teacher the safety to explore bigger questions: How does this lesson relate to long-term goals? What personal meaning do you want students to construct from this? Why is this important to you? (Garmston, Linder, and Whitaker 1993).

Field Dependence

The field-dependent person needs to perceive the whole context (the "field") to make meaning. This person enjoys working with others in collaborative relationships. He/she takes in the overall scheme of something and may have difficulty with individual parts. He/she works from intuition and gut reactions but likes and needs concrete experience. Metaphors, analogies, patterns, and relationships appeal to him because he likes to see things holistically.

Field-dependent persons are often oriented toward relationships, even seeking mentors in their lives. They have a sense of the large picture, and they know where they are going for the semester or the year. They often understand what is occurring in their class through intuition, reading subtleties of body language and voice.

The field-dependent teacher in a faculty meeting appreciates the conversation and interaction about ideas. He finds it important and valuable to share feelings on a topic. He/she is more tolerant of ambiguity than his field-independent peers and more skillful in processing.

Field-dependent teachers want a coach to be warm and show personal interest and support. They want guidance and modeling, but they also want the coach to seek their opinion in decisions. They want the coach to have an open door, practice what she preaches, and use voice tones and body language that support her words.

Christina was a field-dependent teacher in the same school as Jan. Her style of teaching focused heavily on the affective domain. She writes that when she encouraged a focus that was often subjective, students learned through exploration, but specific learnings were often sacrificed in attainment of the broader goal. Jan and Christina both believed that Cognitive Coaching helped them access the "lesser used" sides of their brains. Through self-analysis and "self-remediation" (their words), they searched every corner of their minds, bringing to the surface feelings and ideas that might have otherwise gone untapped. In becoming more bicognitive, they became better thinkers and better teachers (Garmston, Linder, and Whitaker 1993).

Although it is always possible to err by stereotyping with certain characteristics, the traits of field independence and dependence are useful cues for communication in the coaching relationship. A developmental goal for each person, is to become bicognitive—able to switch according to task, attending to detail and big picture with equal ease.

EDUCATIONAL BELIEF SYSTEMS

To be truly flexible and maximally effective, coaches are discerning about, and work congruently with, educational philosophies different from their own. Different educators have different aims in education. Whether they verbalize them or not, educators hold deep convictions about their professional mission, their work, their students, the role of schools in society, the curriculum, and teaching. Furthermore, these beliefs are grounded in and congruent with deep personal philosophies and are powerful predictors of behaviors. They drive the perceptions, decisions, and actions of all players on the education scene.

Beliefs are formed early and tend to self-perpetuate (Pajares 1992). Changing beliefs during adulthood is a relatively rare phenomenon, and beliefs about teaching are well established by the time a student gets to college. Because beliefs strongly influence perception, they can be an unreliable guide to the nature of reality. The filtering effects of belief structures ultimately screen, distort, or reshape thinking and information processing.

People's ideologies significantly influence their educational decisions. The flexible Cognitive Coach recognizes colleagues' prevailing educational ideologies through their use of metaphors, educational goals, instructional methods, approaches to assessing outcomes, and teaching. The coach respects these beliefs (with the obvious exception of destructive orientations like racism and fixed mind-sets) and works with them to help fellow educators articulate and be increasingly effective in achieving their educational goals.

The coach's goal is to maintain trust and communication rather than to change beliefs. When the coach works within a colleague's educational belief system, trust is deepened, rapport is enhanced, coaching services are valued, and the coach becomes more influential. Ultimately, the colleague may become more conditional regarding his or her own educational philosophies and more understanding and appreciative of the educational beliefs of others. This movement, as a step toward greater cognitive complexity, is a developmental goal for all human beings. In their seminal work on educational belief systems, Eisner and Valance (1974) and Eisner (1994) describe at least six predominant ideologies that influence educators' decision making and fuel the ongoing debate among policymakers, politicians, parents, and the public in the 21st century. We do not intend to describe those ideologies fully here, but we present a brief summary of our interpretation of each to illuminate their differences and manifestations in classrooms and schools.

Religious Orthodoxy

Coming from a Christian tradition and with a firm belief in God, the early Pilgrims in the United States provided an educational policy in which schools were legally mandated to teach biblical literacy. Today, numerous religious schools in the United States—such as Roman Catholic, Church of Christ, Evangelical Christian, Buddhist, Seventh Day, Adventist, Muslim, Jewish, Lutheran—have a common goal: to teach the habits and values that will lead students to live in accordance with their faith.

Proponents of religious orthodoxy in the Christian sects include such groups as the Christian Coalition as well as such orders as the Jesuits, the Christian Brothers, and the Sisters of Mercy. Their metaphors lie in the Bible. Others may draw inspiration from the Qur'an, the Torah, Tao Te Ching, or the Veda. They may model their ways after David, Moses, Jesus, Mohammed, Buddha, or Confucius. They wish to transmit the beliefs, laws, commandments, and truths of their interpretation of their scriptures.

Educators with the orientation of religious orthodoxy enable students to learn appropriate norms of morality. They encourage students to conduct their lives in accordance with those norms. Students are assessed, in part, to the degree that they exhibit the teachings of that religion.

In some of these schools, dialogue, debate, and critical thinking are encouraged. In others, questioning, doubt, and the exploration of competing views are often discouraged. Faith in Allah or in God's word may be held to transcend human thought. Fundamentalist Christian parents, for example, may criticize a public school's efforts to teach "critical thinking," questioning, evolution, or other attitudes that might weaken religious commitment. Likewise, such issues as vouchers, school prayer, and dialogues about the separation of church and state often illuminate differing beliefs.

Cognitive Process

With an orientation to cognitive psychology, cognitive processors believe that the central role of schools is to develop rational thought processes, problem-solving abilities, and decision-making capacities. They are drawn to educational theorists and authors such as John Dewey, Jerome Bruner, Hilda Taba, J. Richard Suchmanm, Caleb Gattegno, Robert Sternberg, Jean Piaget, Reuven Feuerstein, Maria Montessori, Howard Gardner, David Perkins, Edward deBono, and Carol Dweck. They believe that the information explosion is occurring at such a rapid rate that it is no longer possible for experts in any field to keep up with new knowledge. Thus, we no longer know what to teach but instead must help students learn how to continuously learn throughout their lifetimes.

Cognitive processors select instructional strategies that involve problem solving and inquiry. They use concepts like Bloom's taxonomy, intellectual development, habits of mind, metacognition, and thinking skills and dispositions. They organize teaching around the resolution of problems and the Socratic method, and they bring in discrepant events for students to explore and analyze. They assess student growth by how well they perform in problem-solving situations. Their metaphorical model of education is that of information processing: Human beings are meaning makers, and schools and teachers mediate those capacities.

Self-Actualization

Those who believe in self-actualization regard the ideal school as child centered. They believe the teacher's role is to liberate each human's inherent capacities for learning. They believe that the purpose of teaching is to bring out the unique qualities, potentials, and creativity in each child. They support multisensory instruction, with many opportunities for auditory, visual, and kinesthetic learning. They value student choice, whether for classroom topics, the nature of assignments, or classroom activities. They value self-directed learning and individualized instruction and support schools of choice and magnet schools so as to better meet the unique interests of each learner.

To provide for students' unique and multiple needs, interests, and developmental tasks, those who believe in self-actualization use learning centers focused around themes. They value student autonomy and look for increases in self-directedness as a central measure of teacher effectiveness. They are drawn to such humanists as Abraham Maslow, Arthur Combs, Carl Rogers, and Sidney Simon, along with George Leonard, Gerald Jampolsky, Deepak Chopra, Parker Palmer, Steven Covey, and Ekhart Tolle.

The self-actualizer's vocabulary includes words referring to the affective domain and concepts such as the whole child, nurturing, peak experiences, choice, democracy, holism, self-esteem, continuous progress, creativity, empathy, and caring. Their metaphoric model of education is grounded in a vision of children's potentials.

Technologism

Influenced by the behavioral psychology of B. F. Skinner, Ivan Pavlov, and Edward Thorndike, technologists may be attracted to such education authors as Robert Mager, Madeline Hunter, Seymour Papert, Arnie Duncan, Raymond Miltenberger, Bill Gates, and Steve Jobs. They place strong emphasis on accountability, test scores, learning specific subskills, and measurable learning. Their metaphor for education is as an "input, throughput, output system" in which data and opportunities to learn skills are provided.

Technologists are skilled at task analysis and are interested in technology-based learning systems and instruction. They value opportunities to diagnose entry levels and prescribe according to what is known and what is yet to be learned. They are probably more field independent and skilled in detail, with great ability to analyze, project, and plan. They talk about accountability, evaluation, reinforcement, task analysis, time on task, mastery, templates, diagnosis and prescription, interrater reliability, and percentiles. When policymaking bodies adopt this orientation, external assessments and high-stakes testing abound.

Academic Rationalism

Keeping company with such philosophic leaders as Robert Maynard Hutchins, Mortimer Adler, E. D. Hirsch, Arthur Bestor, William Bennett, and Chester Finn, academic rationalists are drawn to teacher-centered instruction. They believe that knowledgeable adults have the wisdom and the experience to know what is best for students. Their metaphors for education are the transmission of the major concepts, values, and truths of society, and they consider students as clay to be molded or vessels to be filled. They value and are highly oriented toward increasing the amount and rigor of student learning. They are drawn to the Classics (preferably original sources), Great Books, and traditional values. For them, the most worthwhile learning centers on those enduring ideas and artifacts that have stood the test of time. The works that contain the greatest products of the human mind thus become the canon of the school curriculum. Academic rationalists believe that human nature is unchanging and that there are eternal truths to be discovered in a world outside of human beings. They therefore emphasize those perennial issues of human life as embodied truths. They appreciate basic texts and the teaching strategies of lecture, memorization, demonstration, and drill. They evaluate students through summative examinations, achievement testing, and content mastery. They speak about authority, humanities, basics, the disciplines, scholarship, and standardized tests as being central to effective instruction.

Social Reconstructionism

Social reconstructionists are concerned with problems of society such as the future of the planet, the destruction of the food chain, the hole in the ozone layer, global warming, endangered species, the deforestation of timberland, the protection of wildlife, and the threat of overpopulation. Social reconstructionists view the learner as a social being: a member of a group, a responsible citizen, one who identifies with and is proactive regarding the environmental ills and social injustices of the day.

Social reconstructionists believe that we have gone beyond the age of representative democracies and have moved to a stage of participative democracies. They believe that this is a world in which we must care for our neighbors and take action at the grassroots level. The social reconstructionist teacher engages students in recycling centers, contributing to social issues, community service, cooperative learning, environmental education, and global education. Their metaphor of education is as an instrument of change, and they believe that schools are the only institution in our society charged with the responsibility of bringing about a better future and a better world. They are drawn to such activists as Pablo Freire, Alvin Toffler, Jane Goodall, Margaret Wheatley, John Naisbitt, Ralph Nader, Fritjof Capra, and Al Gore. They employ concepts such as environmentalism, democracy, consumer education, student rights, the 21st century, multiculturalism, futurism, global intellect, pluralism, change, save the Earth, ecology, peace, and love.

All of these belief systems are necessary and valuable aims of education. As Carl Glickman observes, "When one concept of education wins out over all others and we are left with a single definition determined by others for all of us, then we all lose" (2001). We want students to find spirituality and to become effective problem solvers, to be self-actualized, knowledgeable, efficient, and concerned. So the task for coaches is to become at ease with a variety of educational beliefs and to refine their ability to work with people whose beliefs may be different from their own.

Belief systems don't change easily. As we mature, the less likely we are to change. However, change usually occurs in three instances. If the prevailing culture begins to shift its values persistently and pervasively, we may begin to move our thinking. For example, the 1960s saw the pervasive influence of individualized instruction, and many of us began to behave more like self-actualizers. In the 1970s, social reconstructionism came to the forefront as we became aware of globalization and confronted environmental problems. With the onset of the information age in the 1980s, we realized that information overload demanded the exercise of intellectual prowess, and cognitive processing became paramount. The technological paradigm of the 1990s influenced our educational thought with high-stakes accountability and international comparison of various countries' science and mathematics test scores.

Teachers also adapt their belief systems to accommodate new realities. For example, a 12th-grade history teacher may take a position as a kindergarten teacher, or an instructor may move from an affluent school to one of pervasive poverty. In these cases, the teachers' paradigm changes and so do the beliefs they use to explain their role in their new environment. The third case in which beliefs might change is when a shift in identity occurs. See chapter 2 and *Overcoming Resistance to New Ideas* (Powell and Powell 2015).

HUMAN VARIABILITY IN MEANING MAKING

We've examined three perceptual filters: representational systems (the way we represent thoughts in our minds), cognitive style (a biologically informed way of taking in information), and educational belief systems (five orientations toward learning that inform teacher practices). Next, we suggest that while each of the above plays an important role in teacher decisions and behaviors, four more filters lie behind the curtain and exert significant influence on teacher thinking and collegial communications. These four are race, ethnicity, and culture; gender; adult stages of development; and generational distinctions. Culturally proficient coaches are sensitive to differences in each of these areas, and they adapt the coaching process for each person with whom they work. No person can attend to all these variations at once, nor possibly know each of them in a detailed way. Since coaches aim to develop their capacity to work with others, many coaches learn about one of these variables at a time, and then gradually increase knowledge of others. Coaching effectiveness increases with each new perceptual filter held by the coach. We've found that with multiple filters to consider, we can try on various perspectives as we work with others, testing which might apply.

We, as Culturally Proficient Coaches, are not only aware of these differences, but we are also aware of the danger of assigning specific behaviors to groups of people. We know that differences exist within cultures and at the same time most individuals have a need to identify with a larger culture group. As we wrote in the preface, this book is not written to inform coaches about the specific cultural characteristics of individuals or groups of people; rather, we offer you a frame for acknowledging those differences in ways that demonstrate a high value for diverse behaviors, perspectives, beliefs, languages, and cultures. (Lindsey, Martinez, and Lindsey 2007, 40–41)

RACE, ETHNICITY, NATIONALITY, AND CULTURE

The shoe that fits one person pinches another; there is no recipe for living that fits all cases.

—Carl C. Jung

The culture in which each of us is socialized influences the way we communicate. For each of us, there are human qualities that are different from our own that belong to groups to which we do not belong. It is not the purpose of this section, nor would it be possible in the space provided, nor are we knowledgeable enough, to describe how to effectively communicate with the many cultures educators are increasingly encountering. It is, however, our intention to emphasize that awareness of groups other than one's own, appreciation of differences we may not understand, valuing human uniqueness, and learning to negotiate diversity are essential traits of a person embracing interdependence. Wherever one lives, even in a seemingly homogenous group, differences abound. Dimensions of otherness always exist and affect a sense of belonging, motivation, success, and communications, even though we may not be overtly conscious of their existence. Race, age, ethnicity, gender, sexuality, and nationality are usually noticed. What may not be noticed but yet can affect comfort, connection, and understanding are the influences of heritage, mental or physical abilities, being a single parent, a newlywed, an adult whose worldview was shaped by being an orphan, a struggling teacher, a person grieving for fading parents, or those with distinctive work roles, religious convictions, or unstated sexual orientation.

Racism and Language as Conveyer of Stereotypes

The big three stereotypes psychologists talk about are gender, age, and race (Raz 2014b). In a fraction of second, one makes a judgment about others: old, young, male, female, Chinese, black, Arab. This is an unconscious process and they show up as soon as you can test babies. Babies like to hear their own language over other languages. They like to look at faces that are familiar over those that are unfamiliar, and so on. So a white baby raised by white people prefers to look at white faces rather than black faces. Yet if the baby is white but raised by a multiethnic bunch, the baby will show no preferences.

Of the three big stereotypes, evolutionary theorists point out, race is an anomaly because noticing race has only occurred fairly recent in human history; we all looked alike on the African savanna. Sex and age are going to be relevant for any primate. There's no experiment that's ever found that children don't care about sex and age when making decisions. When kids pay attention to race, it's because they've learned that race matters in the world we live in.

> So there's so many social psychology experiments showing—and this is really the pernicious side of stereotypes—that even in cases where we know a stereotype doesn't apply or we know it's mistaken, we're guided by it anyway and these are done for the most part, I think, by well-intentioned people who would never dream that they'd be racist, but we are influenced by factors we don't know about. (Paul Bloom, in Raz 2014b)

Humans have gut feelings, instincts, and emotions, and these affect judgments and actions for good and for evil. Generalizations about race are often flat-out wrong, But, says Bloom, we are also capable of rational deliberation and intelligent planning, and we can use these to, in some cases, accelerate and nourish our emotions and in other cases staunch them; it's in this way that reason helps us create a better world.

Language Conveys Our Prejudices

> In audio 10.1m4a, we are reminded of how language can evoke stereotypes. To listen, access http://www.thinkingcollaborative.com/supporting-audio-videos.

Culture can be seen as a set of rules for living, based on who people think they are, where they think they are, and the resources they perceive are available. We suspect that many of us, like fish in water, are not aware of the effects of our own culture and are much less sensitive to differences of those in groups who might perceive, think, communicate, and value differently from us. Writing about cross-cultural friendships, Pogrebin (1995) writes that when with another person who is different yet same in some respects, double consciousness is required. Gay–straight friendships must cross boundaries of oppression and culture, disability raises issues of independence and vulnerability, and age differences evoke different value systems and life experiences (Du Bois 1903). So, too, do differences in race, ethnicity, gender, and nationality cause challenges to successful relationships. If this is true about friendships, it seems even more so when the context is a work setting or a supervisory or coaching arrangement where additional requirements for interaction may be present. It is a peculiar sensation, says Du Bois, this double-consciousness, this sense of always looking at one's self through the eyes of others.

We often work in overseas schools and have become sensitive to the fact that when we work in Latin America, the Middle East, the Pacific Islands, or Africa, the local people we encounter have a polychromic sense of time, easily integrating tasks with socializing. When individuals from these areas come to live in North America, misunderstandings and judgments easily arise as they encounter the monochromic—scheduled—linear-oriented systems here. We sometimes witness misunderstandings and criticisms arising from these differing perceptions of time in faculty interactions and in parent-teacher conferences or scheduled parent meetings.

Many whites may rarely give thought to color, though people of color may often be conscious of it when in mixed groups. Featherstone (1994, iv) writes, "For women of color remembering ourselves is a daily act of courage, a ritual of survival. We live the 'theater of the absurd,' understand at least two languages, speak in a variety of voices and view the world always from two perspectives." In the anthology edited by her, professional women of color—Asian, Native American, African American, Latino, and Pacific Islanders—write of how color sets them apart, in their own eyes and in the eyes of others.

Immigration and birth-rate dissimilarities are changing the face and cultural texture of North America. As of June 23, 2014, the United States had a total resident population of 318,264,000, making it the third-most populous country in the world, yet still a miniscule part of it all at about 4 percent of the world's total population. It is becoming bigger, older, and more racially and ethnically diverse (Shrestha and Heisler 2011). "Once a mostly biracial society with a large white majority and relatively small African American minority . . . America is now a society composed of multiple racial and ethnic groups" (1). Exact data are impossible to achieve because federal agencies are required to use a minimum of five race categories in data collection and one person could legitimately mark four of them and be accurate.

Data gathered during the 2010 census discloses an increasingly complex picture of diversity in the United States. Non-Latino whites remain the majority, with their share of the population dropping. The United States is estimated to be a "minority-majority" nation within 35 years. Latinos of any race now make up 16.4 percent, up from 12.5 percent of the population in 2000. The racial breakdown of other groups is: African Americans, 12.6 percent; Asians, 4.8 percent, up from 3.6 percent in 2000; American Indians, 0.9 percent; and Native Hawaiians, .02 percent (U.S. Census Bureau 2011). Furthermore, an increasing percentage of the total population is multiracial. Increasingly, Cognitive Coaches will have to draw upon their flexibility as they interact with colleagues whose racial, ethnic, national, and cultural orientations are different from their own. But race and ethnicity are not the only determinants of culture. Persons living in urban and rural settings display distinct differences, as do people from different regions in the United States: the heartland, West and East Coasts, southern and northern states (Markus and Conner 2013). Conditions of poverty complicate the picture for educators working with parents as well as the number of adults in the home. In 2008, 1.7 million babies, or 40.6 percent of all births, were to unmarried women (Shrestha and Heisler 2011).

Communication meanings differ across cultures. The American "OK" sign of thumb and index finger touching means "money" in Japan, "zero" in Russia, and is an insult in Brazil. Greetings in different countries can take the form of a hug, handshake, kiss on the cheek, kiss on both cheeks, and to complicate this further, if on both cheeks, some cultures go first to the right, others first to the left. Rudeness can be interpreted should a North American initiate a handshake on meeting a person from South America rather than kissing him on the cheek. Patterns of personal space differ. Contact cultures like the Middle East, South America, and southern Europe are those in which people tend to stand close together and touch each other frequently. In noncontact cultures like North America, people maintain more space between them and touch less frequently. When working in the Middle East, we have been reminded on a couple of occasions it was not appropriate for a man to touch a woman during conversation. Eye contact for North Americans is a positive, while in some Asian, African, and Latin American cultures it is considered a challenge.

What to do about all this? Perhaps personal action starts in the heart and mind—in the heart, feeling confidant in who you are and automatically presuming and valuing the goodness in others. A good source to learn more is *Culturally Proficient Learning Communities: Confronting Inequity through Collaborative Curiosity* (Lindsey, Jungwirth, Jarvis, and Lindsey 2009). Lindsey and colleagues describe ways of engaging *all* members of the staff, with a deep focusing on student learning that ultimately produces discoveries and action regarding inequities in the school. From the heart, we are reminded of the Hindu greeting *Namaste* spoken with a slight bow and hands pressed together, palms touching and fingers pointing upward, thumbs close to the chest: "I honor the place within you in which the entire universe dwells. I honor the place within you, which is of Love, of Truth, of Light and of Peace. When you are in that place within you and I am in that place within me, we are one." From here, we maintain "double consciousness"—paying attention to the other person, our interactions, and ourselves. We are more aware of our attentiveness traveling outside North America than within, but a similar form of heightened awareness exists at home when meeting new persons of a different race. Should we feel initially threatened, we know our hearts will beat faster with blood rushing to the parts of the brain that prepare to manage perceived challenges. We remember to breathe, two or three deep breaths flooding the neocortex with oxygen. We may pause to have the other initiate greetings and subsequent conversation. We may align our pace and movements to be more like the other person than we otherwise might do, keeping in mind that matching or copying may be perceived as artificial and ingenuous (Glaser 2014, 82). Partially we trust that positive intentions open physical and emotional changes in the brain that can lead to trusting, healthy conversations (Glaser 2014, 83). Further, when conversations are reflective and respectful, the speaker's and listener's brains show the same patterns of activation. Glaser reports research linking the heart with the brain; when our hearts are in synch with another person, signals are sent to the prefrontal cortex and other parts of the brain to trust and to indicate that we will not be harmed.

Educators working in international schools strive, particularly at the start of this 21st century, for the development of intercultural competence in our students—that capacity to move graciously and gracefully between and among cultures, without causing offense, treating those different from ourselves with dignity and respect.

"Cognitive Coaching stands the chance of being cross cultural. For example, 'listening with respect' probably cuts across all cultures and represents a value that I think most people would hold. The challenge, of course, is that such listening may take different form in different cultures. The values of CC, I think, represent cross-cultural ideals" (Powell 2014).

GENDER

There are two main bodies of research on gender differences in communication styles: academic research and popular literature. Academic research points out major differences in neurology, conversation characteristics, and traits across gender, while popular research focuses on major stylistic differences in conversation styles between men and women (Merchant 2012). The problem in work settings is not the presence of different communication styles but the misunderstanding and misinterpretation that each gender has about the other's communication. Awareness of gender differences brings about change.

Neurological and psychological studies reveal very real biological differences between men and women that influence their perceptions of the world in subtly different ways. A number of gender differences are found in the hypothalamus, the portion of the brain associated with sexual behavior. Although some researchers caution that most gender differences are, statistically, quite small (Gorman 1992), recent brain scans from healthy people have confirmed that parts of the corpus callosum are 23 percent wider in women than in men. Some researchers speculate that the greater communication between the two sides of the brain caused by a thicker corpus callosum could enhance a woman's performance in certain communication transactions.

"The brain structures for empathy show important quantitative gender differences. Plus, there is a qualitative divergence between the sexes in how emotional information is integrated to support everyday decision-making processes" (Christov-Moore et al. 2014, 2). Females have a stronger biological basis for empathy, but males can learn this trait. Women recall more details under stress and men do better in recalling the larger picture (Pletzer, Petasis, and Cahill 2014) because memory processing is affected by testosterone and progesterone in the amygdala.

Females are better at reading the emotions of people in photographs; males excel at rotating three-dimensional objects in their heads. For boys, activities are central to relationships. Girls create close relationships by simply talking. Women excel in their ability to simultaneously process emotion and cognition. However, these same abilities may impair a woman's performance in certain specialized visual-spatial tasks. For example, the spatial ability to tell directions on a map without physically rotating it appears strong in those individuals whose brains restrict processing to the right hemisphere. Brain research indicates that men listen with different parts of the brain than women, who listen with their "whole brain." By contrast, more women listen equally with both ears, while men favor the right one.

Women tend to be more expressive, tentative, polite, social, while men are, on average, more assertive and dominant when it comes to communication style (Basow and Rubenfeld 2003). Popular research has also shown gender differences in communication styles, from men being primarily goal oriented and results focused and women being relationship oriented, process oriented, and placing a high value on closeness and intimacy in interactions with other people (Tannen 1990; Gray 1992). Overall, both academic research and popular literature on gender differences in communication styles tend to agree on how men and women differ in the way they communicate. While academic research focuses more on the communication characteristics and traits that men and women exhibit, popular literature makes the connection between psychological gender traits and communication styles and gender differences in terms of basic goals of conversations (Merchant 2012).

The biggest difference between men and women and their styles of communication boils down to the fact that men and women view the purpose of conversations differently. Academic research on psychological gender differences has shown that while women use communication as a tool to enhance social connections and create relationships, men use language to exert dominance and achieve tangible outcomes.

On average, women use more expressive, tentative, and polite language than men do, especially in situations of conflict. Men, on the other hand, are viewed as more likely than women to offer solutions to problems in order to avoid further seemingly unnecessary discussions of interpersonal problems (Basow and Rubenfeld 2003).

Sociolinguist Deborah Tannen (1990) finds that men and women converse for different purposes. Men talk, she says, to establish independence and status and to report information. Women, on the other hand, tend to use talk to establish intimacy and relationship. In a sense, women's talk is about establishing the fact that we are "close and the same," and men's talk is to establish the fact that we are "separate and different."

Men tend to operate in conversations as individuals in a social world in which they are either one-up or one-down. Women tend to operate as individuals in a network of connections. This is not to say that women are not interested in status and avoiding failure, but these are not consistent goals. Tannen summarizes these differences as "men talk to report, women talk to rapport."

In mixed-gender groups, at public gatherings, and in many informal conversations, men spend more time talking than do women. Men are more likely than women to interrupt the speaking of other people. A study of faculty meetings revealed that women are more likely than men to be interrupted. In meetings, men gain the floor more often, and keep the floor for longer periods of time, regardless of their status in the organization. In faculty meetings, men tend to talk more often and for longer periods of time. Because women seek to establish rapport, they are inclined to play down their expertise in meetings rather than display it. Because men place more value on being center stage and feeling knowledgeable, they seek opportunities to gather and disseminate factual information.

Women are more likely to be auditory than men are. Women tend to ask more questions and give more listening responses than men do. In one startling study, Sadker and Sadker (1985) reported that teachers who were shown a film of classroom discussion overwhelmingly thought the girls were talking more. In actuality, the boys were talking three times more than the girls.

Because sex differences have been misused in the past to assert male "superiority" and female "inferiority," many people are suspicious of the notion that physiological differences between the sexes can lead to different abilities or predispositions. Now we know that differences do exist but that they are also culturally reinforced. This does not guarantee that all men or all women will have the same characteristics, but it does reveal patterns.

One of the most important coaching findings from Tannen's work relates to the misunderstandings that sometimes develop out of the different purposes of conversation. For example, many women feel that is natural to consult with a partner before making a decision. Men, however, may view consulting with a partner as tantamount to asking for permission. Because men strive for independence, they may find it more difficult to consult. This drive for independence may also explain their difficulty in asking a stranger for directions (Goldberg 1991). Women, on the other hand, are more comfortable in asking a stranger for directions when they are lost because it literally affords another person an opportunity to be helpful to them.

Although research conclusions sometimes seem to conflict, the bottom line for the coach is to be aware of gender differences that may or may not exist. The most successful coaches tend to be androgynous in their communications, drawing on both the masculine and feminine characteristics reported by Deborah Tannen and others. Tannen (2014) writes that boys and girls are socialized differently, each tending to play with their own gender. Girls use language to negotiate and tend toward closeness, telling secrets and assuring themselves that they are all the same, while boys use language to negotiate status determining who is in, who is out (Tannen 2014).

Successful male coaches can reflect the other person's emotions on their own face (a trait found more frequently in women), and successful female coaches can restrict the display of facial emotion when called for (a trait found more often in men). Skilled coaches of both genders place a premium on both goal attainment and developing and maintaining relationships.

ADULT DEVELOPMENT: DIFFERENT WAYS OF MAKING MEANING

Another important area of human variability relates to the differing ways adults make meaning. There is developmental growth beyond the childhood ages of Piaget's studies, and many parents today will attest that their own children seemed to continue to "grow up" well into their 20s and 30s. Ellie Drago-Severson (2011) observes that because adults go through different stages of development, they need different types of support and challenges in their professional lives. Environments supportive of adult learning both affirm individuals as they are and provide opportunities for continued development.

Based on Harvard professor Robert Kegan's constructive developmental theory (Kegan 1994), Drago-Severson describes the three most common ways of knowing in adulthood. These are not ways of doing but rather ways of being. Most adults will roughly fall into one of these developmental perspectives. A few will occupy a way of knowing beyond these three. These developmental levels represent principles for how adults construct experience and organize thinking, feeling, and social relating. These different ways of knowing are of special interest to the Cognitive Coach because they may operate as containers for all the other ways of making meaning. As Denise Gerhart, a Kegan student and holistic health practitioner, describes it, each system of knowing is how a person organizes his/her understanding and interpretation of his/her experiences and, based on that understanding, devises ways to work effectively within the systems. It is not about progression through the developmental levels. Instead, the Cognitive Coach's "role is to support the person as they are, and to stretch them in ways that they might consider trying for the good of the group, or the good of the situation, or for the good of the cosmos and so forth" (Gerhart 2014).

Some maintain stability at a given way of knowing for most of their adult life, finding value in their current way of being. Progression through these stages can represent a curriculum of sorts for adult development. Movement through the different ways of knowing is a product of mediational environments and the personal drives related to exploring new theories and dismantling old assumptions. Drago-Severson names the three most common developmental stages as *instrumental, socializing,* and *self-authoring.* A small portion of adults may progress to a further developmental stage of being *interindividual knowers.* Progression through these perspectives requires increasingly complex mental operations.

Instrumental Knowers

These adults tend to follow rules and understand their world in concrete terms, in contrast to those who engage in abstract thinking. Adults with this developmental perspective are pragmatic, will look to others for information useful to them, and will provide information they perceive might be useful to others. Conceptually they understand that observable events, processes, and situations have a reality separate from their own point of view, though they frequently have difficulty accepting points of view other than their own. These adults prefer defined parameters, may be prone to have dependency relationships, to seek approval, and to uncritically accept the values of others. They perceive others as either supporters or barriers to their goals. "How am I doing?" this person frequently wants to know from students, colleagues, administrators, or coaches. Metaphorically this way of knowing might be depicted as an island, with the self as the centerpiece of interest or perhaps, more poignantly, as a still pool with all its rich complexities hidden it its depths.

Drago-Severson suggests we can support these adults by using concrete methods: Provide clear goals and expectations. Offer specific protocols for learning social skills related to aspects of being a professional educator. Include structures to engage in dialogue and standards for meetings, attending, listening, and sending skills in conversations and collaborative engagements with others, each of which can be instrumental in appreciating multiple perspectives. We see her recommendations as relating strongly to the state of mind of craftsmanship. In this regard, efficacy and consciousness also represent desirable states of mind to develop.

Socializing Knowers

For a person at this stage of development, a pond, once still, disturbed by a fallen leaf, might express the gratifying connections with others. "I am my relationships" this image intends to evoke. Adults at this developmental level are sensitive to and influenced by the perspectives of others. They seek approval and want to maintain approval in their relationships with coaches and people in authority. They have a need to be accepted by colleagues. In contrast to adults operating at the previous way of knowing, these people have greater capacity for reflection, they can think in abstract terms, and can subordinate personal needs and desires to those of others. Adults at the socializing stage perceive conflict as a threat, tend to feel responsible for the feelings of others, and feel little personal agency about their own feelings. They behave as if others have "caused" their feeling.

To support these adults, employ practices that facilitate developing and voicing their own views to help them be less dependent on the views of others. Engage them in journal writing or in one-to-one or small-group sharing before engaging with the full group. Foster increased capacity to look inward to express their own perspectives.

Consciousness and efficacy are important growth states of mind. Provide noncontingent reinforcement, and keep these persons in an ongoing relationship.

Self-Authoring Knowers

Able to synthesize various points of view, and informed by abstract principles and clear personal values, this way of knowing might appear as multiple ripples in a pond, perhaps with different velocities. These adults, among the three developmental ways of knowing, operate at the most abstract level. They attain greater clarity about personal values, principles, and long-term goals. They are conscious of their own competencies, pursue personal development, and look for the demonstration of the competencies of others. They can be a special asset to collaborative work as they have the ability to synthesize diverse points of view and critique ideas. Developmental challenges for adults operating at this level are that they prefer their own approaches and need encouragement to be open to other approaches. They can also be resistant to the perspectives of others when those views are in opposition to their own.

Engage them in leadership roles that offer opportunities for them to utilize their analytic abilities and engage in dialogue and learn from the perspectives of others. Flexibility and interdependence are two states of mind in which they can develop. Encourage them to chair committees, lead student projects, or engage in curriculum revision.

Interindividual Knowers

An additional way of making meaning, occupied by only a few adults, might be summarized as "There are relationships and I am part of them." Rather than a pond, this way of knowing might appear as a lively ocean, part of a broader universe of tides, moon, and winds. Adults at this way of knowing live the richest expression of the state of mind of interdependence. They see experiences as opportunities to grow. Conflict is regarded not only as a valuable source of bringing resolution to problems but also as an opportunity to learn about and change oneself. People at this way of knowing can achieve alignment, even with seeming opposite individuals. Identity continues to develop through reflection and self-modification. Validation comes through openness to questions, possibilities, conflict, and reconstruction of meaning. Continual inquiry and data from outside oneself shape values and beliefs. System thinking may be a part of their way of understanding challenges and working toward goals.

Engage the individuals in multifaceted projects seeking complex goals or in seeking resolution to issues with no clear answers. Foster the understanding of others with different processing styles and capacities. Encourage entrepreneurial efforts. Invite the initiation of action research studies or projects benefitting the school.

A still pond, another disturbed by a fallen leaf, yet another with multiple ripples and varying velocities, until finally a living ocean, dynamic, interacting with the broader environment—these images are intended to convey the increasing complexities of the inner lives of adults on this developmental continuum. The intricacies, although internal, reveal a way of interacting with the surrounding environment. Sensitive coaches, mindful of the developmental nature of human growth, without judgment and with understanding, seek to meet coachees where they are and support them in their developmental journeys.

CAREER STAGES

Adults learn, retain, and use what they perceive as relevant to their professional needs. As teachers mature in their professional roles, their needs change as well. Early-career teachers and expert teachers view their practice differently. Expert teachers draw largely from their tacit knowledge acquired from years of experience to create classroom conditions that facilitate access to, and successful learning for, diverse groups of students across content areas (Berliner 1988). Experts possess a special kind of knowledge about learners that is different from novices, knowledge that influences classroom management and is the basis for the transference and application of subject matter. Richly developed appreciation of and strategies for accommodating student differences and learning variables result in experts designing more-effective learning environments and solving problems at a much higher conceptual level. Expert, experienced teachers tend to perceive problems from a much broader context and are able to draw upon their tacit knowledge to determine effective interventions. These conceptual strategies do not automatically appear in a teacher's mind; they are constructed through years of experience.

Teachers' developmental levels have a direct correlation to their performance in the classroom (Garmston, Lipton, and Kaiser 1998). These levels also correlate to their interactions with other adults in the school environment. Teachers who function at higher conceptual levels are capable of greater degrees of complexity in the classroom and are more effective with students. Teachers with more advanced conceptual levels are more flexible, stress tolerant, and adaptive in their teaching styles. Thus, they are able to assume multiple perspectives, employ a variety of coping strategies, and apply a wide repertoire of teaching models. In addition, they are responsive to a wider range of learning styles and culturally diverse classes. Teachers at higher levels of conceptual development can take an allocentric perspective, employing the learner's frame of reference in instructional planning, teaching, and evaluation. Understanding and being sensitive to these dimensions of expert teacher thinking can inform coaching practices with accomplished teachers.

In addition, expert teachers seem not only to perform better than novices do, they also seem to do so with less effort. "Cognitive resources are limited in human beings but experts seem at times to stretch these limits; they seem to do more at a given level of expenditure of resources" (Sternberg and Horvath 1995).

Early-career teachers, on the other hand, tend to focus their attention on the immediate situation, on practical teaching techniques, on classroom management, and on relationships with students (Leat 1995). (As with all factors of human variability, there are always exceptions. Some early-career teachers focus on the broader level and some highly experienced teachers focus on the more technical level, probably as an influence of cognitive style.)

Beginning teachers move through several stages of emotional relationships to their job, which influences what they need from a coach. When they have completed a credentialing program, *anticipation* characterizes their emotional state. They are eager to engage and have big plans. *Survival* is the next stage, as reality hits. They are overwhelmed with aspects of the job they could not anticipate. *Disillusionment* often follows. They feel they are working hard but not getting anywhere. Perhaps this is not the right profession for them, they think. *Rejuvenation* often follows the winter break. They have completed half a year, they have rested, the end is in sight, and they have gained some coping strategies to manage the problems they encounter. Toward the end of the year, *reflection* begins as they think about changes that they want to make for the next year. Finally, a second stage of anticipation begins as they look forward to the next year. Beginning teachers in year-round environments may not experience the extremes of the lows reported by teachers in traditional programs. Any educator working with new teachers knows how profound these differences are between the novice and the expert teacher. Career stage is one factor that calls for another form of coaching flexibility, adjusting the functions within a supporting relationship to match teachers' developmental needs without compromising Cognitive Coaching.

Regarding the mediation of individuation and psychological differentiation,

> It is no less important to recognize and promote individual identity—hence creating and reinforcing what we have labeled individuation and psychological differentiation. While the fusion that needs to occur through the mediation of sharing behavior is important, there is also the companion need for individuation, to value one's uniqueness. The individual has to be given the opportunity to learn that whatever is done, the way it is done, the way one thinks or expresses oneself, is absolutely acceptable and appropriate (when it is in fact so). (Feuerstein et al. 2015, 90)

CONCLUSION

Each person constitutes a distinct system of mental, emotional, cultural, and physical interplays, characterized by the unique ways in which all humans hold and process information, communicate, relate to others, learn, problem solve, and become stressed. All of these ways of being are of equal value. Any combination may be more or less intelligent, compassionate, or capable of communicating meaning. As Robert Kegan noted, "The most powerful driver for behavioral change is a change in how one understands the world. If you want powerful ongoing changes in teaching or leadership, you have to get to the underlying beliefs and conceptions that give rise to behaviors" (cited in Sparks 2002, 70).

Although this topic goes far beyond the scope of this book, we offer several general assumptions that are useful guidelines for achieving communication free of gender, racial, or ethnic bias. The following guidelines also apply to gender sensitivity (Kochman 1983):

1. If you are a different race or gender than I, you carry experiences, perceptions, and meanings that I cannot know directly.
2. The origins of your perceptions, processing, and communication styles emerge from personal experience. Because they are ecologically sound, they become persevering patterns.
3. To the degree that our personal histories are different, my communication may be misinterpreted by you and yours by me.
4. When we have misunderstandings about communication style differences, it doesn't make them go away, but understanding the source of the differences can diminish mutual mystification and blame. This makes the world a more familiar and comfortable territory.
5. Our communication and the mutual interests that bring us together provide valuable opportunities to grow and learn from each other. Our differences enrich us both.
6. We all have unexamined prejudices and biases. We can work respectfully together and accomplish tasks important to us to the degree that we can become conscious of and set aside these thoughts and feelings.
7. My most useful personal attributes in communicating with you are integrity, consciousness, flexibility, and interdependence, manifested through respect, openness, curiosity, and inquiry. From these sources, you and I can continue to learn more about each other.
8. As a result of these assumptions, I strive to be free of ethnic, racial, and gender bias in my communication.

Cognitive Coaches use the following guidelines in their private and public communication (Bettman 1984):

• They guard against making generalizations that suggest all members of a racial, ethnic, or gender group are the same. Generalizations and stereotypes may lead to insupportable or offensive assumptions while ignoring the fact that all attributes may be found in all groups.

- They avoid qualifiers that categorize others and call attention to racial, ethnic, and gender stereotypes. A qualifier is additional information that suggests an exception to a rule (e.g., a "sensitive" man, a "mathematically gifted" girl).
- They identify others by race or ethnic origin only when relevant—and few situations require such identification.
- They consciously avoid language that to some people has questionable racial, ethnic, or sexual connotations. Although a word or phrase may not be personally offensive to you, it may be to others.
- They are aware of the possible negative implications of color-symbolic words. They choose language and usage that does not offend people or reinforce bias.

Because the making of meaning is an individual process, Cognitive Coaches are conscious of human variability. They not only promote diversity but also are effective in working with others who possess a wide range of human variables. They recognize, respect, and understand them so they can make better connections, work together more effectively, learn and teach successfully, leverage their own and others' diverse gifts and affinities, and consciously foster individual and collective development. This is critically important to developing trust, to accelerating the learning of the coach and the person being coached, and to enhancing interdependence.

The opportunities for coaches to assist others in the process of meaning making demand continual learning about human differences and developing their capacities of consciousness and flexibility. The person with the greatest flexibility has the greatest influence.

I ka noho pu ana—a'ike i ke aloha.
It is living together that teaches the meaning of love.

In Hawaii, the sugar planters mixed ethnic groups on purpose, their idea being that different folks would not naturally mix and that dissension among the workers would make them easier to manage and less likely to band together against their bosses, the Lunas. What actually happened, however, is a tribute to the adaptability and resiliency of the human spirit. The workers not only got along, but of necessity they shared the very elements that made them different, even as they maintained their ethnic integrity.

First they had to adapt to their new situation. There were things they were used to in their home cultures that they simply couldn't get in Hawaii. Second, they adopted the good things from other cultures they were exposed to, but they retained some of the important cultural elements of their own ethnic groups. Finally, they made special contributions to their new multicultural society on the plantation, a working-class culture made up of many minorities who recognized the different capacities inherit in each person and developed skills to work together. This is more than an organizational need. It is a human need: to enhance the quality of life that people individually and collectively express.

11

The Maps of the Planning and Reflecting Conversations

In this chapter we relate recent findings in cognitive neurology as we describe the basic structures and variations of Cognitive Coaching processes, and we elaborate on two of the four mental maps that guide the coach's interactions. Examples of conversations are presented in videos to illustrate the process in action. We also offer suggestions on how to get started and how to find time to implement the coaching process.

THE MENTAL MAPS OF COACHING

One of the most important capabilities that a mediator possesses is the ability to navigate between and within coaching maps and across support functions as described in chapters 1, 5, and 12, to guide mediational interactions. A map can be thought of as an internal representation, or scaffold, that displays the territory involved in a goal-oriented conversation. Knowing the territory, in this sense, allows the coach to make a conscious choice from the various routes that may be taken. The four mental maps, and their variations, guide the Cognitive Coach. They are as follows:

1. The *planning conversation* occurs before a colleague conducts or participates in an event or attempts a task. The purpose of the planning conversation is to bring to consciousness in the colleague the importance, the mental processes, and the benefits of planning. The coach may or may not be present during the lesson or event being planned or available for a follow-up conversation.
2. The *reflecting conversation* occurs after a colleague conducts or participates in an event or completes a task. The purpose of the reflecting conversation is to bring to consciousness in the colleague the importance, the mental processes, and the benefits of reflection. The coach may or may not have been present at or participated in the event.

3. The *problem-resolving conversation* occurs when a colleague feels stuck, helpless, unclear, emotionally flooded, or lacking in resourcefulness; experiences
a crisis; or requests external assistance from a mediator. This is explored in
chapter 12.

Finally we offer a special-purpose map, the fourth, developed in response to the need
for self-directed learning in reference to the pervasive use of teaching standards during this decade. It is:

4. The *calibrating conversation* engages teachers in assessing their own performance related to a set of standards or external criteria (Marzano 2007;
Marzano, Frontier, and Livingston 2011; Marzano, Pickering, and Pollock,
2001; Danielson 2007; Hattie 2003; Silver et al. 2014). The aim of these conversations is not about external evaluation but rather to support the teacher in
self-assessment and personal goal setting. To calibrate is to set a standard for
performance by assessing the congruence with, or deviation from, that standard
and then to modify the performance to become closer to the standard. The calibrating conversation map is explored in chapter 13.

The coaching cycle is a combination of these planning and reflecting maps, with
some slight variations. It is conducted both before and after an event for which
the coach may or may not be present and gather data. They can also be used in
circumstances where no one serves in the role of an observer, yet a coach might
engage a colleague either before or after an event (the coaching cycle is discussed
later in this chapter).

These mental maps are not recipes, prescriptions, or step-by-step instructions. As
maps, they simply display the territory, and travelers choose which route they want
to take to reach their destination. Skilled Cognitive Coaches know these maps back-
ward and forward. They use them spontaneously and know exactly where in the map
they are during the coaching processes.

Alert to cues, the coach decides to shift from one map to another depending on the
needs of the person being coached, the time available, and the context. For example,
toward the end of a reflecting conversation, a colleague may begin talking about
plans for future activities. The coach might decide to shift into the planning map as a
conversational guide. Or, during a planning or reflecting conversation, the colleague
may seem frustrated, confused, or vague. The coach may decide to use the problem-
resolving conversation to amplify and liberate the colleague's internal resources in
relation to a problem that seems overwhelming. Coaches may use the maps flexibly,
but these maps are always the basic structure of the coaching process.

The ultimate goal of Cognitive Coaching is for the colleague to self-manage,
self-monitor, and self-modify. Therefore, the coach will bring these maps to the

consciousness of the colleague to explore the benefits of using these maps and, over time, to transfer to the colleague increasing degrees of self-mediation. The mediator's greatest achievement would be to put him/herself "out of business"!

THE PLANNING CONVERSATION

Before a colleague participates in an event or attempts a task, the Cognitive Coach "mediates" by illuminating and facilitating the refinement of the colleague's cognitive processes of planning. Holding the planning map in mind, the coach engages the colleague's cognitive processes that will maximize the significance, success, and meaning of the event. A map of the planning conversation is displayed in figure 11.1

This planning map displays the territory of a planning conversation, the regions within it, and the possible routes to reach a destination. The coach and the colleague visit each of the regions of the map with specific intentions. Within the territory of planning, there are five dominant regions, which the coach mediates by having the planner do the following:

- *Clarify goals.* In the region of "clarify goals," the intent is to decide what purposes or outcomes the person being coached wants for the event that is being planned.
- *Specify success indicators and a plan for collecting evidence.* In "specify success indicators," the intent is to envision and specify those observable indicators

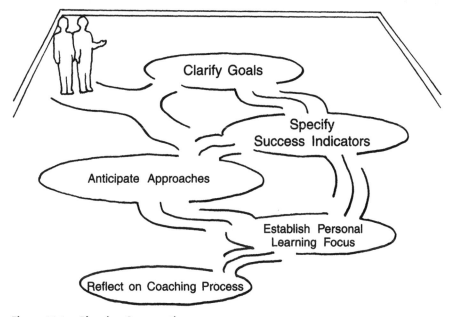

Figure 11.1. Planning Conversation

of success: What would participants (e.g., students) in the event be doing or saying that would indicate that the approach is achieving its purposes? Also in this region, the coach and the colleague devise a plan for monitoring (recording and/or collecting evidence) for specific success indicators.

• *Anticipate approaches, strategies, decisions, and how to monitor them.* Within the "anticipate approaches" region, the intent is to describe those strategies or activities that are intended to achieve the outcomes.

• *Establish personal learning focuses and processes for self-assessment.* Because Cognitive Coaches consider each event as an opportunity for meaning making, the intent of the "establish personal learnings" region is for the partner to establish a basis for self-directed learning. Inviting the teacher to establish a personal learning focus is unique to Cognitive Coaching and is founded in the values related to supporting self-directed learning.

• *Reflect on the coaching process and explore refinements.* Finally, the coach may invite the partner to reflect on the entire coaching conversation, to explore its effects on thinking and decision making, and to recommend modifications that could enhance future planning conversations.

When we've asked teachers what information they would like observers to have about their class prior to an observation, we have gotten an extensive list of requests. These are situational, of course, and depend on many environmental and personal factors. However, teachers often want coaches to know the following:

• Where this lesson fits into the teacher's overall, long-range plan for the students and what has happened previously on this topic
• Information about the social dynamics of the class
• Behavioral information about specific students
• Aspects of the lesson about which the teacher is unclear
• Concerns about student behavior
• Concerns related to trying a new teaching technique
• Why this lesson concept is important to students
• Events beyond the classroom experience affecting students

Although the teacher or the coach might introduce any of the above topics into a planning conversation, the planning conversation items presented earlier represent the five cornerstones of any and all Cognitive Coaching planning conversations.

The regions of a planning conversation exist in the coach's mind in the above, logical order. However, the regions may be visited in any sequence, depending on the interests and thinking processes of the colleague. At any time during a planning conversation, either the coach or the colleague may decide to revisit a region for more specificity or to examine its influence on other regions.

You may now wish to access the 12-minute video of a planning conversation, using the link www.thinkingcollaborative.com/supporting-audio-videos/11-1. In this video, James, the teacher, is being coached by Alejandro. Viewers are invited to focus on the steps in the planning conversation map. You may wish to use the planning map in figure 11.1 to observe the journey.

THE REFLECTING CONVERSATION

After the colleague participates in an event, resolves a challenge, or completes some task, the Cognitive Coach may seize this opportunity to mediate the cognitive processes of reflection. Reflection is a process of self-discovery, pattern finding, and analysis. It is an exercise in comparative thinking, recalling what was planned and finding similarities and differences with what actually occurred. The coach's intent is to habituate the process of reflection. It reaches the colleague on both an affective level and a cognitive level and is designed around a set of questions that the person thinks through and reacts to with concrete responses (York-Barr, Sommers, Ghere, and Montie 2001; Sanford 1995a, 1995b; Feuerstein, Falik, and Feuerstein 2015).

Sometimes, only the opportunity of engaging in reflection is available, even though the coach was not present during the event. Holding the reflecting conversation map in mind, the coach engages the colleague's cognitive processes of reflection so as to maximize the construction of significant meanings from the experience and to apply those insights to other settings and events.

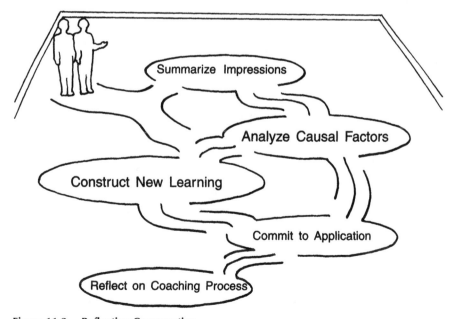

Figure 11.2. Reflecting Conversation

There are five regions in the reflecting conversation. The coach and the colleague visit each of the regions of the map with specific intentions. The coach begins to mediate by inviting the reflector to summarize his or her impressions of the event and recall data that support those assessments.

- *Summarize impressions and recall supporting information.* In summarizing impressions and recalling supporting information, the intent is to elicit the overall thoughts and feelings of the coachee about the event and to ground those in specific data. Because sensitive coaches realize that there is a tremendous emotional investment in any event for which a person has had the responsibility of planning and conducting, the summarize and recall region provides an opportunity to summon those feelings and to recollect the events, conditions, or actions that produced them.

  After eliciting impressions, the coach always prompts the reflector's memory of events (data) before providing data she/he may have collected in observing the event. It is important to understand and hear the internal data (what the reflector knows and is aware of) before providing external data. The coach offers external data and puts this information into the hands of the reflector, pausing to allow the person time to scan and examine the data. Providing this time allows the coachee an opportunity to reflect. The coachee analyzes and generates insights by comparing internal and external data. The intent of this is to heighten consciousness about possible congruencies and discrepancies.

- *Analyze causal factors; compare, analyze, infer, and determine cause-and-effect relationships.* To analyze involves deconstructing an event into its parts and considering the relationship between the parts. The intent of the "analyze" region is to compare the planned event with what actually happened, to identify and interpret causal factors that produced the results, to explain and give reasons for the "in action" decisions that were made, and to make inferences from the information that has been recalled.

- *Construct new learning and applications.* The intent of the "construct" region is to make meaning from the analysis, to draw insights and patterns, and to synthesize the personal learnings that were described in the planning conversation.

- *Commit to applications.* Because Cognitive Coaches value self-directed learning, the purpose of the "commit" region is to transcend this event and to make applications of the learnings to future events, to bridge to other life situations, to transfer such learnings, to self-prescribe, and to take actions to modify personal behaviors. This helps to override what Feuerstein terms *episodic thinking* (Feuerstein and Feuerstein 1991).

- *Reflect on the coaching process and explore refinements.* Upon completion of the learning journey, the coach invites the partner to reflect on the entire conversation, to explore its effects on thinking and decision making, and to recommend modifications that could enhance future reflecting conversations. This is important as it reinforces the intent to communicate that this is a collaborative relationship, addressing an ever-present subconscious concern about status (Rock 2009). Additionally, reflecting on the process brings it to the awareness level so that the process is more likely to be performed autonomously by the coachee.

Although the steps in the reflecting process may be visited in any order, exploring the "summarize impressions" region is usually a first destination and committing to action is often the last before reflecting on the coaching process.

THE COACHING CYCLE

This is a combination of the planning and the reflecting maps, and it is conducted before, during, and after some goal-directed event or while attempting some task. Figure 11.3 presents a diagram of a coaching cycle. It depicts a continuous process

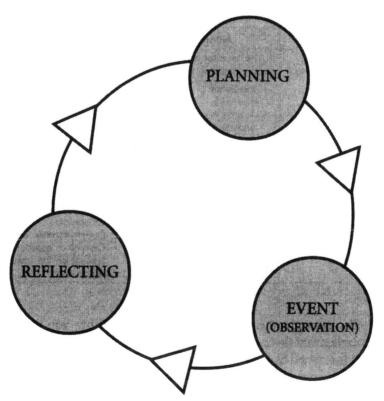

Figure 11.3.   A Coaching Cycle

Table 11.1. Planning Conversation

Planning	Monitoring	Reflecting
I. Planning Coaches mediate by having the planner: • Clarify goals • Specify success indicators and a plan for collecting evidence • Anticipate approaches, strategies, decisions, and how to monitor them. • Establish personal learning focus and processes for self-assessment	**II. The teacher and the coach observe for:** • Indicators of student success • Approaches, strategies, and decisions	**III. Analyzing:** Coaches mediate by having the reflector: • Summarize impressions and recall supporting information • Analyze casual factors; compare, infer, and determine cause-and-effect relationships **IV. Applying:** Coaches mediate by having the reflector: • Construct new learning • Commit to application • Reflect on the coaching process

of learning in which goals are set, actions are taken, and data are monitored, collected, and interpreted, leading to a revision in actions from which new goals are set. It is intended to illustrate that coaching is a continual process and, though mediated by a coach, is intended to help another person internalize this cycle of learning and become self-coaching.

Table 11.1 is a basic map of the steps in a typical coaching cycle: the planning conversation, observation of the event, and the reflecting conversation. A planning conversation differs slightly when the coach will not be present for the event. The coach may serve as a data collector for the planner by serving as another set of eyes monitoring the performance of either the participants (students) in the event or the planner. The coach may collect data and mediate the planner's thinking by establishing mental mechanisms for self-observation or self-monitoring as well.

For example, the coach might say, "Okay. Now, you mentioned earlier that there was something you'd like me to look for. What is it that you'd like me to observe today when I'm in your classroom? How would you like me to collect and record that data?" The goal is to have the planner identify a personal learning focus and processes for self-assessment. The coach might ask, "In addition to paying attention to evidence of student learning, how might you use this lesson for your own growth? What might you focus on, and gather data about, to support your own continuing growth as a teacher?"

The Planning Conversation in a Coaching Cycle

The planning conversation is both powerful and essential to the coaching process for six reasons.

1. It Is a Trust-Building Opportunity

Learning cannot occur without a foundation of trust, and establishing and maintaining this trust is one of the coach's primary goals. To this end, the colleague, or teacher, controls the planning conversation agenda. In the planning conversation, the teacher suggests the time for a classroom visit, specifies which data should be collected during the observation, describes how the data should be recorded, and even chooses where the coach will sit, stand, or move about in the classroom. These specifics are especially important in the early stages of a coaching relationship. (The significance and skills of building trust are expanded in chapter 6.) Suffice it to say here that conversations trigger physical and emotional changes in the brain that either open you up or close you down (Glaser 2014). Grimm, Kaufman, and Doty (2014) note that typical approaches to observation often serve the observer, not the teacher. Ideally the teacher identifies the focus of the observation and develops a question about which she/he wants to know more.

2. It Focuses the Coach's Attention on the Teacher's Goals

Art was once leaving a school when he peeked into a classroom where the students were being especially loud. He saw children jumping on the desks, running around the room, and yelling and screaming. Alarmed, he hurried to the principal's office: "You'd better get down to Room 14! Those kids are going wild and the teacher has lost control!" The principal calmly assured him, "Don't worry. They're practicing the school play. That's the riot scene." Without knowing a teacher's objectives or plan, an observer can make entirely incorrect inferences. Coaches cannot know what to look for in an observation unless they have met with the teacher before a classroom visit.

3. It Provides for a Detailed Mental Rehearsal of the Lesson

As teachers talk with coaches about their lessons, they refine strategies, discover potential flaws in their original thinking, and anticipate decisions they may need to make in the heat of the moment. Specific questions in the planning conversation can spark this mental rehearsal, promote metacognition, and prepare teachers with a repertoire of strategies for the lesson ahead—for example, "How will you know that your strategy is working?" or "How will you know when it is time to move into the activity portion of the lesson?"

4. It Establishes the Parameters of the Reflecting Conversation

Agreements in the planning conversation establish the coach's role and the data to be collected during the lesson. The teacher sets this agenda, and it provides the context for the reflecting conversation. Without a planning conversation, teachers can evaluate their lesson only in terms of what happened rather than in terms of what their intentions were.

5. It Promotes Self-Coaching

The planning conversation may ultimately be the most important, as it relates to a long-range goal of teacher automaticity in instructional thought. The planning map represents a way of thinking about all instruction—in fact, all goal-directed activities. After experiencing a number of planning conversations, teachers adopt this way of thinking about most lessons. They internalize the conversation questions, automatically asking themselves, "What are my objectives?" "What are my plans?" "How will I know students are learning?" They are activating the first step in self-coaching.

Bob discovered this to be true in an evaluation of a Professor's Peer Coaching program at California State University, Sacramento, California (Garmston and Hyerle 1988). Previously, a number of professors lectured until the bell rung, then picked up where they left off for the lecture in the next class period. After several sessions with a Cognitive Coach in which they deliberately identified goals, teaching strategies, and success indicators for their lessons, they reported increases in mental activity, improvements in student learning, and increased satisfaction in teaching.

We have discovered that, as the coach probes and clarifies in an attempt to better understand the teacher's plan, the teacher also becomes clearer about the lesson. The coach engages the teacher in a process of mental rehearsal similar to what athletes do before competition.

6. It Accelerates Growth toward More Sophisticated Instructional Thinking in Teachers

Novice teachers tend to focus on the event. As teachers become more experienced, they tend to focus on goals and success indicators. (We have heard of teachers who, when being evaluated, asked their principal what indicators supported their conclusions about their performance!)

The coach also invites the teacher to describe which strategies will be used to accomplish the goals. The coach leads the teacher to anticipate what students will be doing if they are, indeed, successfully performing the goals and objectives of the lesson. The coach helps the teacher to specify what will be seen or heard within or by the end of the lesson to indicate student learning. Throughout the conversation, the coach clarifies her role in the process, the kind of data she is to collect, and the format of data collection.

We have found that in addition to the items previously listed for a planning conversation, there are two other areas of a planning conversation that are most frequently

Table 11.2. Four Phases of Thought in a Coaching Cycle

	Dialogue	Regions of the Reflecting Map
Coach	Hey, Raul! So how was the curriculum meeting? I'm sorry I wasn't there. Tell me about it.	Summarize impressions . . .
Raul	Well, it went pretty well. Our task was to review and adopt standards of learning for our primary grade social studies curriculum. We made a lot of progress.	
Coach	What exactly did the group accomplish?	. . . and recall supporting information.
Raul	We didn't adopt the standards yet, only reviewed them. We argued a lot about what primary kids could accomplish in a project-based curriculum. I must say, I've had to revise my thinking about my youngsters' abilities.	
Coach	What contributed to your change of heart?	Analyze causal factors, compare, infer, and determine cause-and-effect relationships.
Raul	Well, some of the teachers brought in samples of their kids' extended projects. I was amazed at how much those second and third graders knew about Australia and the problems the Australians were having with conservation efforts.	
Coach	So what insights are you gaining?	Construct new learnings and commit to applications.
Raul	Well, two things. One, I'm going to go back and review those standards to see if they are too simple for primary students. And second, I'm going to expect more from my kids. I think this project stuff is the way to go.	
Coach	So you've been resisting project-based learning for your primary students because you thought they couldn't handle those types of assignments, and now you are seeing things differently?	
Raul	Yeah. Thanks for asking. This conversation helped me realize I've got some expectations to change!	Reflect on process.

useful to a coach or a teacher: (1) information regarding the relationship of this lesson to the broader curriculum picture for the class, and (2) information about teacher concerns. The coach may ask the teacher, "Any concerns?" This question allows teachers to say no or to discuss anything that might be troubling them. Even though the description of the planning conversation described above appears to be focused on the lesson, the Cognitive Coach is actually focused on more long-range outcomes: developing and automating these intellectual patterns of effective instruction.

A tip: Don't ask teachers to bring a written response to planning conversation. For busy teachers, this often feels like one more burden, and it robs both of you of the spontaneity related to deepening thought as you talk.

Monitoring the Event

Coaches do not specify data-gathering instruments for their observations. Instead, they assist the planner in designing the instrument in the planning conversation and in evaluating the instrument's usefulness in the reflecting conversation. The intent is to cast the colleague in the role of experimenter and researcher, and the coach in the role of data collector. In some settings, video might be used for the teacher to see his/her own instruction and student responses before a reflecting conversation (Knight 2014).

It is important for the coach to strive for specificity of what should be recorded. Data gathering during coaching is not the place for subjective judgments. Pursuing definitions and observable indicators of what behavior will look like or sound like will determine the objectivity and therefore the usefulness of the data. For example, the coach might say, "You want me to determine if students are on task as they are working on their projects. What will you hear them saying and see them doing when they are engaged?" Or the coach might say, "So your goal is to have the students comprehend the symbolism of this story. What data might I collect that would help you to assess your students' comprehension of the symbolism?"

It is also important that the coach invite the planner to construct the system for collecting the data. Data must make sense to the teacher during the reflecting conversation, which is the time when meaning will be constructed from the data. The coach might say, "So my job is to collect evidence of students' engagement. How could I record that for you? When during the lesson should I record it? How often should I collect the data? Do you want me to record it on a seating chart? A checklist? Are there specific students you want me to observe?" Or the coach might say, "In what form might I collect evidence of the students' understanding the symbolism of the story? Should I record all their comments? At what point in the lesson should I start recording? Do you want me to specify which student made which comment?"

Having clarified the instructional goals and how the teacher and coach will collect evidence of their achievement—and having determined with the teacher exactly what data should be collected for teacher growth and how the data should be recorded—the coach is now prepared for the observation. During the classroom observation, the coach simply monitors for and collects data regarding the teaching behaviors and student learning as decided during the planning conversation. The coach may employ a variety of data-collection strategies, including classroom maps of teacher movement, audio and video recordings, verbal interaction patterns, verbatim recording of what teachers say, student participation, on-task counts, or frequency counts of certain teacher behaviors. Of more importance, however, is the teacher's perception of the data and the format in which the data are collected. Both must be meaningful and relevant to the teacher's self-improvement efforts.

As the reflecting conversation begins, the coach encourages the teacher to share his impressions of the lesson and to recall specific events that support those impressions. We have found that it is important for teachers to summarize their own impressions at the outset of a conversation. This way, the teacher is the only participant who is judging his own performance or effectiveness. The question "How did you feel about the lesson" is often a useful opener, as emotion is a primary sorting mechanism for people and this question is open-ended enough to allow the teacher to reflect about the areas of greatest interest to him.

The coach also invites the teacher to make comparisons between what he remembers from the lesson and what was desired (as determined in the planning phase). The coach facilitates the teacher's analysis of the lesson goals by using reflective questioning. The coach also shares the data collected during the observation and invites the teacher to make inferences from the data. The aim is to support the teacher's ability to draw causal relationships between his actions and student outcomes. Drawing forth specific data and employing a variety of linguistic tools are important coaching skills in supporting the teacher as he makes inferences regarding instructional decisions, teaching behaviors, and the success of the lesson.

As the reflecting conversation continues, the coach will encourage the teacher to project how future lessons might be rearranged based on new learnings, discoveries, and insights. The coach also invites the teacher to reflect on what has been learned from the coaching experience itself. The coach invites the teacher to give feedback about the coaching process and to suggest any refinements or changes that will make the relationship more productive.

You may now wish to access the 12-minute video of a reflecting conversation using the link www.thinkingcollaborative.com/supporting-audio-videos/11-2/. In this video, Carolee coaches Ochan through a reflective conversation. You may wish to use the reflecting map in figure 11.2 to follow the journey.

Navigating within and between the Coaching Maps

The mental maps of cognitive coaching are not necessarily intended to be used separately. As the coach interacts with a colleague, the coach may detect that it is an opportunity to switch from a reflecting mode to a planning conversation. For example, in the reflecting conversation with Raul, the coach might seize the opportunity to begin planning: "So, Raul, you'd like to implement project-based learning with your primary students. What might you do to get started?" Or, during a planning conversation, the vigilant coach might, based on the comments of the planner, find it profitable to switch to a reflecting conversation: "What was it about some of those projects you saw that you might use in your planning for your students?"

By knowing these mental maps deeply, skillful coaches have the flexibility to maneuver within each map and to draw forth components of each of the maps as the situation demands.

INNER COACHING

The success of an intervention depends on the inner condition of the intervener.

—Bill O'Brien

Mark Treadwell (2014, 12) states: "Our 'inner voice' is what we use to reflect on what we do, how and why we behave in the way we do, how we critique ourselves and how we connect the knowledge, ideas, concepts, and concept frameworks developed using each of our four learning systems. It is the voice that challenges us to strive further and the voice that condemns our foolishness."

Before being a spectator of others, you need to be a spectator of yourself. Before any visit to classrooms or conversations with a teacher, you must engage your "inner coach." You need to engage your own internal dialogue—a metacognitive rehearsal inside your own mind. The questions you ask yourself and the answers you provide yourself help to clarify and direct your skills and competencies as an observer and coach. They also influence the success of the observation and the conversation for the teacher. Interrogative self-talk—questions—elicit in your mind an active response and give you an opportunity to prepare and rehearse (Pink 2014). For examples of inner dialogues in preparation for a variety of coaching situations, please refer to appendix B.

Getting Coaching Started

We are often asked questions like the following:

- How do I get started?
- Where do I begin?
- How do I find the time to coach?
- How much time should I allow?
- How long will it take to implement Cognitive Coaching?
- What should I do about reluctant teachers?

Although we have no prescriptions for each reader's unique situation, we have some general suggestions gleaned from working with numerous educators and school districts:

- As a novice Cognitive Coach, begin with a colleague who is relatively secure and with whom you already have a trusting relationship. Preview your intentions and procedures. As your skills become more automatic, you may wish to become more venturesome with new acquaintances and less-experienced teachers.
- For administrators whose staff has experienced a more traditional evaluative form of supervision, a demonstration in which staff members observe the differences between Cognitive Coaching and their previous experiences is a highly successful strategy. Communicate clearly at the outset that the purpose of Cognitive Coaching is to refine and make automatic the intellectual skills associated with effective instruction. Be equally clear that the coaching process is not an evaluation. Explain that your coaching behaviors will include pausing, paraphrasing, clarifying, and posing questions.
- Explain why you remain nonjudgmental and do not give advice. Some Cognitive Coaches have begun by switching roles: having the uninitiated teacher follow the planning conversation map. While the coach teaches a lesson or conducts a meeting, the colleague observes. Thus, the teacher becomes acquainted with the process by actually coaching the coach. Variations of this process have been used with student teachers and first-year teachers to model for them the basics of instructional thought.

Coaching and Time

When new to Cognitive Coaching, some participants show concern about time to coach. When one has an identity as a mediator of thinking, every interaction is an impromptu opportunity to coach. Thus coaching does not require adding time to a person's already busy day. Coaching can be scheduled as a formal sit-down conversation, a hallway conversation, or as part of a planned coaching cycle or "walk-through." Often coaching occurs in the normal course of conversations throughout the day. As the identity of a mediator is internalized, Cognitive Coaches discover that coaching is something they use automatically. When approached by a colleague about something coming up in the future, the coach begins an informal planning conversation. When someone strikes up a conversation about a recent event or experience, the coach engages the conversation using the reflecting conversation map. Coaching is an investment in another person's capacity for self-directedness. It replaces less-productive patterns that may enable or create dependency.

Time, that precious and rare commodity, must be allocated for conducting the coaching cycle. Teachers engaged in peer coaching often establish times for the segments of the full conversation cycle before or after school, during planning periods, or even over lunch. Often, substitutes are hired or other resource teachers and administrators take the coach's class to free the teacher to conversation and observe other teachers. Hiring a substitute teacher for a day, to spend one hour in each grade level to free each teacher, is a powerful and inexpensive way to make some time for coaching.

The planning, observation, and reflecting conversations may consume extra time initially, because both parties are learning the process. However, that time diminishes as the expectations of the teacher are better understood and the coach becomes more skillful. It has been found that highly proficient coaches working with teachers experienced in the process can conduct a planning conversation in about 8 to 12 minutes. This does not mean that speed is valued, however. It only points out that as experiences are gained, there is a greater economy of time for coaching.

Administrators and supervisors find time for coaching by incorporating it into their regular duties. They block out time on their calendars and inform their secretaries, staff, central office, and community that they will be in classrooms during these hours.

Arranging for coaching by peers is most often done informally, with teachers agreeing to meet at times convenient to them. Sometimes, however, substitutes have to be hired and teachers released. One school keeps a sign-up sheet in the faculty lounge to arrange for the substitute to take a class at a certain time. Often, the school administrator initially sets up a coaching session with the teacher, both of them suggesting and agreeing on a suitable time. As trust develops, the teacher will request the conversation with the administrator or other resourceful colleague.

Planning and reflecting conversations can be conducted in informal, unscheduled settings: in the teacher's lounge, in the car while driving, in the hallway on the way to or from a class or meeting, and even on the telephone or over the Internet.

Deepening Skills

Skillfulness in Cognitive Coaching takes time and practice beyond reading this book. Other opportunities can help coaches acquire the skills and understanding of this complex process. Several visual aids are available as an introduction to Cognitive Coaching and in skill refinement. Numerous resources and learning opportunities are available and are listed in the preface of this book and in the section in appendix F.

CONCLUSION

In chapter 3, The Mediator's Skills, we described the specific kinds of knowledge, techniques, and skills that coaches need to achieve the goals described in this chapter. You may wish to examine the videos presented in this chapter to see how the principles of Cognitive Coaching are applied.

This chapter introduced the basic structures and variations of the Cognitive Coaching model. Two of the four mental maps that guide a coach's interactions, along with video examples of conversations intended to illustrate the process in action, were presented.

Skillful coaches are consistently conscious of their intention, which is self-directed learning. Mediating reflection on the planning and reflecting conversations brings the entire process to the colleague's conscious level with the intent of habituating (interiorizing) the cognitive processes of the coaching/learning cycle. It means "I will increasingly coach myself" (Feuerstein et al. 2015, 45).

12

Resolving Problems: The Third Coaching Map

Our deepest struggles are in effect our greatest spiritual and creative assets and the doors to whatever creativity we might possess. It seems to be a learned wisdom to share them with others only when they have the possibility of meeting them with some maturity. We learn to remain attentive to the mood and outlook of the listener even before we begin to speak about the darker side of our existence.

—David Whyte

Astonishing advances in understanding the human brain have been made since the second edition of this book. Our earlier observations, intuitions, and studies about the role of the nervous system in problem resolving are being borne out in current findings from the neurosciences. Obviously more learning is yet to come. In this chapter, we explore what we now know about the most complex of the coaching maps and explain why the use of this map is effective, based on previous and new research. One dilemma in describing these responsively interactive processes in print is that the context for this map is problem solving, but not in the usual sense. Problem solving is often construed as a linear sequence: tell me a problem and together we can search out possible causes, and then envision a desired state with some probable alternative approaches to problem resolution. This lends itself to linear problem solving, in which both parties have the data and internal resources necessary for problem analysis and solution generation. This process is not effective, however, for the majority of problems that, by definition, are "wicked"— problems that are nonlinear and recursive. These are the types of problems that engage conflicts inherent in holonomous dichotomies and in which the protagonist often approaches the problem situation already feeling defeated, confused, overwhelmed, or simply nonresourceful.

The approach to problems that we present is both simple and sophisticated. On the one hand, it is a search for what the person wants as a desired state. On the other

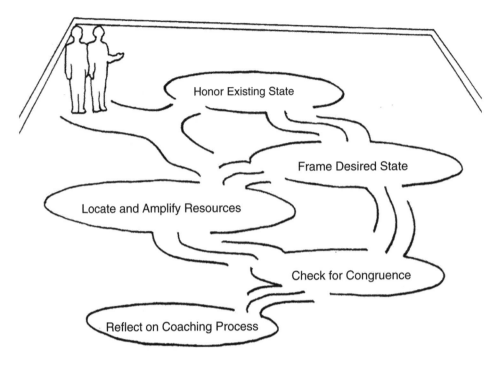

Figure 12.1. Problem-Resolving Conversation

hand, it is the identification and amplification of the internal resources necessary to generate and apply the strategies required to achieve the desired state. In concept it is simple; in execution it is complex. As illustrated in figure 12.1, it is a coaching conservation that moves one from an existing state of limitations to framing a state the choachee desires and locating resources through which it can be attained.

PACING AND LEADING

Imagine that your friend says the following: "I feel awful. I lost my temper during seventh period today. I was exhausted and, when Alex acted up, I just lost it. I know I embarrassed him in front of his peers, but I had just had it."

Our well-intentioned human tendency is to respond in the following ways—each of them not helpful and, indeed, counterproductive if the intention is to assist a colleague in feeling better about, learning from, and being able to resolve the problem with Alex:

- Provide sympathy and comfort
- Share similar experiences to let your friend know this happens to others
- Identify a silver lining
- Blame the student

There is a more effective way through a conversation that can actually trigger physical and emotional changes in the brain that open one up to optimism, resource-fulness, and creative energy (Glaser 2014). Cognitive Coaches, working with another person who has temporally lost access to his or her resourcefulness, use a pattern of pacing and leading. The problem-resolving conversation uses this pattern to medi-ate the other person's capacity to liberate and use the efficacy, craftsmanship, con-sciousness, flexibility, or interdependence that has been temporarily obscured. The long-range intent of the problem-resolving map is to help the other person become self-accessing of internal resources, even when the coach is not present. To be clear, pacing and leading is not the map, rather it is a structure on which the map is built.

When to Use the Problem-Resolving Map

The problem-resolving map is used to assist a person who, when facing adversity, feels stuck, uncertain of what to do, or so emotionally flooded that it is difficult to make good choices. When people feel trapped in a situation without alternatives, their hearts beat faster, driving them into self-protective states rather than resourceful ones. In these instances, they need two lifelines: states of resourcefulness and think-ing strategies to address the specific issue. Yet one without the other is not enough. When people are resourceful, electrochemical interactions occur that make available their inventiveness, their decision-making capacities, and a repertoire of steps they might take. Thus, a coach helps a colleague become more resourceful by accessing the colleague's internal resources, creating chemical adjustments in the bloodstream and thereby liberating reflective thinking.

Most often, pacing and leading is used in conversations specifically focused on resolving a problem about which the person feels stuck. However, if the coach senses the person is stuck and unable to think clearly during a planning or reflective conversation, the coach may transition to this Cognitive Coaching map. Pacing and leading conversations are sometimes brief. More often they are extended, focused conversations in an environment without distractions. A coach paces to honor the existing state and create awareness of a possible desired state. This is done because any effort to "resist" the existing emotional state by placating, advising, or denying causes the resistance and the person is driven deeper into "stuckness." This is a necessary first step before the coach "leads" a person to locate and amplify his or her own internal resources—states of mind or knowledge—necessary to deal ef-fectively with a problem situation. When nonverbal cues are congruent with verbal descriptions of the desired state and necessary resources, the coach and colleague reflect together about the coaching process. Matching or mirroring of perceptions might be synonyms for pacing. The goal of pacing is to let the other person know that his or her experience is understood without judgment. Pacing first reflects what is and then makes visible what is possible. In the following example, the response to the speaker might be called a pace and jerk because the speaker's experience was never really acknowledged.

Speaker: "I'm feeling bad."

Friend: "Ohhh. You shouldn't feel that way. Look at the bright side!"

By denying the speaker's feelings, it is more likely that the feelings stay unchanged. To pace, the coach should empathically paraphrase to reflect the emotional state of the speaker and give assurance that the speaker's condition is recognized and acknowledged. After several interchanges with repeated paraphrasing, the coach can then offer a description of what conditions the speaker might like to see instead of the existing state.

Leading begins after the speaker has indicated that the description of a desired state is accurate. By now, oxytocin has been released into the bloodstream. The listener and speaker are showing the same patterns of activation (Hasson, Ghazanfar, Galantucci, Garrod, and Keysers 2012). Then the coach begins a series of questions intended to help the speaker identify the states of mind or other internal resources necessary to achieve the goal. Leading locates, elicits, and amplifies the internal resources necessary to achieve what is possible.

It may be useful to keep in mind that the limbic system and the prefrontal cortex are sources for responses to coaching. Awareness of the structure and functions of these parts of the brain allows the coach to be intentional in managing their interrelationships. The limbic system (hippocampus, amygdala, hypothalamus) regulates emotions and memory. It protects us by asking whether we are safe and triggers fight-or-flight responses if we are not. A limbic response is reflexive and short-circuits capacity in the prefrontal cortex (Sylwester 2005). The prefrontal cortex allows for reflective responses, including problem solving and decision making, thus moving us to action. These two structures work best together under conditions of relaxed alertness (Caine and Caine 1991)—that is, low stress and high challenge. Attention to emotional safety is the gateway to cognitive complexity, underscoring our emphasis on rapport in coaching interactions (Dolcemascolo, Miori-Merola, and Ellison 2014). Mirror neurons relate to learning social skills and cultural behaviors, and may be closely linked to empathy and rapport, essential in pacing (Rizzolatti and Sinigaglia 2008; Iacoboni 2008; Ramachandran 2011). Iacoboni (2008) speculates that there is linkage between the mirror neurons and the limbic system at the insula, which accounts for the human capacity for empathy. Figure 12.1 illustrates the problem-resolving conversation.

Learning the Pattern of Pacing and Leading

The processes of pacing and leading are simultaneously simple and complex. They can be learned from this book, but one must study, practice, test, and reflect upon

them in order to have the patterns become internalized. For this reason we have organized the information in several ways. First, we have just offered a brief definition
of pacing and leading. Second, we offer a more detailed account of the structures and
skills of pacing, then leading. Next, we explore theories about the changes in body
chemistry during pacing and leading, along with responses of the autonomic nervous
system that the coach can observe and use to guide the interaction. We explore more
about leading after that, and finally we describe a set of principles of intervention.

The Problem-Resolving Conversation

Pacing is first in this conversation. By doing so the coach validates the experience of
the colleague, acknowledging in a sense, that whatever feelings and conceptions of
reality the colleague is experiencing are true for that person. Pacing the experience of
another requires dedicated attention to rapport. Humans appear to be hard-wired for social connection and relationship (Lewis, Amini, and Lannon 2000; Rock 2009). Without this wiring for connectivity, our brains would be unduly influenced by the release
of cortisol, a stress hormone, which would inhibit thinking. The value of this synchron-
icity is that during pacing, the coachee's brain, in a sense, releases defensive barriers
leading to relaxation and openness. Oxytocin is released, leading to feeling connected.
During this physiological dance, the coach attends to the four verbal phases of the pace:
expressing empathy, reflecting content, inferring and stating a goal, and presupposing a
search for a way to get started working toward the goal (i.e., a pathway).

   Everything we learned about rapport in chapter 6 becomes extremely important
in the problem-resolving conversation. Rapport means setting aside our desire to
"report," the tendency to be consultants rather than coaches, not to be experts and
fixers. Instead, a coach enters the world of the coachee with humility, empathy, and
compassion. In pacing, the primary verbal tools of the coach are pausing to provide
time for reflection and paraphrasing to deepen thought and develop shared understanding, along with the nonverbal tools of physical alignment (Dolcemascolo et al.
2014). When the limbic system perceives threat, incorrect links of thinking increase
(Bolte Taylor 2008). Heart rhythm patterns become erratic and block our ability to
think clearly. In pacing, these patterns are reversed. The coach's posture, voice qualities, gesture, facial expressions, breathing rates, and other nonverbal manifestations
of human experience communicate acceptance of the other's emotional state. Pacing
reflects and honors the other person's experience, which fosters the convergence of
heart patterns and high levels of trust (Glaser 2014, 82).

   As illustrated in the list that follows, the first two verbal steps are embodied in any
well-formed paraphrase. The third is another paraphrase about the coach's inference
regarding a goal the colleague desires. The fourth element paraphrases what has been
unsaid verbally and presumes a readiness to begin exploring options.

Table 12.1.   The Neurobiological Effects of Pacing

Phase	Neurological Effects
Empathy "You're worried . . ."	Name the emotion to decrease limbic activity (Rock 2009). The emotion is named, not talked about, because talking about it increases limbic activity and decreases cognitive resourcefulness. Peptides produced in the limbic brain have stimulated a chemistry of distress. Lung muscles constrict, making breathing shallow and reducing oxygen to the neocortex; heart contractions constrict coronary vessels that supply blood to the heart muscle cells; changes in skeletal muscles and visceral organs produce changes in skin color and muscle tone. During the empathy and content phases of pacing, the person often shows signs of distress through muscle tautness and shallow breathing.
Content " . . . because others are perceiving you as negative."	Distress: After a few interchanges in which the coach continues to paraphrase, blood chemistry aborts its downward decline and hypothetically reaches a state of balance. The coach paraphrases for goals, values, beliefs, and assumptions rather than for specific details. Stating the content in a few words decreases the load on the prefrontal cortex, and that alone may lead to new insights.
"What you want is to be a good team member . . ."	Eustress: This is "good stress." The coach observes shifts in posture, voice qualities, and animation. The coach also notices deeper breathing or other signals that the goal statement, worded as a paraphrase, has shifted the body–mind system into the chemistry of resourcefulness. Reducing the goal statement to a few words reduces overload on the prefrontal cortex. A brief statement, says Rock (2009), can foster insight. This phase of the conversation deliberately avoids delving into "the problem" because focusing on the problem activates emotions and increases activity in the limbic system, blocking prefrontal work. Reframing and relief occurs from focusing on a more global desired state.
Pathway " . . . so you're looking for a way to make that happen."	Eustress: The physical manifestations of resourcefulness remain strong. The coach knows the person is no longer at the affect stage of limbic system processes and is capable of neocortical thinking, which will occur in the lead stage.

1. Express empathy by matching intonation and accurately naming the person's feeling: "You're frustrated . . ."
2. Accurately reflect the speaker's content: " . . . because you have so many things on your plate you can't keep up . . ."
3. State the goal that you infer the speaker is trying to achieve: "What you want is to be in control of your time and your work . . ."
4. Presuppose readiness to find a pathway to attain the goal: "So you're searching for a way to make that happen."

The above scaffold is simple yet often demanding to learn because it may require unlearning some long-held communication patterns. It is also composed entirely of paraphrases. Any question posed at this stage interrupts the colleague's focus in the story and is experienced as a distraction. Skillful and empathic pacing is essential to effective leading. Repeated and guided practice in pacing is usually necessary for pacing responses to come easily and naturally. Table 12.1 illustrates the neurological effects of the process of pacing.

One reason we recommend making the pace automatic is that pacing is a complex, multifaceted intellectual task. The coach listens to the feelings and context of what is being expressed, infers an authentic goal, and decides how to begin the lead. The five states of mind serve the coach as a template to initiate a lead statement or question. When verbal and nonverbal messages are congruent and related to both goal and pathway, the coach knows she/he can start the lead.

Crafting a Goal Statement

Leading without pacing is ineffective. Most people can describe what they don't like. It is only when we are resourceful that we can describe what we want instead. Pacing "listens below the story." During the goal stage, pacing tests the coach's understanding of a desired state. The goal statement helps the person see the problem differently. Of the four elements in the pace, formulating an appropriate goal statement is the most complex. A well-formed goal statement does the following:

- It is stated with artful vagueness. ("You want to be in control, be confident, have satisfaction.") If the goal is stated with too much specificity, it narrows the solution possibilities and increases the likelihood that it does not represent the person's goal. Vague language here allows the coach to engage the energy of the person being coached to move forward. It does not matter yet if the pictures formed by "confident" mean different things to the coach and the partner.
- It uses verb forms of be, have, or feel, but never do. The latter is used in reference to an action or a plan, not a goal. A useful goal statement is always about a destination (what the person will be or have when satisfied), not a journey (what he or she might do to get there).
- It is stated in positive form (what the person wants, rather than doesn't want).
- It is brief and stated with simple language.
- It acknowledges feelings. (This is sometimes communicated through posture, facial expression, or tone of voice.)
- It assumes nobility of intention on the part of the person being paced.
- It may identify two or more goals that are seemingly incompatible: "You want to be accountable for test results but at the same time be true to your responsibilities to students."
- It is about the person being coached, not a third party.

To craft a goal statement, the coach listens for an unexpressed yearning to resolve the tension. What might the person feel, be, or have if the challenging situation were resolved? Search for the tension arising from the drive for self-assertion. Another way to listen is to search for contrasting emotions. If the person is sad, perhaps he or she wants to be happy. If she is feeling disconnected, then she may want to be connected. If he is feeling powerless, he may want to have control.

Naming Universal Goals

Another tip for crafting goal statements is to listen for the expression of three universal goals. After the basic survival needs of food, water, shelter, and sex, humans desire three goals: identity, connectedness, and potency. Coaches may use words in the goal statement that relate to these concepts. If, for example, circumstances threaten a colleague's ability to act congruently with his sense of self, the coach may choose goal words that relate to identity. For example, the coach may use phrases such as "and you want to be true to yourself" or "who you are is a person of integrity and you want to be that with others" or "you want to be who you really are in this situation."

If a situation is challenging one's sense of efficacy, words related to potency may express that desire. For example: "and what you want is to be a catalyst" or "you want to be a person of influence" or "you want to be resourceful in this situation."

If a person feels isolated, words that convey connectedness might be appropriate. For example: "so your desire is be a team player," "you want a sense of family," or "you want to be a friend."

Table 12.2 contains a bank of words and phrases that the coach may want to use with this verbal skill.

> All behavior is communication.
>
> —P. Watzlawick, J. Beavin, and D. Jackson (1967)

Table 12.2.   A Word Bank for Universal Goals

Identity	Connectedness	Potency
Integrity	Community	Effective
Valued	Bonded	Influence
Creative	Team	Successful
Trusted	Family	Resourceful
Balanced	Unified	Compelling
In Charge	Relationship	Powerful
Self	Unit	Grounded
Unique	Integrated	Strength

Body–Mind Connections

Humans have an integrated body–brain system. What is seen on the outside is testament to what is happening internally. In this section, we explore some signals emitting from visceral, vestibular (which maintains balance and spatial orientation), and musculoskeletal neurochemical systems coaches use to interpret in-the-moment experiences of a coachee. For example, transitions from distress to eustress are consistently observed during pacing, as we saw in table 12.1. *Dis* is a prefix denoting, in general, separation or negation. Distress is "bad" stress, separating us from our cognitive resources, leaving our body–mind system primed to protect itself in ancient ways. Eustress is "good stress"; *eu* means "good," as in eulogy (good words) or euphoria (good feelings). Often, a colleague begins a pacing conversation with muscle tautness, shallow breathing, and other nonverbal indicators associated with emotional flooding and a shutdown from the neocortex. At the exact moment the coach makes a goal statement congruent with the teacher's desires, an immediate shift into eustress can be observed in breathing, posture, animation, and voice qualities. Given our understanding of the methods of information processing, our current hypothesis is that at the moment eustress is achieved, electrochemical energy is released to the body–brain system, stimulating two reactions. One reaction relates to neurotransmitters like serotonin, which move messages from neuron to neuron, make the brain more efficient, and metaphorically allow access.

A second reaction is that peptides, produced in the limbic brain and routed through the bloodstream to the entire body, including the brain, carry chemical signals of hopefulness that are manifested as "gut" reactions and changes throughout the autonomic nervous system.

Changes in the visceral, vestibular, and musculoskeletal systems can be observed in pupil dilation, breathing changes, pulse changes, and changes in skin color and muscle tone as skeletal muscles and visceral organs are affected. Even glands respond with discharges of tears, perspiration, or saliva. Hair follicles can become erect. The coach is highly attentive to these signals as feedback about his/her work (Ledoux 1996).

Emotions decide what is worth paying attention to, and one's sensing of the outer world "is filtered along peptide-receptor-rich sensory way stations, each with a different emotional tone" (Pert 1997, 146). These are one of three information delivery systems in the body that produce the behavioral manifestations to which the coach attends. The other two are neurotransmitters (like acetylcholine, norepinephrine, dopamine, histamine, and serotonin) and steroids (like testosterone, estrogen, and progesterone). Oxytocin is one example of a peptide. It releases the chemistry of "tend and befriend," which is reported to be a feminine response to stress. Peptides regulate most of life processes and carry 95 percent of information transmission through the bloodstream to every cell in the body and brain. Less than 2 percent of neuronal communication actually occurs at synapses. Pacing releases oxytocin, and the coach recognizes when this occurs when changes in the coachee occur. These changes—we call them BMIRS (behavioral manifestations of internal states) can include full unlabored breathing, relaxed muscles, postural shifts, a softening of facial features, and speech-pattern changes apparent from a previously stressed state. All this information is important because it assures the coach that cognitive, emotional changes in the coachee are not "skin deep," so to say, but the product of fundamental, dynamical alterations occurring with body systems, assisting the coach's recognition of when a cognitive shift occurs. These behavioral manifestations of internal change will be elaborated below.

COGNITIVE SHIFT

What occurs during a cognitive shift is what Kuhn (1962) might have labeled a *paradigm shift*—a discontinuous and often radical shift in point of view. But there is more. What we observe are *cognitive* shifts in which we experience a change in how our conscious and unconscious minds communicate with one another. Insight is often the stimulus for shift and is thought to represent a break in mind-set registering unusual activity in the anterior cingulate cortex and may be related to detection of conflict between old and new cognitive modes at the moment of insight (Xiao-Qin et al. 2004). A cognitive shift may also suggest that regions of the brain–body system not previously engaged are now active, possibly related to an abrupt change from certainty to not knowing; visual to auditory processing; or egocentric thought to considering another's perspective. The shift is usually sudden and is noticeable by an observer. Sighs, laughter, changes in eye accessing, extended silence, postural shifts, unconscious movement of legs or arms, or other physical manifestations of neurochemical changes appear. In Cognitive Coaching, we see cognitive shifts often and most frequently in pacing at the goal statement stage and in an "Aha!" moment during the lead. Euphoria often results.

   Cognitive Coaches who facilitate cognitive shifts for their partner bring a valuable added quality to teachers in the conversation. Without such shifts, the teacher may experience a planning conversation, for example, as an opportunity to talk about what he or she has already decided, to gain greater clarity and precision about certain portions of the lesson, and to instruct the coach on how to serve the teacher as a data gatherer. When cognitive shift is facilitated, the coach has added a dimension totally unavailable to the teacher on his or her own. One teacher described this metaphorically by saying that without the shift, you have the equivalent of a good physical exercise; with the shift, it is akin to an aerobic exercise in which the very core and capacity of the being is strengthened.

Metaphorically, then, we regard the state of distress as being related to the chemistry of depression or frustration, and the state of eustress as a shifting to a chemistry of hopefulness or resourcefulness. We believe the paraphrase to be the heroine of this unfolding neurological drama. Pacing is paraphrasing, and its first function is to interrupt the decline of a chemical-emotional state and restore or refresh the chemistry of hopefulness. Skilled pacing honors and validates the emotions of the existing state without increasing the chemistry of defeat or frustration. Questions during pacing counteract this movement and reverse it to plunge the person into deeper negative chemistry. Exploring the speaker's frustration reinforces and often amplifies the body chemistry of negative emotion. Paraphrasing is so consistently associated with the physical signals of these changes because it often clarifies a person's feelings and conflicts, it is accepting of the existing condition (hence lessening physiological "resistance" to it), and it leads the person to feel understood.

Bruce Wellman and Diane Zimmerman (2001) offer the following metaphor, illustrated in figure 12.2. The paraphrase acts as the fulcrum of a lever assisting the increase of chemical resourcefulness. As paraphrasing continues, levering action floods the bloodstream (and every cell in the body) with better chemistry. Envision a fulcrum under a lever. Paraphrases that acknowledge and clarify are like a fulcrum placed at about the middle in which some movement will occur. This will balance the chemistry. Paraphrases that summarize and organize move the fulcrum closer to the "load" to be lifted and add more resourceful chemistry. Abstraction paraphrases that shift the logical level, which is the form of paraphrase offered in a goal statement, move the fulcrum to its most potent spot, flooding the body–brain system with resources manifested throughout the whole body, including oxygen and glucose, which the brain needs for reasoning. A shift from limbic system dominance to neocortical dominance has been completed. Positioning the fulcrum closer to "the load" reduces the amount of "force" required to lift it.

Focused inquiry (the lead) is now possible for exploring solutions within the newly identified desired state. (See descriptions of these forms of paraphrase in chapter 3.) During distress, the limbic system is in command, focusing inward, which is its function. The neocortex looks outward and can attend to situations and oneself from a neutral observation point. (See "How Do I Know When I Am Finished?" later in this chapter for a description of other physical signals stimulated by the autonomic nervous system.)

The Structure and Skills of Leading

If pacing is paraphrasing, then leading is questioning. When pacing is complete, questioning begins. Your partner has indicated, often through the autonomic nervous system, that the stated goal is indeed congruent with his or her desires. One cue rela-

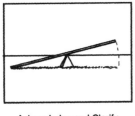 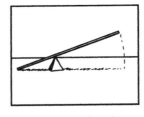 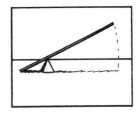

Acknowledge and Clarify   Summarize or Organize   Shift Logical Level

Figure 12.2.   How Paraphrasing "Lightens the Load"

Table 12.3. A Macro Map for Leading

Since leading another to essential internal resources required for problem resolving is as much art as science, there can be no prescription for how a Cognitive Coach achieves this.There do, however, seem to be at least three stages in the conversation with touch points as follows.

1. Where do I start?	2. Where do I go from there?	3. How do I know when I am finished?
• Initiate a lead question intended to access and illuminate a state of mind in chapter 7 • If the first question unproductive, explore other states of mind • Continue questioning guided by the colleague's responses • Maintain empathy and rapport	• Emphasize questions that liberate internal resources – feelings, values, sub-goals amplify states of mind • Ask questions to create better goal specificity if necessary • Paraphrase to shift conceptual levels • Ask questions that shift the focus from a third party to the person being coached	• Observe major and sudden shifts in voice qualities, posture, breathing, gesture, muscular tone and or skin tone. • Detect congruence between speech content and non verbal messages • Observe manifestations of cognitive dissonance as in shifts from certainty to uncertainty or sudden insights • Obsere indications of ccongruence such as bi-lateral gesturring

tively easy to detect is the slowing and deepening of breathing. Now the coach has tacit permission to inquire (Zoller 2007).

Questions become the coach's primary language tool, supported by paraphrasing and pausing. The map for leading a colleague to internal resources is less clear and less standard than the map for pacing.

The pathway completes the pace and is an on ramp to the lead: "So you are looking for a way to do that." Now the coach enters more uncharted water, guided by three overarching questions and the moment-to-moment verbal and nonverbal responses of the person being coached, as shown in table 12.3. These questions are: Where do I start the lead? Where do I go from there? How do I know I am done?

Since leading another to essential internal resources required for problem resolving is as much art as science, there can be no prescription for how a Cognitive Coach achieves this. There do, however, seem to be at least three stages in the conversation, beginning with "Where do I start?"

- Initiate a lead question intended to access and illuminate a state of mind described in chapter 7.
- Emphasize questions that liberate internal resources—feelings, values, subgoals.
- Observe major and sudden shifts in voice qualities, posture, breathing, gesture, muscular tone, and/or skin tone.
- If the first question is unproductive, explore other states of mind.
- Ask questions to create better goal specificity, if necessary.
- Detect congruence between speech content and nonverbal messages.

- Continue questioning guided by the colleague's responses.
- Use abstracting paraphrases to shift conceptual levels.
- Observe manifestation of cognitive dissonance, as in shifts from certainty to uncertainty or sudden insights.
- Maintain empathy and rapport.
- Ask questions that shift the focus from a third party to the person being coached.

Where Do I Start?

During pacing, the coach asks himself which of the five states of mind appears to be unawakened in the colleague's experience. If liberated, which of them might convey the necessary resources? The lead to internal resources may begin with a question designed to illuminate that state of mind and its related resources. Although none of the questions listed here would be the first question asked in the lead, as they are the central focus of the conversation occurring later, they illustrate inquiries that might activate a particular state of mind within the context of an interaction. For example, if efficacy is unawakened, the coach might ask the following questions:

- Self-prescribing: "How do you go about motivating yourself?"
- Choice making: "So you can think of several approaches. How will you know which one to choose?"
- Correcting fate control: "What was it in your behavior that might have caused that?"
- Shifting toward an internal locus of control: "What do you need to do?"
- Identifying resources: "What knowledge, skills, or attitudes might you need to accomplish that?"
- Content reframing: "What might be the positive value of that response?"

If the colleague needs to access flexibility, the coach might ask questions like these:

- Entering other perspectives: "Considering their history, what might they be feeling?"
- Enlarging frames of reference: "What additional values might there be?"
- Predicting consequences: "What might be the result of . . .?"
- Considering intentions: "What do you think they want to achieve?"
- Testing outcomes: "What's the worst [or best] possible results of achieving that outcome?"
- Style checking: "Given your knowledge of learning styles, what might be going on?"
- Changing time horizons: "If you had twice the time [or half the time], what difference might that make?"
- Context reframing: "If there were an upside to this misfortune, what might it be?"

A coach who wants to help a colleague access craftsmanship might ask questions like the following:

- Communicating with specificity: "How do you define . . .?"
- Managing time: "How will you know it is time to move on?"
- Defining criteria for judgment: "What criteria will you use to . . .?"
- Striving for refinement: "How will you know when you are really satisfied?"
- Exploring standards: "What are the connections with [curriculum, teaching, district] standards?"

To access consciousness, the coach might ask these kinds of questions:

- Metacogitating: "What was your thinking that led to that decision?"
- Mental rehearsing: "As you picture this event, what will you be doing?"
- Mental editing: "Looking back, what might you change?"
- Owning intentions: "What are your intentions?"
- Considering effects: "How might your choice influence you and others?"
- Checking assumptions: "So you are assuming . . . . How did you arrive at that?"
- Values searching: "What is it that makes this so important to you?"

In the area of interdependence, a coach might ask the following:

- Collaborating: "Who else is concerned about this?"
- Talent searching: "What useful skills do other group members have?"
- Resource banking: "How might others assist you?"
- Group supporting: "In what ways does this relate to staff concerns?"
- Envisioning potential: "As this group continues to develop, what other ways might it handle this?"
- Coordinating: "Where else in the system does this value of yours reside?"

Where Do I Go from There?

As table 12.3 illustrates, the coach begins the lead with questions that illuminate a particular state of mind. If that seems unproductive, the coach may explore other states of mind. Other mediational strategies include paraphrasing to shift the conceptual level of abstraction and asking questions that illuminate feelings, values, or more details about the goals. The coach might also ask the person to compose self-generated questions or use a series of questions that move from focusing on a third party to themselves. For example:

- What behaviors do you want from the group?
- What knowledge, skills, or attitudes will they need to perform those behaviors?

- What might you do to help them develop these resources?
- What internal resources do you need in order to do that?

Language offers cues to thinking. Language also abides by certain conventions. One thing we want to avoid is detail in which every nuance of thought is expressed. When people talk or write, generalizations, deletions, and distortions are quite common. At times, however, they blur understanding and hide opportunities for choice. Coaches are sensitive to these language forms at all times, but most particularly when leading colleagues to resources. When coaches suspect that one of these forms of language is hiding important information from thinking processes, they may respond in the following ways.

Generalization

Generalization eliminates the necessity to relearn a concept or behavior every time we are confronted with a variation of the original. However, generalizations can carry imprecision and inaccuracies that limit thought and choices. The coach might ask, "Under what conditions might that be true?"

A complex equivalent is a statement that ties two separate ideas together as if they were inseparable: "He is failing; he doesn't study enough." In that case, the coach might ask, "Have you ever known someone to study hard and still fail?" or "Can one not study at all and still pass?"

A lost performative is a statement in which the performance criteria for a value judgment are missing: "That was a good paper." The coach might then ask, "Good in what way?" or "What criteria are you considering when you rate it as good?"

A modal operator (derived from mode of operation or modus operandi) is a rule word like *can't* or *impossible*: "I can't do this." Then the coach might ask, "What is stopping you?" or "What would happen if you did?"

In all of these situations, the coach's intention is to illuminate possible inaccuracies represented by the generalizations.

Distortion

Distortion is the process by which we alter our perceptions; it is the source of creativity, play, humor, and multiple versions of remembering or interpreting an event. Three linguistic cues may signal distortions that are limiting to the speaker.

Cause-and-effect statements inaccurately name the source of an event. This type of distortion may be a cue to low efficacy, placing the power outside oneself for responses: "They make me mad." In that case, a coach might respond, "They have that much control over you?" or "When do you first notice that you are becoming angry?"

Mind reading is the expression of interpreting another person's intentions, thinking, or feelings. It differs from the paraphrases in a pace in that it is expressed as an explanation of another person's behavior rather than an attempt to understand when someone says something such as: "He is doing that to annoy me." In that situation, the coach might respond, "What other possible reasons might he have?" or "Can you think of a time when you have done something similar for a different purpose?" or "If he had a positive intention, what might it be?"

A nominalization names a process as if it were a thing. It describes a verb as if it were a noun. The words *relate*, *love*, *learn*, *communicate*, and *trust* are all nominalizations. In the sentence "They don't respect each other," the word *respect* is used as an abstraction, or nominalization. Respect can mean a number of things. The question "What might they be doing when they are respecting each other?" returns the conversation to a description of behaviors. These can be understood by both parties and therefore addressed.

Coaches seek to return active status to nominalizations because we are doomed to frustration and failure whenever we try to solve an abstraction. For example, nothing can be done about "resistance," because the language form characterizes it as a nominalization, a thing, an immovable object. "Resisting," however, has form: a beginning, middle, and an end with which a person can observe, interact, and intervene. Denominalized words allow participation and efficacy.

Nominalized forms suggest choicelessness and low efficacy. When someone says, "Their resistance is the problem," the coach might respond, "What are they doing when they resist?" or, "How does the resisting begin?" or, "What would they be doing if they were not resisting?"

Deletions

Deletions are information left out of statements. Deleted information occurs more frequently in speech than either distortions or generalizations because if we were to speak inclusively of all details, our conversations would take a century: "The man— that is, the tall, mustached man with the gray suit, green shoes, and white tie (and holding a cane), standing three feet from and to the north side of . . ."

The sensitive coach is selective about the frequency of challenges to deletions. Paraphrasing gives permission to ask any type of probing questions. For example, when a person says, "They say it will get better," the coach can respond, "So, they believe it will improve. Who are 'they'?" Given the choice to clarify a vague referential index (vague noun or pronoun) or vague verb, we recommend clarifying first the subject of the conversation—who is being referenced. If the colleague says, "We want them to think," the coach can respond, after paraphrasing, "Think how, specifically?" Unlike a conversation in which the parties are working to solve a problem and a solution has emerged worthy of trying, parties in a problem-resolving conversation need different signals about when to move on.

How Do I Know When I Am Finished?

> Conversation does not have to reach conclusions in order to be of value.
>
> —Parker Palmer

New connections are not always consciously formed. Many insights involve unconscious processing (Rock 2009). Think of times when you have a new way of approaching a problem suddenly come to you in the middle of the night. One thing a coach can do is to pose questions to think about at the end of a conversation, not asking for an immediate response. Some examples might be: "What are some questions you are still asking yourself?" or "What are some of the variables still in your control?" Pondering those "walk-away" questions assumes the person has the capacity to make advances in his/her thinking, which has a very different effect than making a suggestion or offering your own personal observations.

The problem-resolving conversation releases a person from the trap of impotency about how to approach a problem. This is accomplished by accessing and amplifying the internal resources necessary for clarity, courage, and right action. Most often these require amplifying states of mind, as depicted in figure 12.3. In consulting conversations, where the goal is to solve problems, you know you are finished when the person has articulated a plan of action. When the goal, however, is to liberate sufficient internal resources to restore states of mind, different indicators of completion are needed.

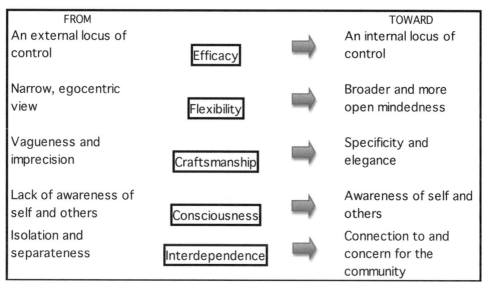

Figure 12.3. **From Limitations to Resourcefulness**

Observations of physiological changes provide cues as to when a cognitive shift has occurred, signaling internal resolution, insight, fresh determination, or an "Aha!" moment. These shifts occur in systems outside the coachee's control; breathing, postural changes, gestures, voice qualities, and blushing are some examples. Our sense is that these physical alterations may mark abrupt changes in mental functioning, perhaps activating portions of the mind not previously engaged. An observant coach sees and hears and infers meaning about these BMIRS.

During pacing they may be interpreted as movement from distress to eustress. During the leading portion of the conversation, the coach takes these as signals that the internal work has been done and the coachee is in touch with the necessary interior resources. Emotions play a commanding role in being stuck or becoming resourceful and are registered in the body. Behavioral manifestations of internal resource states (BMIRS) will appear involuntarily as they are informed by chemical shifts affecting the autonomic (self-regulating) system.

To observe a cognitive shift (BMIRS), please access two videos: www.thinkingcollaborative .com/support-audio-videos/12-1/ and www.thinkingcollaborative.com/video-12-2/. In the first video, Jane, the coach, paraphrases Dana. Watch Dana's response, noting ways her behaviors reveal what is happening internally. In the second video, Carolee asks a question that triggers an observable cognitive shift in Ochan.

RESEARCH: EMOTIONS MAPPED IN THE BODY

When people are emotionally flooded, feeling stuck, discouraged, experience low efficacy, or feel angry or bad about an event, these emotions play out in the body as changes in the facial expression, skin tone, posture, muscle tone, upper and lower limb activity, and gut sensations as discovered in the bodily mapping of emotions (Nummenmaa, Glerean, Hari, and Hietanen 2013). Psychoacoustics research has also detected changes in voice quality (Cook 2002, 4). Emotions are trigged by perceptions of events in the moment or in the past. Body mapping research was conducted with West Europeans and East Asians speaking four different languages. The researchers conclude that emotions are displayed in a culturally universal pattern across distinct body regions and that positive emotions (happiness, love, and pride) formed one cluster in the body, and negative emotions diverged into four clusters (anger and fear; anxiety and shame; sadness and depression; and disgust, contempt, and envy) in different body regions.

Most basic emotions were associated with sensations of elevated activity in the upper chest area, likely corresponding to changes in breathing and heart rate. Similarly, sensations in the head area were shared across all emotions, reflecting probably both physiological changes in the facial area (i.e., facial musculature activation, skin temperature, tearing) as well as the felt changes in the contents of mind triggered by the emotional events. Sensations in the upper limbs were most prominent in approach-oriented emotions, anger and happiness; whereas sensations of decreased limb activity were a defining feature of sadness. Sensations in the digestive system and around the throat region were mainly found in disgust. In contrast with all of the other emotions, happiness was associated with enhanced sensations all over the body. The nonbasic emotions showed a much smaller degree of bodily sensations and spatial independence, with the exception of a high degree of similarity across the emotional states of fear and sadness, and their respective prolonged, clinical variants of anxiety and depression. (Nummenmaa 2014)

In a typical emotion, regions of the brain send commands to other regions of the brain and to almost every area of the body. Two routes send the commands. One route is the bloodstream, where chemical molecules act on cell receptors. The other route consists of neural pathways, and the commands along this route take the form of electrochemical signals that act on other neurons, on muscular fibers, or on organs. Organs (such as the adrenal gland) in turn can release chemicals of their own into the bloodstream. Each emotion shares a common biological core described by neurologist Antonio Damasio (1999, 2010):

- Emotion travels via the bloodstream and neural pathways; it acts on cell receptors, neurons or muscular fibers, or organs. Organs release chemicals into bloodstream.
- Emotions are collections of chemical and neural responses. All emotions contribute in some way to the advantage of the organism. The role of emotions is to assist the organism in maintaining life.
- Emotions are biologically determined processes, depending on innately set brain devices, established by a long evolutionary history. Learning and culture may alter the expression of emotions.
- Emotions both regulate and represent body states. They are produced in the brain stem and move up to the higher brain.
- All the triggering devices for emotions can be engaged automatically, without conscious deliberation. While individual variation and culture play a role in bringing on emotion, there is still a fundamental stereotypic, automatic, and regulatory purpose of the emotions.

All emotions use the body as their theater, but emotions also affect numerous brain circuits. The coach is attentive to bodily changes that signify involuntary changes in the autonomic nervous system. Although the coach can never know the exact meaning of each behavioral change, the coach can combine observations of BMIRS with her knowledge of the person, the conversation taking place, and her intuition to presume that a problem resolving conversation goal has been met. Voice, posture, breathing, gesture, muscular tone, and skin-tone color shifts are major sources of data. Specifically, the coach will look for the following:

- Major and sudden shifts in any of systems outside the conscious control of the coachee, as well as congruence in content of speech, voice, posture, and gesture.
- Manifestations of cognitive dissonance, as in shifts from certainty to uncertainty or in sudden insights. Bilateral body movements—gestures with both hands—allow the coach to infer that right- and left-hemisphere integration is present regarding a topic.

THE COACH'S INVISIBLE SKILLS

In chapter 4 we discussed three listening set-asides useful to coaches: setting aside autobiographical listing, solution listening, and inquisitive listening. Within the problem-resolving conversation, three additional metacognitive skills are useful; coaches should set aside their own need for closure, comfort, and comprehension.

Closure

Humans have a natural need for closure: to understand the end of a story, to be satisfied that we know how it turns out, to put an end to a dilemma. Two internal questions regarding closure inform the coach's decisions during problem-solving conversations: What are my intentions? How will I know when my work is finished?

In the problem-resolving conversation, the coach's intention is to help a colleague illuminate and accept tensions, clarify goals, and locate the internal resources needed to meet whatever challenge the coachee is addressing. This is different from help-ing the colleague to identify a solution and devise a plan. The latter is a legitimate consulting goal, but most often it is attainable only when the teacher already has the confidence and emotional resources for that level of thinking. Thus, in the problem-resolving conversation, the coach sets aside his or her own natural interest in know-ing "the end of the story" or what specific action the teacher will take. Instead, the coach looks for the involuntary nonverbal cues that signify the colleague has accessed the necessary resources to proceed. Internal resources may include knowl-edge, skills, and states of mind.

Comfort

Because problem resolving is the most complex of the maps and depends on moment-to-moment sensory acuity, the coach may not always know what to say next. Being comfortable with this dimension of ambiguity is a critical resource to both parties. For the coach, living with this ambiguity helps to prevent premature and inappropriate comments. For the colleague, it is reassuring to have someone fully present and not discomforted as he works through his uncertainties.

If the coach becomes uncomfortable by being lost in her own emotional reactions to the issues being discussed, two strategies can be employed. First, she can remind herself to set aside focus on self. Second, she can use a paraphrase. Paraphrasing helps to focus the coach's emotional attention on the colleague, not on the coach's discomfort. If the coach becomes lost in the process and a way out is not apparent, living with the discomfort at these times is often the best action. In doing so, the coach communicates her trust in the colleague and in the process.

Comprehension

The only time one needs to know extensive details about a problem is when one intends to be the problem solver. Since the coach's intentions are to support the other person in clarifying goals and to locate the internal resources to meet those goals, the coach doesn't need to know the history, background, context, and sources of the problem. Process knowledge is paramount for the coach; specific information about the situation is secondary.

NINE PRINCIPLES OF INTERVENTION

To support colleagues in getting "unstuck" and back to resourcefulness, a coach uses the tool clusters of pacing and leading. This strategy has been described in two stages: pacing, with four major verbal phases, then leading, with a more generalized map to guide choices. In addition, larger ideas than these are useful guides in promoting thinking and development. These ideas are principles of intervention we have adapted from the work of Stephen and Carol Lankton (1983).

Although theories of personality or human behavior are useful, they have shortcomings because resolutions are sought within the frame offered by the theoretical model. Since all models illuminate only selected views of human behaviors and place others in shadow, interventions based on theories of personality will always be doomed to partial success. In contrast, coaches working with the principles of intervention listed here will find themselves limited by only their own flexibility.

1. People act on their internal maps of reality and not on sensory experience. Each person perceives the world from his own unique vantage point, seeing it through frames of personal history, belief systems, representation systems, and cognitive styles. This principle reminds the coach to learn how the world appears to the other person and to sensitively gather data to understand the other person's maps and how they are constructed.

2. People make the best choice for themselves at any given moment, essentially doing what they think is best. This principle does not suggest that people make the best choice possible, only the best choice available to themselves at the moment. These choices are not necessarily good choices or the most effective ones; they are what occurs to the individual in the heat of the moment. High emotionality will limit choices, as will strongly held positions or points of view.

3. Respect all messages from the other person. Empathy and respect are critical resources the coach brings to the coaching relationship. Skilled coaches nonjudgmentally attend to and respect both verbal and nonverbal messages. This requires being attuned to the more subtle elements of communication—voice, tone, gestures, facial expressions, eye shifts, and speech metaphors.

4. Provide choice; never take choice away. In several cases, Milton Erickson worked with suicidal persons. His approach was, in effect, "Yes, that's one choice. What are others?" When we attempt to limit a person's choices, the person is often drawn even more stubbornly to the choice we've removed. Effective coaches exercise options.

5. The resources each person needs lie within his or her own personal history. Erickson would frequently tell students or clients that their "unconscious contains a vast storehouse of learning, memories, and resources." Coaches need to be mindful that they are there to facilitate the teacher's access to these inner resources, not to offer a solution.

6. Meet the other person in her own model of the world. The skills of rapport are especially helpful in connecting with another person psychologically. However, this does not imply that the coach must stay in the other person's model.

7. The person with the most flexibility or choices will be the most effective. The greater a repertoire that teachers have, the greater flexibility and choice they have in terms of instructional strategies and classroom management. The same is true for the coach or any person in a helping relationship. When interventions fail, it is frequently due to the coach's inability to exercise the necessary flexibility.

8. A person cannot not communicate. Even if a person is not communicating verbally, the person is still sending nonverbal messages. Some behaviors are very subtle, such as breathing shifts or a slight nod of the head. These behaviors are very important for the coach to attend to when coaching.

9. Outcomes are achieved at the psychological level. This principle is based on the reality that several levels of communication operate simultaneously.

One of these is the social level message carried in words. Another is the psychological message usually reflected in the voice tone, gesture, or emphasis. When these two levels of communication are incongruent, the psychological message will determine the outcome of communication (Garfield 1986). This principle reminds coaches to be conscious of and clearly intentional about their own levels of communication.

CONCLUSION

The danger in writing about a subject like accessing internal resources is that the examples can too easily be seen as mechanical formulas for given problems. Nothing could be further from the truth. For example, on these last pages, we have displayed primarily what Lankton and Lankton (1983) call the "social message" carried in words. We are unable to experience on these pages the postural shifts, inflections, breathing changes, and all the many ideomotor responses that send the psychological message.

The skillful coach attends to these states not by mechanically observing and interpreting nonverbal signals but by caringly, intuitively, with the coach's entire presence being committed interdependently, one human being to another, entering the other's model of the world, accessing resources, and illuminating choice.

Skillful coaching does not mean applying formulas. Rather, it means continually designing, testing, and refining an ever-expanding repertoire of new and more efficient ways to be catalytic in a colleague's growth. As coaches become increasingly skillful in working with the map of problem resolving, there is a reciprocal effect on themselves. In recognizing the five states of mind in others, the coach recognizes these states in himself. The external attention and skill of coaching others becomes internal in the coach, who thus mediates his self to higher states of holonomy. The model of Cognitive Coaching, then, becomes self-mediating, self-transforming, and self-modifying for coaches themselves. As Erich Fromm (1956) wrote:

> In thus giving of his life, the mentor enriches the other's sense of aliveness. He does not give in order to receive; giving is in itself exquisite joy. But in giving he cannot help bringing something to life in the other person, and this, which is brought to life, reflects back to him; in truly giving, he cannot help receiving that which is given back.

13

The Calibrating Conversation

Since the publication of the second edition of this book, a fourth map supporting teachers' self-directed learning was conceived, developed, and tested extensively by a number of Cognitive Coaches. The calibrating conversation uses an externally generated, locally adopted description of teaching excellence with the purpose of supporting the capacities and practice of self-directed learning and instructional improvement (Dolcemascolo, Miori-Merola, and Ellison 2014; Ellison and Hayes 2012). Calibrating is based on the assumption that a teacher is a continuing learner interested in refining her/his craft (craftsmanship) and as a result turns to other sources and research to do so.

Recently there has been a proliferation of documents proclaiming to describe elements of teacher excellence. (See appendix E: Sources of Standards). Many schools and school districts have adopted these standards and rubrics as part of their teacher evaluation practices (although some documents were not intended to be evaluative). Since evaluation, in itself, rarely leads to the improvement of performance while coaching does, it was deemed valuable to find ways to utilize these exemplars of excellent teaching with conversations in which teachers could be supported in self-assessing and self-prescribing improvements for themselves based on these external standards. There is, therefore, a need for providing an additional structure for rich, rigorous, and reflective professional conversations between principals, or other support personnel as they converse with teachers. Effective instructional leadership does matter according to Ebmeier (2003), who reports research indicating that skillful forms of supervision influence teacher commitment and personal efficacy. This chapter adds a fourth map to the Cognitive Coach's repertoire: the map of the calibrating conversation.

To *calibrate* means to measure and attune performance against an established standard. This is done by determining the deviation from a standard so as to develop proper correction factors. Calibrating conversations, therefore, are designed to assist a staff member to calibrate his or her progress against a standard by understanding and engaging with a locally adopted and agreed-upon standard and by determining where his or her skill level falls. Many descriptions of teaching excellence are available; all draw upon the rich knowledge base in development about instruction. All address six domains of inquiry, though frequently with different terminology:

- What's worth learning? (Content knowledge)
- What works in teaching? (Pedagogy)
- What factors influence student learning? (Knowledge of students and how they learn)
- Who am I and who am I becoming? (Self-knowledge)
- How does the brain learn? (Knowledge of cognitive processes of instruction) (see chapter 8)
- How are collegial interactions continually strengthened and enhanced? (Knowledge of collegial interactions)

Districts develop their own, as is the case with the American School of Japan, or are state adopted, or they use descriptions of excellence or rubrics developed by Marzano, Danielson, Hattie, Saphier, Silver, Leithwood, or others (see appendix E: Sources of Standards).

The greatest distinction in the way the calibrating conversation uses instruments that might otherwise be used for evaluation is the focus on cognitive processes and on liberating internal resources for self-assessment, goal setting, and planning. Earlier we described self-directed learning as incorporating self-managing, self-monitoring, and self-modifying behaviors. The calibrating conversation serves these goals well. In these conversations, the standards are less important than the quality of the dialogue fostering professional autonomy and self-directed learning. In order for the coachee to calibrate his/her progress toward the standard, the coach first supports the coachee in making meaning of the standards. Inviting the coachee to make meaning through dialogue is the first step that the coach takes to mediate the thinking of the coachee. The conversation becomes one of reflecting, analyzing, and objectively evaluating the results rather than a judgment of the person.

A THIRD-POINT CONVERSATION

The document being used in the calibrating conversation becomes what is called a third point in the communication. The third point serves as a focus separate from each of the parties in the conversation. A conversation between two people looking at each other is called two-point conversation. In a third-point conversation, the data—

in this case, the standards rubric—is set in a position that both parties can look at it, rather than each other. The value of designating a third point is that both parties can refer to it in an impersonal way. The third point does not belong to either party; it is simply a reference point for the conversation. Physically referencing the third point in a space off to the side between the parties provides a psychologically safe place for information and depersonalizes ideas. Thus, placement of the conversational focus creates a triangle, either literally or referentially, keeping the conversational container psychologically safe (Lipton and Wellman 2011). As can be seen in video 13.1, establishing a visible "third point" for the conversation increases psychological safety for the teacher by shifting the focus to the data and promotes conversations about the factors producing positive results and what may be causing any perceived performance gaps (Grinder 1997).

HOW TEACHERS OWN STANDARDS

Standards, test scores, and rubrics—which propose to define teaching quality but which are developed and imposed without the teacher's involvement, comprehension, and commitment—lead to short-term, shallow results and ultimately to failure (Danielson and McGreal 2000). Sanford (1995a) suggests that feedback must be conducted within a system the person being coached has helped create. For insights to be useful, they need to be generated from within, not given to individuals as conclusions. This is true for several reasons. Teachers will experience the adrenaline-like rush of insight only if they go through the process of making connections themselves. Defining the practice as actions creates a more vivid picture inside the minds of the teachers as to what they will be doing, saying, or feeling if they are performing the behavior. It is more likely that we can agree on actions than on definitions. Teachers, therefore, would be invited to envision what they would see themselves doing or hear themselves saying when they, for example, use powerful questions. Thus they would translate the labels of the criteria for excellence and categories of performance into actions and tactics.

Teachers, who have the opportunity to decide what is good or poor, appropriate or inappropriate, effective or ineffective, are more likely to transfer these decisions into practice. This powerful approach to improving instructional practice focuses on the intellectual skills, perceptions, and decisions that underlie effective teaching.

Involving teachers in the development and application of these practices is another way for them to assess their own performance. The purpose is self-mastery. Teacher's self-authoring of descriptions and indicators of what they will be doing and saying if they are using the practice effectively promotes self-managing, self-monitoring, and self-evaluating. It also provides a mental rehearsal prior to performance. The intent is for teachers to describe the categories of behaviors, hold them in their head as they apply them (self-monitoring), and then self-evaluate their performance and make plans for improvement. Each category should be sufficiently clear so that teachers can learn from self- and other-observed feedback about their behavior and to seek ways to improve.

This would be true for breaking an old habit as well. When teachers envision what a behavior looks like and sounds like, it makes possible the replacement of undesirable habits. Change requires observing the pattern that we presently have and then making a conscious decision to break that pattern. We can put our attention to what was missing. We can begin to attend to changing our behaviors and seeing the benefit when we do so. For example, some teachers are unaware that their questions are posed in such a way as can be answered with a simple "yes" or "no," requiring minimal thought by the students. Hearing, analyzing the structure of, and composing questions that require deeper, more complex thinking helps teachers compare their typical questions with these more thoughtful questions—which initiates a journey of replacing of old habits.

The categories of excellence can be derived from many sources (see appendix E: Sources of Standards). Any set of criteria may be used that is based on research, values, and staff agreements. The five states of mind described in chapter 7 could be used as categories to derive performance. For example, if school goals include student efficacy, consciousness, craftsmanship, flexibility, and interdependence, what would learning look like? What would teachers be doing if they use these as outcomes?

As an alternative to adopting external standards, school leaders could build common agreements among staff and provide time and practice as a group in translating them into observable behaviors. School mission statements can also serve as starting points. For example:

> City High School is a home for the active mind—a cooperative community promoting knowledge, self-understanding, mutual respect, global understanding, adaptability to change, and a love for lifelong learning.

Based on this mission statement, conversations would center on what a classroom might look like if minds were active: What would we as teachers need to do to keep minds active? Or adaptable? Or cooperative? And so forth. Such endeavors would draw upon frameworks and research studies as informative sources rather than evaluative criteria.

Another reason for promoting self-evaluation for teachers is that it models the same values that we hold for students. We spend far too much time and resources evaluating students and thus we rob them of the opportunities to evaluate themselves. Self-monitoring and self-evaluating models a cascading quality that pervades the culture of the school and becomes a norm for all the school's inhabitants—for the leadership personnel, the teachers, and the students. Standards are not standards until they are owned. As long as they remain someone else's expectations, teachers will simply comply or conform rather than make real, authentic changes. Because systems can mandate compliance but not excellence, schools need ways in which the faculty and individual teachers can own externally generated standards. For achieving the goal of teacher ownership of standards, see suggestions in appendix A.

REGIONS OF THE CALIBRATING MAP

Within the territory of the calibrating conversation map, there are six regions that the coach navigates to mediate the coachee's thinking (table 13.1). Each region has a specific purpose, designed to support the coachee in reaching objectives or goal(s) reflected by the document. The coach uses the basic tools of Cognitive

Table 13.1.   Regions of the Calibrating Conversation

Map

Coach mediates by having the coachee:

- Select a focus

- Identify existing level of performance or placement on a rubric and give supporting evidence

- Specify desired placement and explore values, beliefs, and identify congruence with desired placement

- Establish behavioral indicators for new placement on rubric or level of performance

- Describe support needed to get to a higher level of performance and commit to action

- Reflect on the coaching process, explore refinements, and explore ways of using this process on your own

Tools

Coaches navigate the map using the tools below:

- Pause to allow you and your partner to think.

- Summarize by paraphrasing your partner's thoughts from time to time by saying, "So. . . ."

- Pose questions to explore thinking, by asking, for example, "Specifically which area might you want to focus on?"

- Pose questions to explore thinking, by asking, for example, "What are you aware of in your students that is causing you to move to a higher level of performance?"

- Pay close attention to your partner, attend with your mind and you body.

Coaching (rapport, pausing, paraphrasing, and posing questions) to mediate the coachee's thinking.

- *Select a focus.* In this region, the coach asks the coachee to decide on what aspect of the document she/he wants to focus. This is important, given that most documents contain a great deal of information. The selected focus should be one that can be addressed in the amount of time scheduled for the conversation. The coach might ask, "On what aspect of the document would you like to focus today?" or "What aspect of the document is of interest to you for today's conversation?"

- *Identify existing level of performance or placement on a rubric and give supporting evidence.* In this region, the coach is interested in finding out where the coachee sees himself on the document. The coach poses questions to specify thinking in order for the coachee to be clear about the data that supports his self-assessment. The coach might ask, "Where do you see yourself currently?" and "What might be some examples of how that plays out for you?" "What do you see in your students' performance that leads you to see yourself here?"

- *Specify desired placement and explore values, beliefs, and identify congruence with desired placement.* In this region, the coach asks the coachee where she would like to be. This supports the coachee in establishing a goal or objective for herself, toward which she wants to move. The coach might ask, "So at what level of competence would you like to be on this behavior?" This region is also designed to go to the deep structure of the coachee's thinking to validate the importance of the desired placement. The coach is interested in raising the consciousness of the coachee about the importance of the desired placement. The coach might ask, "What might be some of the values motivating you to reach this level?" "What makes this important to you?" "How do you want to see your students performing when you reach this level?" or "What would you need to tell yourself . . .?).

- *Establish behavioral indicators for new placement.* In this region, the coach is interested in having the coachee envision himself doing what it is he aspires to do. The coachee should be specific in identifying what it looks, feels, and sounds like to achieve the level he desires. The coach might ask, "What might students notice that's different about you when you are performing at this level?" "What might this change cause students to do differently?" "What might it look and sound like when you reach that level?" "Please describe some examples." "By when do you want to achieve that?"

- *Describe support needed to get to a higher level of performance and commit to action.* In this region, the coach is interested in having the coachee draw on her resources to determine what it's going to take to reach the goal/desired placement. The coachee should identify what support she will need to reach the goal.

This support might be in the form of strategies, materials, or the support of other people. Once support is described, the coachee should state what she will do to implement the plan and the data collection tool(s) that might be used. The coach might ask, "What might be some resources you will need to reach this level?" "What might it take for you to apply these strategies?" "What kind of help might be useful to you?" "What is the most powerful step you might take?" "As you implement your plan, what will you be aware of to know it is working?" or "What data collection tool(s) might be helpful to you?"

- *Reflect on the coaching process, explore refinements, and explore ways of using this process on your own.* This is the same region that concludes the other Cognitive Coaching conversations. In this region, the coach asks the coachee to reflect on the conversation in which he just engaged. The intent of this region is to give the coachee the opportunity to identify what was helpful and what supported thinking and to raise to consciousness the process of self-calibrating. The coach might ask, "How has this conversation been helpful to you?" "How has this conversation supported your thinking?" "Where are you now in your thinking compared to where you were when we started?" or "Given your desire for continuing improvement, how might the process that we engaged in today assist you in doing this on your own?"

Although the calibrating conversation is grounded in the Cognitive Coaching support function, it can also be used in each of the other support functions. Appendix D is an example of a teacher and coach in a conversation about a professional development leadership role in which the teacher is about to make a presentation to staff.

> You may now wish to access the 15-minute video of a calibrating conversation. Using the link www.thinkingcollaborative.com/supporting-audio-videos/, select video 13.1 *Calibrating Conversation*. In this video, Sue coaches Carolyn through a calibrating conversation. Also note the use of a third point—the list of standards—in the conversation.

CONCLUSION

As a high school English teacher at the American School in Japan puts it, "In terms of ownership of teaching quality standards, the calibrating conversation is a meaningful and nonthreatening way for these standards to become 'real' for me—the conversation helps generate clear connections between the abstract descriptions and actual practice" (Gotterson 2014). Ed Ladd, head of the school where Gotterson works, observes: "I think this is the best process for evaluation and growth that I have ever used and it is well respected by our teachers and actually viewed as a helpful process by most" (2014).

Supporting self-directed learning, assisting teachers in gauging current performance with aspirations as noted on locally developed statements of excellence, adopted standards, or rubrics from other sources, the calibrating conversation embodies the values and goals of Cognitive Coaching.

Why is it important to use Cognitive Coaching to support professional performance review? The answers are clear: Cognitive Coaching provides the procedural skills to focus on the thinking that produces behavior, it is a research-based growth model congruent with current neuroscience, and it's grounded in a solid foundation of trust.

The evaluator's primary role, then, is to engage thinking that results in self-modification that will sustain longer-lasting change. Far more important than simply telling people what to do are developing rapport, listening, and posing questions that support thinking. Cognitive Coaching provides evaluators with the conversation structures and skills to engage teachers' thinking so they can be self-managing, self-monitoring, and self-modifying (Dolcemascolo, Miori-Merola, and Ellison 2014).

Part 4

THE IMPACT OF COGNITIVE COACHING

In this final section, a synthesis of research on the effects of Cognitive Coaching is presented by Dr. Jenny Edwards who, for over 20 years, has been monitoring work on this topic.

In this section, we restate the higher goals of Cognitive Coaching in systems and acknowledge the difficulties and stresses teachers and schools experience. We comment on the long-held myth that schools can be run from the top down and describe the attributes of natural systems that are more characteristic of schools. We close with the attributes of an "agile organization" in which Cognitive Coaching would most likely take root and flourish.

14

Research on Cognitive Coaching

Dr. Jenny Edwards

As you have read the other chapters in this book, you might have wondered, "Does Cognitive Coaching achieve its goals? What is the evidence that Cognitive Coaching is making a difference?" Numerous researchers have investigated the impact of Cognitive Coaching since it was first developed in 1984. First, Garmston and Hyerle (1988) examined the influence of Cognitive Coaching on professors' thought processes. Next, Foster (1989) studied the influence of Cognitive Coaching on teachers' thought processes. In 1993, Edwards investigated its impact on first-year teachers' conceptual development and reflective thinking; Garmston, Linder, and Whitaker (1993) explored its effects on middle school teachers' practices; Geltner (1993) investigated its impact on the reflection of graduate students; Liebmann (1993) explored the perceptions of human resource developers of the five states of mind; Lipton (1993) examined the reflection of administrators; and Sommers and Costa (1993) investigated the teaching practices of senior high teachers. Since 1993, researchers have studied the effects of Cognitive Coaching in a variety of areas and reported multiple findings.

This chapter begins with findings from studies that have shown increases in self-directed learning and the five states of mind. It continues with information about the benefits of Cognitive Coaching for students, for teachers and principals professionally, for supervisory relationships, for school staffs as teams, for mentoring and teacher preparation programs, and for individual teachers personally. Then, 12 research-based recommendations for implementing Cognitive Coaching are presented. The chapter closes with suggestions for further research. Earlier studies are presented first, followed by subsequent studies.

SELF-DIRECTED LEARNING

One of the major goals of Cognitive Coaching is to create self-directed learners, and research evidence suggests that it achieves this goal. Garmston and Hyerle (1988) found that Cognitive Coaching increased growth in self-directed learning in university professors. Professors trained in Cognitive Coaching showed substantial improvement in using language more precisely in their teaching, as well as in expanding their teaching repertoires. In Edwards's (1993) study, first-year teachers who received Cognitive Coaching grew significantly on the conceptual level question "When I am told what to do . . . ." They moved from "black and white" thinking to having more "shades of gray" in their thinking and to thinking more on their own.

Cognitive Coaching has been associated with teachers expanding their repertoire of teaching strategies with their students. In one study, two middle-school English teachers who received Cognitive Coaching expanded their teaching repertoires, requested greater student accountability, exhibited more power as they planned lessons, and became more conscious of their behaviors and options as they worked with students (Garmston et al. 1993). They also began to use the coaching behaviors of paraphrasing, probing, and gathering data with their students as they internalized the coaching process.

Administrators trained in Cognitive Coaching were asked to reflect on an event in which they used their Cognitive Coaching skills to obtain an outcome in their administrative positions (Lipton 1993). They made conscious choices and used paraphrasing, pacing and leading, and elements of rapport. They kept the five states of mind in their awareness, identified goals for their own growth, related their positive results to specific coaching maps and tools, and reflected on their growth in coaching. They also identified the next steps for their own growth, recognized their own growth as well as the growth of the person they were coaching, reflected on the importance of coaching for their staff, and expressed appreciation for being asked to reflect in writing on the event.

Teachers have indicated that an advantage of being involved in Cognitive Coaching is having the opportunity to learn and grow (Edwards and Newton 1994a). Teachers who had taken Cognitive Coaching training mentioned the following sources of satisfaction with teaching as a profession more often than those who had not taken the training: learning and growth, the ever-changing nature of the profession, the opportunity to make a difference, the opportunity to be creative, and relationships with school staff. In another study, when principals commented on the portfolios of teachers who were using Cognitive Coaching to help them reach their goals, they emphasized their teachers' increased desire to learn and grow professionally (Edwards and Newton 1994c).

Awakuni (1995) found that teachers who were involved in a yearlong Cognitive Coaching program took on more leadership positions during that year, such as giving presentations to the faculty, increasing their involvement in state activities, and joining the school leadership team. In her study, teachers expanded their teaching practices over the course of a year, adopting such practices as team teaching, use of learning logs, peer tutoring, marketing strategies, student choices, multiple forms of assessment, exhibitions, interviews, and developing lessons based on learning styles. Both interview data and observations confirmed that they increased the use of inquiry methods, including asking more higher-level questions, varying their explanations and feedback to students, and involving students more in discussions. They also made changes in their classroom management, using more strategies to work with misbehavior, organizing the classroom to prevent behavior problems from occurring, motivating students with grades, and rethinking how curriculum was aligned. In addition, they adopted new strategies for teaching and assessing students, such as using cooperative learning, helping students to clarify their thinking in writing projects, and attempting to accommodate different learning styles through art, video, music, projects, and demonstrations.

Training in Cognitive Coaching has been associated with increases in the ability to solve problems. Teacher interns who received mentoring for a year by experienced teachers who had been trained in Cognitive Coaching wrote that they felt confident that they could solve problems, and they wrote evaluations of how they solved the various problems that they encountered (Burk, Ford, Guffy, and Mann 1996). They were able to draw their own conclusions rather than looking outside themselves for answers.

Edwards and Green (1997) found that teachers who had taken Cognitive Coaching training, when compared with a control group, spent more hours in workshops, both during school time and outside of school time. In addition, they had implemented more new teaching practices in the previous two years than teachers who had not taken the training. They also had more positive attitudes toward Cognitive Coaching and toward Professional Growth Planning, which was a method of self-evaluation that their district was using.

Poole (1987) and Poole and Okeafor (1989) reported that higher levels of efficacy, when coupled with more frequent task-focused interactions among teachers, predicted higher levels of curricular change. In a three-year project in which Cognitive Coaching was utilized to increase teachers' implementation of standards-based education, teachers who coached each other increased significantly in their reported levels of use of standards-based education than did teachers in a matched control group (Hull, Edwards, Rogers, and Swords 1998).

Coy (2004) investigated the yearlong Transition to Teaching Program, of which Cognitive Coaching was a key component. In her study, protégés who were receiving Cognitive Coaching moved from being concerned about surviving to being focused on "the success of students" (159). "The data provided instances from the various focus groups, case-profiles, and field notes of examples of self-directed behavioral changes both in terms of the mentor and protégés" (146). In "the final mentor focus group," mentors pointed out a "number of instances" of "the protégés' self-management, self-monitoring, and self-modification" (140). "Self-directed learning was evident for mentors and protégés in that they recognized their strengths and weaknesses and either sought solutions to their dilemmas or modified their behaviors as they taught and learned" (148).

Batt (2010) conducted a study in which 15 teacher leaders were trained in the Sheltered Instruction Observation Protocol (SIOP) for working with English language learners. After participating in the training, 80 percent of the teachers indicated that they were committed to implementing the model. In reality, only 53 percent of them implemented it. Then, they received Cognitive Coaching around implementing the model. After that, 100 percent of the teachers were fully implementing it, which was an indication of the teachers' self-directedness. This was also evident in the qualitative data. In addition, nearly all of the teachers said that students were benefiting from the SIOP model after they had received Cognitive Coaching.

THE FIVE STATES OF MIND

A goal of Cognitive Coaching is to help teachers increase their awareness of and skill with the five states of mind: efficacy, flexibility, consciousness, craftsmanship, and interdependence. A number of researchers have measured teacher growth in those areas. Ushijima (1996a) and Alseike (1997) developed instruments to measure the five states of mind. Ellison and Hayes (1998) developed the Energy Sources Team Self-Assessment Survey to measure the five states of mind in a group setting. González (2009) developed the "Relaciones con Dios y con el Prójimo" (Relationships with God and with Others) survey to measure the five states of mind related to the spiritual life. Studies that were focused on measuring all of the five states of mind are presented first, followed by studies on specific states of mind.

Studies on All Five States of Mind

Liebmann (1993) investigated how human resource developers from product and service organizations perceived the five states of mind. She found that they identified the states of mind of consciousness and interdependence, followed by flexibility, as critical attributes for all employees to have. They also placed a high value on holonomy for a democratic society.

Ushijima (1996b) studied teachers who had used Cognitive Coaching for a year and a half. They showed a 40 percent increase in efficacy, a 33 percent increase in flexibility, a 27 percent increase in consciousness, and a 37 percent increase in craftsmanship. Quantitative measures were supported by qualitative measures.

When teachers used the coaching process at least four times, they showed gains in the states of mind.

Alseike (1997) focused specifically on measuring the five states of mind in her study, and she investigated factors that could have been associated with teacher growth in them. She found that teachers who had received Cognitive Coaching from experienced coaches scored significantly higher than teachers who had not received Cognitive Coaching on measures of efficacy, flexibility, consciousness, interdependence, and overall holonomy. No significant differences were found between males and females who received Cognitive Coaching in their receptivity to coaching, nor in their states of mind of efficacy, flexibility, consciousness, interdependence, and overall holonomy. Males scored higher than females on craftsmanship.

In Alseike's (1997) study, no differences were found in the impact of Cognitive Coaching on the five states of mind between teachers who had been coached by the principal, a building resource teacher, another teacher, or a combination of the three. In addition, no differences were reported by number of years of teaching experience for the states of mind of flexibility, craftsmanship, consciousness, and interdependence. Teachers with 16 to 20 years of teaching experience had higher overall states of mind, and teachers with 6 to 20 years of experience had higher efficacy than did teachers with under six years of experience or teachers with 21 years or more of experience.

Ellison (2003) investigated the impact of weekly coaching sessions with a trained Cognitive Coach on 12 principals and four assistant principals over four months. They received six to eight hours of coaching in 10 to 13 sessions lasting between 20 and 60 minutes. They also responded to two reflective questions via e-mail after each coaching session. Data included the coach's observations and a 40-question survey designed to measure the five states of mind filled out by the administrator, three staff members, and his/her supervisor. The data from the principals, assistant principals, and their supervisors indicated that the participants had grown in the five states of mind. Participants grew the most in the states of mind of consciousness, craftsmanship, and flexibility. Each of the assistant principals told the Cognitive Coach "that this was one of the most valuable professional development experiences in which they had ever engaged" (23). Five of the participants requested coaching after the study had ended.

Slinger (2004) coached five first-grade teachers for nine weeks. As a result of being coached, the teachers reported that they were able to think on deeper levels and with greater clarity (consciousness), become more accountable for their actions (interdependence), solve problems (flexibility), feel the support of others (interdependence), feel as though they were challenged (efficacy), and target their teaching (craftsmanship).

González (2009) worked with priests, nuns, and lay people who were participating in a three-year course in Spiritual Accompaniment, as well as those earning their master's and doctoral degrees in spiritual theology or in Christian anthropology at a university in Rome, Italy. He developed the survey "Relaciones con Dios y con el Prójimo" (Relationships with God and with Others) to measure the five states of mind related to the spiritual life. In the factor analysis of the instrument, flexibility was contained in efficacy. In addition, the state of mind of interdependence contained two separate subscales—Collaboration and Altruism. In other words, one type of interdependence involved working *with* others (collaboration), and the other type involved working *for* others (altruism).

González (2009) used four groups in his study. Participants in Group 1, who received the full eight-day training in Cognitive Coaching and received coaching twice a month from October through May, grew the most in the five states of mind from pretest to posttest (71.8 points overall, with significant growth from pretest to posttest in consciousness, craftsmanship, efficacy, collaboration, and altruism). Group 2, who only received coaching twice a month, grew the next most (56.8 points, with significant growth in consciousness, collaboration, and altruism). Group 3, who only took the eight-day Cognitive Coaching training, grew the next most (40.2 points, with significant growth in consciousness and efficacy). Group 4, who did not receive coaching or Cognitive Coaching training, either regressed or stayed the same (−7.3 points, with no significant growth in any of the five states of mind).

Avant (2012) found that four effective instructional coaches demonstrated the five states of mind of flexibility, efficacy, consciousness, craftsmanship, and interdependence when talking about their practice, as well as elements of emotional intelligence (managing emotions, understanding emotions, perceiving emotions, and facilitating thought).

Henry (2012) compared the weekly reflective journals (13–15 weeks) of teacher candidates who had been exposed to Cognitive Coaching with the journals of teacher candidates who had not been exposed to Cognitive Coaching. She selected 18 journals from each group (every fifth journal), beginning with the journal entries from Week 4. She found that the teacher candidates who had experienced Cognitive Coaching "were able to learn more from their experiences and used more words indicating higher levels of the five states of mind in their reflective journals" (Abstract).

Rich (2013) compared the experiences of "two novice alternatively certified reading teachers positioned at two chronically underperforming and high poverty schools" (Abstract). One teacher received Cognitive Coaching from her, and the other teacher did not. She gathered data through interviews, observations, and four coaching cycles that included planning and reflecting conversations. She found that "Cognitive Coaching substantially influenced her ability to self-monitor and self-modify her teaching behaviors" (124). The teacher became more efficacious, as demonstrated by her "increased capacity to make informed instructional decisions resulting in improved learning outcomes for her students" (124). She showed an internal locus of control when she made errors. In addition, she became more flexible, as evidenced by her changing her instruction when she found out her students would learn more. She became more craftsmanlike, as evidenced by her continual quest to improve her teaching craft. She improved in asking questions and differentiating instruction. As a result, her students grew in their ability "to think critically and to engage in complex thought" (124). She also grew in consciousness as evidenced by changing the decisions she made, as well as her behaviors. Rich concluded that:

> all four states of mind played an important influence on Participant 01B's ability to make changes to her practice. However, if Participant 01B had not taken the time to reflect and to become conscious of her own teaching decisions and behaviors, she would not have been able to make the modifications that she made in the study. (168)

Thus, for Rich, consciousness appeared to be the state of mind that drove the other states of mind.

Studies on Efficacy

Perhaps the most frequently investigated area in Cognitive Coaching research has been teacher efficacy. Numerous researchers have found benefits for students that were associated with higher levels of efficacy in teachers. These included teacher use of fewer control tactics with students (Ashton, Webb, and Doda 1983); more time in large-group instruction (Gibson and Dembo 1984); student achievement in reading, language, and mathematics (Tracz and Gibson 1986); more parental involvement and support (Hoover-Dempsey, Bassler, and Brissie 1987); greater use of cooperative learning by teachers (Dutton 1990); less teacher anger for student misbehavior (Glenn 1993); more teacher enthusiasm and higher student grades at the middle-school level (Newman 1993); and fewer referrals of children from low socioeconomic status to special education (Podell and Soodak 1993). In a study by da Costa and Riordan (1996), teachers with higher levels of efficacy were more willing to have other teachers observe them. Betoret (2006) found that teachers with higher levels of efficacy and more resources to help them cope had less stress and burnout than teachers who had lower levels of efficacy and fewer resources. In addition, Caprara, Barbaranelli, Steca, and Malone (2006) found a relationship between teacher personal efficacy, job satisfaction, and student academic achievement using structural equation modeling.

Edwards and Newton (1994a, 1994b, 1995), Edwards and Green (1997), and Edwards, Green, Lyons, Rogers, and Swords (1998) used the Teacher Efficacy Scale, developed by Gibson and Dembo (1984), to investigate the relationship between Cognitive Coaching and teacher efficacy. Krpan (1997) and Smith (1997) used the Teacher Efficacy Scale developed by Guskey and Passaro (1993), while Alseike (1997) used a self-developed measure of efficacy. Baker (2008) used Bandura's

Teacher Self-Efficacy Scale (2006), and Maginnis (2009) used the Teacher Sense of Efficacy Scale (TSES) (Tschannen-Moran and Woolfolk Hoy 2001). Skytt, Hauserman, Rogers, and Johnson (2014) used the Principal Sense of Self-Efficacy Scale (Tschannen-Moran and Gareis 2004). Wooten Burnett (2015) used the Ohio State Teacher Efficacy Scale (OSTES) (Tschannen-Moran and Woolfolk Hoy 2001) and the Physical Education Teacher Efficacy Scale (PETES) (Humphries, Hebert, Daigle, and Martin 2012).

Edwards and Newton (1994a, 1995) found that teachers who had been trained in Cognitive Coaching showed higher levels of teaching efficacy as measured by the Teacher Efficacy Scale (Gibson and Dembo 1984) than did teachers in a matched control group. In addition, teachers who frequently used Cognitive Coaching scored significantly higher in teaching efficacy than did teachers who used it less often (Edwards and Newton 1994a, 1995). In a follow-up study in the same school district, teachers who were trained in Cognitive Coaching had grown significantly in teaching efficacy between 1993 and 1996 (Edwards and Green 1997). Growth in teaching efficacy was correlated with length of time in the district, more positive attitudes toward Cognitive Coaching, implementation of a larger number of teaching practices in the previous two years, and more positive attitudes toward Professional Growth Planning.

Alseike (1997) investigated whether teachers who were coached by experienced Cognitive Coaches would score higher on a measure of the state of mind of efficacy that she developed for the study. She found that they did score significantly higher on efficacy than did a matched control group. She also found that teachers who had attended seven or more trainings given by experienced Cognitive Coaching trainers scored significantly higher on the measure of efficacy than did teachers who had not attended any trainings or who had attended one or two trainings, even though they had received coaching from experienced coaches. In addition, teachers who received Cognitive Coaching from experienced coaches and had never coached another teacher scored significantly lower in efficacy than did teachers who had coached another teacher seven or more times.

Edwards and Green (1997) examined teacher growth in empowerment as a result of Cognitive Coaching. Teachers who participated in Cognitive Coaching scored higher than a control group on all subscales of the Vincenz Empowerment Scale (Vincenz 1990) and on total empowerment. Teachers who grew more in total empowerment on the Vincenz Empowerment Scale indicated that they informally coached their colleagues more frequently, built rapport with colleagues more frequently, paraphrased colleagues more often, asked questions of their colleagues more often, and used pacing and leading more frequently. In another study, higher levels of empowerment were also associated with more frequent coaching conversations (Edwards and Newton 1995).

Second-year, third-year, and fourth-year teachers who participated in a seven-month program of Cognitive Coaching grew significantly on the Teacher Efficacy Scale developed by Guskey and Passaro (1993), compared to a matched control group (Krpan 1997). They also grew significantly over the control group during the same period of time on written expressions of efficacy. In addition to scoring higher on the measure, teachers in the seven-month study grew significantly over a matched control group in their perceptions of their abilities to bring about purposeful change (Smith 1997).

Teachers who participated in a three-year project utilizing Cognitive Coaching to help them implement standards-based education grew significantly more in teaching efficacy on the Gibson and Dembo (1984) Teacher Efficacy Scale than did teachers in a matched control group (Edwards et al. 1998). Teachers with higher levels of teaching efficacy indicated that they used paraphrasing more frequently, asked questions more often, coached students and parents more, and generally used coaching skills more often than did teachers with lower levels of teaching efficacy.

Baker (2008) assessed the content knowledge related to Cognitive Coaching of 11 mentors and correlated it with the level of self-efficacy of 15 initially licensed teachers whom they were mentoring. She found that mentors who had a high level of content knowledge had mentees with high levels of self-efficacy as measured by Bandura's Teacher Self-Efficacy Scale (2006), and mentors who had low levels of content knowledge of Cognitive Coaching had mentees with lower self-efficacy.

Student teachers who received mentoring during a semester from clinical faculty who were trained in Cognitive Coaching grew more in teaching efficacy on the Teacher Sense of Efficacy Scale (TSES) (Tschannen-Moran and Woolfolk Hoy 2001) than did teachers who received mentoring from clinical teachers who were not trained in Cognitive Coaching (Maginnis 2009). Formative assessment and the language of support were most important in helping them grow in efficacy. Mutual trust, positive relationships, and formal feedback were also important in their growth.

Skytt et al. (2014) used the Principal Sense of Self-Efficacy Scale (Tschannen-Moran and Gareis 2004) to measure the efficacy of 15 experienced principals who coached 23 beginning principals. At the beginning of the study, the coaches' efficacy was significantly higher than the principals' efficacy on two of the three subscales (Instructional Leadership and Management). Their scores did not differ on the Moral Leadership scale. After two years, no differences were found between the efficacy of the coaches and the efficacy of the principals.

Wooten Burnett (2015) used the Ohio State Teacher Efficacy Scale (OSTES) (Tschannen-Moran and Woolfolk Hoy 2001) and the Physical Education Teaching Efficacy Scale (PETES) (Humphries et al. 2012) to measure the efficacy of seven physical education teacher candidates in a master's program as a result of receiving three cycles of Cognitive Coaching (three planning conversations and three reflecting conversations) over a period of six weeks. Seven teachers participated in the control group. She also collected data from semistructured interviews, an open-ended survey, and the three cycles of planning and reflecting conversations. She found that "Cognitive Coaching had a statistically significant impact on physical education teacher candidates' teacher efficacy measured by the PETES and OSTES" (121).

Studies on Flexibility

Researchers have also found indications that Cognitive Coaching was related to increases in educator flexibility. Garmston et al. (1993) conducted a study in which Garmston used Cognitive Coaching with Linder and Whitaker, who were teaching. As a result of the process, Linder and Whitaker became more flexible in their thinking and teaching styles as they began to use the side of their brains that they formerly had not used as much. They became more balanced in using analytic and intuitive styles.

When asked about their sources of satisfaction in their teaching positions, teachers trained in Cognitive Coaching were more likely to mention autonomy and flexibility, as well as the opportunity to contribute to students (Edwards and Newton 1994c). Thus, they valued the flexibility that comes with teaching. In another study, teachers reported that they increased in creativity and flexibility as a result of being involved with Cognitive Coaching (McLymont 2000; McLymont and da Costa 1998).

Dougherty (2000) found that teachers who participated in Cognitive Coaching training liked the training, learned how to use Cognitive Coaching, changed their behavior, and obtained results from using their new communication skills. These were all elements of Kirkpatrick's (1998) model for evaluating the effects of training.

When 64 experienced third- and fourth-grade teachers from low-performing schools who were in a master's-level program coached each other for four months focusing on using rubrics for teaching strategies and wrote reflections, they became more comfortable with peer assessment, more tolerant of the lack of structure that comes with working in groups at the college level, and more open to receiving feedback and ideas about their teaching (Fine and Kossack 2002). In addition, they developed more appreciation for Cognitive Coaching, used it in their classrooms, and shared it with others.

As Evans (2005) used Cognitive Coaching skills to coach two middle-school teachers in using research-based methods for teaching at the middle-school level, he moved from believing that he needed to tell teachers what to do to realizing the importance of listening to them. He also grew in his coaching skills the more he coached them. He believed Cognitive Coaching to be a way of improving instruction at the middle-school level. According to Evans, "I can report to a teacher what I have seen during a three-minute visit; however, I have learned from this study that reporting on what I have observed will not lead to improved instruction. The teachers want to engage in professional conversations that will help their performance" (122).

"When teachers engaged in coaching conversations with the instructional coach and other teachers, they had opportunities to create new mental models and attempt new strategies and techniques they might not have otherwise attempted without support" (Reed 2007, 230). "While teachers struggled to identify the process of Cognitive Coaching, they acknowledged changes in their instructional practice over time as they engaged in individual conversations with the instructional coach and met in cross-grade level groups with the instructional coach" (212).

Beltman (2009) found that teachers who participated in Cognitive Coaching training enjoyed the training, used the skills, grew in their understanding of coaching over the eight days of training, and refined their skills. In addition, they no longer felt like they needed to solve problems for others, and they continued practicing their skills. After eight days of training, teachers had a mean over 3.0 on a scale of 1 to 5, indicating that they felt that their training had made a difference, they consciously used the skills they had learned, and they were looking forward to more experiences in Cognitive Coaching. Wooten Burnett (2015) found that seven physical education teacher candidates in a master's program who received three planning conversations and three reflecting conversations over a period of six weeks became more flexible.

Studies on Consciousness

Teachers in numerous studies became more conscious of their teaching as they reflected on their classroom practices. In a peer coaching program in which professors received 42 hours of training in Cognitive Coaching, they showed a maximum improvement in their ability to analyze and evaluate themselves. They also grew in self-perception and the ability to autonomously perform cognitive activities (Garmston and Hyerle 1988).

In a study of first-year teachers who received Cognitive Coaching from trained coaches, higher numbers of formal and informal interactions with coaches were correlated with increased growth on the Reflective Pedagogical Thinking Instrument (RPT) (Simmons et al. 1989), a measure of reflective thinking about teaching (Edwards 1993). In addition, the first-year teachers who filled out more interaction sheets, which were their reflections on their coaching conferences, grew more on the RPT than did those who filled out fewer interaction sheets.

Graduate students in educational administration and leadership who used formative portfolio assessment, reflective practice, and Cognitive Coaching indicated that they deepened in their understanding, had opportunities to create meaning, and were able to engage in metacognitive analysis. They also said that their ability to link theory and practice was enhanced, that they had become more open to exploring complex problems because they were in a safe environment, and that they had redefined and reaffirmed themselves as developing leaders (Geltner 1993).

Edwards and Newton (1994c) analyzed the portfolios of teachers who were using Cognitive Coaching to help them reach their goals. In their comments on the portfolios, principals emphasized their teachers' growth in self-reflection. In a study by Clinard, Ariav, Beeson, Minor, and Dwyer (1995), teachers who used Cognitive Coaching with student teachers became more reflective as they reexamined their teaching methods and classroom management strategies, revisited strategies they had been using for many years, and brainstormed teaching ideas with their colleagues.

More training in Cognitive Coaching correlated with higher levels of consciousness in one study (Alseike 1997). Teachers who participated in seven or more training sessions scored significantly higher in the state of mind of consciousness than did teachers who had not attended any trainings or had attended one or two trainings, even though they had received coaching from experienced coaches.

Teachers who were supervising student teachers indicated that using Cognitive Coaching with their student teachers had impacted their own teaching (Clinard et al. 1997). In addition, they found themselves using nonjudgmental feedback with their students and reflecting more on their lessons.

In a study of second-, third-, and fourth-year teachers, teachers indicated that they grew in awareness of their teaching practices as a result of Cognitive Coaching (Krpan 1997). They also indicated that they had become more conscious of their teaching practices.

Over an eight-month period, special education interns in a master's program who met monthly to learn and practice Cognitive Coaching skills grew significantly in awareness, skill development, and application of those skills (McMahon 1997). At the end, they had also grown in their ability to think reflectively, to self-analyze and self-evaluate, and to apply the coaching skills in their teaching.

Second-year, third-year, and fourth-year teachers who participated in Cognitive Coaching over a seven-month period increased their ability to think reflectively (Smith 1997). They grew in awareness about their teaching practices, became more observant, and gained greater insights into their teaching. This self-reflection helped them to redefine their perceptions about their teaching roles. They also grew in their appreciation for the value of self-reflection.

Schlosser (1998) found that teachers who participated in Cognitive Coaching training and coached each other reported that Cognitive Coaching assisted them in thinking more precisely about their teaching. They hypothesized that they had become more reflective as a result. In another study, teachers using Cognitive Coaching reflected more deeply on their practice at the end of the project than at the beginning (McLymont 2000; McLymont and da Costa 1998).

In Uzat's (1999) study, teachers who had been trained in Cognitive Coaching scored above the midpoint on the subscales of analyzing and evaluating, and applying on the Teacher Thought Processes questionnaire (Foster 1989). This indicated that Cognitive Coaching had impacted those two areas of their thinking about instruction, suggesting that they had grown in higher-order thinking skills (Uzat 1999). In addition, teachers scored above the midpoint on thinking more while they were teaching, "using clear and precise language during teaching," evaluating their actions while they were planning their lessons and teaching them, thinking about the behaviors they used while they taught, and thinking about possible alternatives while they were teaching (Uzat 1999, 61).

Jewish day-school teachers who received three cycles of Cognitive Coaching over a period of seven months were compared with a group of teachers who received traditional evaluation, which included observations by administrators and letters containing suggestions after the observations (Moche 1999, 2000). They were also compared with a group of teachers who participated in informal discussions about their classroom instruction. Although all of the teachers grew on the Reflective Pedagogical Thinking Instrument (Simmons et al. 1989), teachers who received Cognitive Coaching grew significantly more than did the control groups.

During exit interviews after receiving Cognitive Coaching for nine weeks, first-grade teachers said positive things about Cognitive Coaching (Slinger 2004). They reported that the experience had impacted their thought processes. They spent more time reflecting about their practice at the end of the study, and they thought in different ways about their craftsmanship as teachers. In addition, they said that they felt supported and stronger as a result of having a coach, and they valued having the opportunity to be coached.

Robinson (2011) found that five teachers who received monthly Cognitive Coaching sessions from August to November and participated in a Community of Practice to support them in taking the National Board certification "became more self-reflective in their teaching practice throughout the semester of study" (34). The teachers indicated that "they more often analyzed why they teach, what they teach, and what the benefits on student learning might have been" (36). In addition, the teachers indicated "that they believed [the Cognitive Coaching conversations] helped them through the National Board process" (37). Teachers spent approximately 75 percent of their time in their Communities of Practice reflecting on their practice.

Loeschen (2012) conducted a phenomenological study of four mentor-teachers who had received four days of Cognitive Coaching training to assist their mentees. (Typically, the training lasts for eight days.) The teachers reported that Cognitive Coaching had influenced their own practice by supporting the way they reflected, helping them to experience cognitive shifts in the way they thought, and influencing the way they taught their own students. They increasingly asked themselves questions as they were teaching and afterward. They realized the need to reflect, and they connected their new thinking with what they already knew. They increasingly believed that asking themselves questions was important, and they used their questions to help them refine what they were doing with their students in their own classrooms.

Bjerken (2013) studied teachers at the elementary, middle, and high school levels in a school district to determine their thoughts about how receiving Cognitive Coaching for four years had impacted their teaching. They had participated in three coaching cycles per year for four years with a certified Cognitive Coach. They filled out surveys and participated in focus groups. They indicated that they had increased in reflection and had decreased in their sense of isolation. Rather than focusing on the faults in their past lessons, they were able to focus on the positive aspects of lessons they had taught.

Chang, Lee, and Wang (2014) compared 117 elementary and secondary teachers who had participated in three days of Cognitive Coaching training and used the skills in their schools for a year with 117 teachers in the comparison group. They found that the teachers who used Cognitive Coaching improved significantly more than the comparison group in their ability to reflect on their practice, both in the strategies that they used and the content of their reflections.

Jaede, Brosnan, Leigh, and Stroot (2014) examined the influence of Cognitive Coaching on 28 middle school and high school mentor-teachers in an urban setting. They found that the mentor-teachers increased in their ability to reflect on their practice, and they were able to assist their interns in reflecting on their practice.

Studies on Craftsmanship

Teaching is a craft, and another goal of Cognitive Coaching is to help teachers grow in their craftsmanship. As with the other states of mind, research evidence suggests that teachers do, in fact, grow in craftsmanship as a result of Cognitive Coaching. Bal and Demir (2011) developed the Cognitive Coaching Questionnaire (CCQ) to measure the growth of 180 preservice teachers in metacognitive skills as a result of receiving Cognitive Coaching. The teachers were studying in the Science and Technology Education Department at Çukurova University in Turkey. The instrument, which contains 34 items related to planning (five items), thinking (23 items), and evaluating (six items), had a reliability of .94.

According to Costa and Garmston (2002), teaching consists of planning, teaching, analyzing, evaluating, and applying. Foster (1989) conducted a study in which she examined the impact of Cognitive Coaching on teacher perceptions of their thought processes in those areas. She found that teachers who participated in seven or more coaching conferences perceived that Cognitive Coaching had a high level of impact on their thought processes in the areas of planning, teaching, analyzing, evaluating, and applying. Teachers who participated in four to six coaching conferences reported that Cognitive Coaching had an average impact on planning, teaching, and applying, and a high impact on analyzing and evaluating. Teachers who participated in one to three coaching conferences reported that Cognitive Coaching had an average impact on their thought processes in these areas, and teachers who participated in no coaching conferences reported that Cognitive Coaching had a low impact on their thought processes in these areas. No differences were found between elementary and secondary teachers in their perceptions of the impact that Cognitive Coaching had on their thought processes. Furthermore, no differences were found in the perceptions of the impact on thought processes between teachers who were coached by an administrator and teachers who were coached by another teacher.

As a result of being coached for a year, senior high teachers reported that the coaching facilitated the achievement of their goals because they changed the strategies they used in teaching their students (Sommers and Costa 1993). In other words, they grew in craftsmanship.

Teachers who used Cognitive Coaching more frequently obtained significantly higher scores on the Self-Reflection Survey: Cognitive Coaching Rating Scale (Schuman 1991) than did those who used it less (Edwards and Newton 1994b). Subscales included Planning, Teaching, Analyzing and Evaluating, and Applying. Edwards and Green (1999a) found that practice in using Cognitive Coaching skills is essential to bring about the growth that is possible through Cognitive Coaching.

Alseike (1997) built on Foster's (1989) study and examined the impact of Cognitive Coaching on teacher thought processes. Teachers who participated in seven or more Cognitive Coaching training sessions scored significantly higher on teaching, applying, and overall instructional processes than those who had not attended any sessions or had only attended one or two sessions. All of the teachers received Cognitive Coaching from experienced trainers.

Alseike (1997) also examined the effects of Cognitive Coaching in a variety of coaching situations. She found that teachers who had been coached by experienced coaches scored significantly higher on measures of planning, teaching, analyzing, and applying than did teachers who had not received Cognitive Coaching. No differences were found according to years of experience in how teachers reported that Cognitive Coaching impacted their instructional processes in the areas of planning, teaching, analyzing, applying, or overall instructional process. Furthermore, teachers who received Cognitive Coaching from experienced coaches yet had never coached another teacher scored significantly lower in teaching and applying than did teachers who had coached another teacher seven or more times.

Whether teachers were coached formally—including planning conversations, classroom observations, and reflecting conversations—or informally, they identified Cognitive Coaching as having a positive impact on their teaching (Alseike 1997). No significant differences were found between males and females who received Cognitive Coaching in their report of its impact on their planning, analyzing, and applying. Females scored higher than males on teaching. No differences were found in the impact of Cognitive Coaching on the instructional processes of planning, teaching, analyzing, and applying between teachers who had been coached by the principal, a building resource teacher, another teacher, or a combination of the three.

In their reflection logs, first-grade teachers who were coached during a nine-week period of time reflected continually about their teaching processes (Slinger 2004). This reflection resulted in the teachers changing their lesson plans and refining their instruction based on the data. "Conversation topics shifted from relating that students 'had needs' to specifically identifying those needs and discussing what could be done about it" (164). The teachers felt affirmed from reflecting and became more conscious about how they delivered instruction in the classroom. In addition, the teachers reflected between the times when they were coached. They reported that being coached impacted their reading instruction "in the following positive ways: (a) instruction became more focused, (b) more thoughtful planning occurred, (c) teachers increased their craftsmanship in particular areas of instruction, and (d) the status quo was questioned" (Slinger 2004, 153).

Evans (2005) coached two middle-school teachers on their implementation of research-based recommended practices for teaching middle-school students. During the course of five coaching cycles (including five planning conversations, five observations, and five reflecting conversations), the teachers moved from describing their lessons and reflecting on their lessons in general terms to being more specific about their lessons and reflecting in more depth.

Eger (2006) found that for veteran teachers who had received training in Cognitive Coaching who were mentoring second- or third-year teachers using Cognitive Coaching and had chosen to be supervised with the Cognitive Coaching model, "Cognitive coaching has at least 'some' to a 'great' extent of impact on the four phases of teacher thinking, with the greatest impact occurring at the evaluation and analysis phase" (50). When Eger (2006) gathered data to determine the effects of Cognitive Coaching on the four phases of teacher thinking, she found

> a significant difference between those [veteran] teachers that had three or more courses in cognitive coaching to those only having one course in cognitive coaching. Teachers with three or more courses in cognitive coaching had a greater ability when it came to clear and precise language, lesson planning, and evaluation of lessons. These same teachers reported that cognitive coaching helped them to monitor their progress when it came to implementing their lessons more so than teachers who had only taken one course in cognitive coaching. (54)

In Eger's (2006) study, veteran teachers "discussed how cognitive coaching's reflective practice resulted in higher levels of thinking and more critical analysis of goals, lesson, plans, and teaching behaviors, as well as evaluation of their own teaching and student performance" (67). In addition, second- and third-year teachers who were coached by mentor-teachers who had been trained in Cognitive Coaching reported that Cognitive Coaching had the most impact on supporting their emotional needs and helping them think about their behaviors in teaching students.

Fifteen teachers who had taken instruction in the Sheltered Instruction Observation Protocol (SIOP) received Cognitive Coaching as they were implementing what they had learned with English-language learners (Batt 2010). The teachers in the study believed that they had developed higher expectations for their culturally diverse students as a result of participating in Cognitive Coaching.

Robinson (2011) found that five teachers who received monthly Cognitive Coaching sessions from August to November and participated in a Community of Practice to support them in taking the National Board certification improved significantly in "perceived knowledge of their students" (31). They "were more aware of setting high, meaningful instructional goals based on their knowledge of students and their needs" (31). "They became even more consistently aware of who their students were and even more attentive to how they might meet their students' individual needs" (41). As a result, "they felt better prepared to meet those student needs" (42). In their Cognitive Coaching conversations, they "discussed students and their learning needs approximately 90% of the time" (43). In addition, participants spent approximately 75 percent of their time in their Communities of Practice "discussing [their] students and their learning" (44).

Robinson (2011) also found that "after the intervention, the participants were more likely to agree that their assessment provided evidence of student learning" (32). The intervention "provided structured opportunities for the participants to analyze and discuss their students' instructional needs, how they approached these needs through instructional strategies and the impact on student learning" (32). "They were more aware of their students' prior knowledge and understood students' learning needs at a deeper level" (45). During their Cognitive Coaching conversations, they "discussed their grade level curriculum and their instructional approaches approximately 80% of the time" (47). The teachers also "began to align their teaching practice to the NBPTS standards" (49).

Lin (2012) studied the growth of 28 mathematics teachers in their instructional conversations over a year as a result of receiving Cognitive Coaching. She used multilevel modeling to analyze the data. Two coders analyzed the teachers' lesson plans for eight sessions. Additional data included the teachers' reflection journals, feedback from the coaches, and the observation notes from the coaches. Lin also conducted interviews with the coaches to clarify questions she had while analyzing the data. Lin found that the coached teachers grew consistently as a result of receiving Cognitive Coaching, noting that they had low scores on instructional planning prior to receiving Cognitive Coaching. Teachers grew the most after the first coaching conversation, the next most after the second conversation, and the next most after the third conversation. Teachers who taught at the elementary level grew more quickly than secondary teachers in their instructional conversations.

Bjerken (2013) studied teachers at the elementary, middle, and high school levels in a school district to determine their thoughts about how receiving Cognitive Coaching for four years had impacted their teaching. They had participated in three coaching cycles per year for four years with a certified Cognitive Coach. They filled out surveys and participated in focus groups. Participants focused more on the details of the lessons when they were being observed and coached than when they were not being coached. Teachers also focused more on planning for specific groups of students, as well as individuals. They became aware of how students were engaged while they were teaching, how they were interacting with their students, and how the students were learning. They expressed the desire for more specific feedback and ideas for improving their lessons. They wanted the coach to have taught their grade level and subject. Some participants indicated that they were more able to know when students were achieving, and they were better able to measure student achievement. They believed that student learning had increased for the unit they were teaching and could not say whether student achievement in general had increased.

Diaz (2013) conducted a study in which teachers (grades 2–5) who were not associated with the National Board Certification process had the opportunity to experience the elements of the process. She compared three groups—Group One: those who received elements of the National Board Certification process, including Cognitive Coaching; Group Two: those who participated in the National Board Certification process; and Group Three: a control group. Each group included four teachers. Those in Group One received 8–10 coaching sessions in four months. As a result of reflecting, collaborating with one another, and working with a coach, the teachers in Group One grew in their ability to understand content knowledge and the needs of their students. They indicated that they were reflecting more, and that they were more focused on meeting their students' needs. They refined what they did as a result of reflecting on their practice. They talked about what they believed as teachers, and they explored the positive dispositions that they held as professionals.

Chang et al. (2014) compared 117 elementary and secondary teachers who had participated in three days of Cognitive Coaching training and used the skills in their schools for a year with 117 teachers in the comparison group. They found that the teachers who used Cognitive Coaching improved significantly more than the comparison group in their teaching effectiveness in the areas of communicating, grasping teaching strategies, and applying teaching strategies dynamically.

Donahue-Barrett (2014) investigated the effects of Cognitive Coaching with six elementary teachers in a writing workshop. She found that with four to six cycles of coaching, the teachers increased in their knowledge of writing instruction, as well as in their instructional practices. She also found that the planning conferences, demonstration lessons, and co-teaching lessons were most effective in helping to increase the teachers' knowledge and instructional practices. The teachers said that they would have preferred to have had more time with their coach.

Jaede, Brosnan, Leigh, and Stroot (2014) examined the influence of Cognitive Coaching on 28 middle school and high school mentor-teachers in an urban setting.

Their "thinking and talking about teaching and learning moved from the generic question of 'How do students learn?' to the fine-grained and nuanced questions, 'How will I ensure these students, in this classroom, in this school, in this community learn?' and 'How does who I am in the context of my classroom impact learning?'" (27).

Wooten Burnett (2015) studied seven physical education teacher candidates in a master's program who received three cycles of Cognitive Coaching (three planning conversations and three reflecting conversations) over a period of six weeks. Seven teachers participated in the control group. She also collected data from semistructured interviews, an open-ended survey, and the three cycles of planning and reflecting conversations. She found from the qualitative data that "the planning conversation helped them become more aware of student needs" (96). The participants in the treatment group indicated that "the planning conversation helped them (a) think critically—category 1, (b) set realistic expectations—category 2, and (c) be more flexible—category 3" (97). Finally, the participants indicated that they had "become more aware of their lesson planning development" as a result of participating in the planning conversations (98). Data from the open-ended surveys indicated that the teachers in the treatment group had grown in their ability to plan lessons. They planned lessons to help students be successful.

Studies on Interdependence

Another goal of Cognitive Coaching is to assist educators in growing in the state of mind of interdependence (Costa and Garmston 2002). In a study by McDonough (1991, 1992), principals identified their collaboration with supervisors on work goals, frequent interaction with and observation by supervisors, responsive practices, and trusting relationships as contributing to their growth. Sommers and Costa (1993) found that as a result of being coached for a year, senior high teachers talked more with their colleagues about their teaching and ceased being concerned about the extra time involved in coaching.

Teachers involved in a yearlong Cognitive Coaching program indicated that they were satisfied with their positions because of the support they gave to one another (Awakuni 1995). In a study by Alseike (1997), teachers who were coached formally—including the planning conversation, observation, and reflecting conversation—scored significantly higher on a measure of interdependence. She also found that teachers who were coached by experienced Cognitive Coaches scored significantly higher on interdependence than did a matched control group of teachers who had never experienced Cognitive Coaching.

Teachers who participated in Cognitive Coaching scored significantly higher than a control group on the Relatedness subscale of the Vincenz Empowerment Scale (1990), which was equivalent to the state of mind of interdependence (Edwards and Green 1997).

In Eger's (2006) study, which was done at the secondary level, "there was a strong conviction that cognitive coaching was responsible for developing deeper and stronger relationships with their peers, as well as with their students" (57). Teachers said that Cognitive Coaching created more "collaboration, more conversations, and improved relationships more so now than in the past" (58). They reported that they were able to listen more effectively, "become more patient with their colleagues and students" (60), and improved in their relationships with others. In addition, Cognitive Coaching "increased teachers' appreciation and awareness of what other teachers did" (60). They enjoyed having coaching partners who were in different departments.

Robinson (2011) found that five teachers who received monthly Cognitive Coaching sessions from August to November and participated in a Community of Practice to support them in taking the National Board Certification improved significantly in the way they "viewed themselves as a community of practice" (31). Robinson suggested "that the intervention may have had a positive impact as it provided structured time devoted to supporting the participants' needs as they navigated through the board certification process" (32).

Diaz (2013) conducted a study in which teachers (grades 2–5) who were not associated with the National Board Certification process had the opportunity to experience the elements of the process. She compared three groups—Group One: those who received elements of the National Board Certification process, including Cognitive Coaching; Group Two: those who participated in the National Board Certification process; and Group Three: a control group. Each group included four teachers. Those in Group One received 8–10 coaching sessions in four months. They talked about the ways in which they collaborated with parents and colleagues. They enjoyed having time to collaborate with their coach, as well as to watch videotapes of themselves and discuss what they were seeing with a coach. Teachers in Group One saw value in being coached.

THE IMPACT OF COGNITIVE COACHING ON STUDENTS

The goal of every school initiative is to benefit students, and research evidence exists to suggest that Cognitive Coaching achieves this goal as well. Researchers have investigated the effects of Cognitive Coaching on student learning and achievement, teachers' encouragement of student thinking, and other benefits to students.

Increases in Student Learning and Achievement

Researchers have investigated the impact of Cognitive Coaching on student learning and achievement. In a three-year project utilizing Cognitive Coaching, monthly dialogue groups, and nonverbal classroom management (Grinder 1996) to assist teachers in implementing standards-based education, differences in changes for the treatment and control schools were found on the Iowa Test of Basic Skills (ITBS) for the Total ITBS Score and the Integrated Writing Total Score between year 1 and year 3, for Math Advanced Skills and Integrated Writing Advanced Skills between years 1 and 2 and between years 1 and 3, and for the Math Total Score between years 1 and 2. No marked differences were found for Reading Advanced Skills or for the Reading Total Score. Although the scores for control schools improved over time, the improvements found for treatment schools exceeded the changes for control schools (Hull et al. 1998).

When nine experienced third- and fourth-grade teachers from low-performing schools who were in a master's-level program coached each other for four months focusing on using rubrics for teaching strategies and wrote reflections, their students grew significantly over a comparison group on the Degrees of Reading Power (DRP) reading comprehension test (Fine and Kossack 2002).

Kindergarten students' test scores on the Gates-MacGinitie Reading Test, Level BR, were compared for teachers who received Cognitive Coaching for a year after participating in a two-week lecture course in implementing a balanced literacy program, teachers who only attended the two-week course, and teachers who received no training (Rennick 2002). Student test scores for the Cognitive Coaching group were significantly higher, and students of teachers in the group receiving no training scored significantly higher than students of teachers in the group that received the two-week lecture course.

The lower-achieving kindergarten students' test scores on An Observation Survey (Clay 1993) were compared for teachers who received Cognitive Coaching for a year after participating in a two-week lecture course in implementing a balanced literacy program, teachers who only attended the two-week course, and teachers who received no training (Rennick 2002). Student test scores for the Cognitive Coaching group were significantly higher on the subtests of word identification, and hearing and recording sounds in words. No significant differences were found for the subtests of letter identification and text reading. When the four subtests of An Observation Survey (Clay 1993) were correlated for lower-achieving students, the correlations for the teachers who received Cognitive Coaching for a year after participating in a two-week lecture course in implementing a balanced literacy program were greater than the correlations for teachers who only attended the two-week course and teachers who received no training (Rennick 2002).

Students of five first-grade teachers who received Cognitive Coaching in a nine-week study grew in guided reading level, dictation, and word identification more than students in a control group, although not significantly (Slinger 2004). Small sample size, length of the study (only nine weeks), the control group consisting of students from higher socioeconomic status homes, the fact that the control group teachers had more experience and the students began at a higher level, and the fact that not all student progress may have been measured by the instruments that were used may have influenced the outcome. In addition, one treatment group teacher had significantly lower mean gains for students in the low group than the other teachers. This could possibly be because of having higher numbers or English as a second language (ESL) and special education students and other factors.

When 14 eighth-grade students were taught to use Cognitive Coaching to coach each other in writing using the planning conversation and employing pausing, paraphrasing, and probing, all except one student grew two stanines on the International Educational Research Bureau (ERB) Writing Assessment Program (WrAP) (Powell and Kusuma-Powell 2007).

Third-grade students' test scores improved over a three-year period as a result of teachers receiving the Read to Achieve Grant and receiving Cognitive Coaching from an instructional coach as they were implementing the grant (Reed 2007). "Teachers attributed the success of the third grade students to the collaboration of the second and third grade team and the instructional coach" (232).

Diaz (2013) conducted a study in which teachers (grades 2–5) who were not associated with the National Board Certification process had the opportunity to experience the elements of the process. She compared three groups—Group One: those who received elements of the National Board Certification process, including Cognitive Coaching; Group Two: those who participated in the National Board Certification process; and Group Three: a control group. Each group included four teachers. Those in Group One received 8–10 coaching sessions in four months. Diaz found increases in student achievement on the district benchmark reading assessment in Group One. Student achievement in teachers' classes in Group One improved more than in Group Two and Group Three.

Rinaldi (2013) asked third-, fourth-, and fifth-grade teachers who had received Cognitive Coaching training to teach their students about the five states of mind and invite them individually to develop a Question Bank for each of the states of mind. They asked themselves and each other the questions to help them solve problems in mathematics. She found that when the students used the Question Banks for 10 weeks, they grew significantly from pretest to posttest on their overall scores on the Mathematics Constructed Response Rubric (Los Angeles Unified School District, n.d.). Third-grade students grew significantly on both of the subscales, fourth-grade students grew significantly on four of the five subscales, and fifth-grade students grew significantly on three of the four subscales.

Rinaldi (2013) also found that when the students developed and used Question Banks based on the five states of mind for 10 weeks, they grew significantly on all five of the subscales of the five states of mind scale (Consciousness, Craftsmanship, Interdependence, Flexibility, and Efficacy) (Ushijima 1996a) in a repeated measures analysis of variance. In addition, Rinaldi (2013) found that the third-grade students grew significantly from pretest to posttest on the Implicit Theories of Intelligence for Children—Self-Form (Dweck 1976) in a paired-samples *t* test.

Teacher Encouragement of Student Thinking

Cognitive Coaching has also resulted in teachers encouraging student thinking. As a result of being coached for a year, senior high teachers taught more thinking skills to their students and modeled the skills of coaching (Sommers and Costa 1993). In their comments on the portfolios of teachers who were using Cognitive Coaching to help them reach their goals, principals emphasized their teachers' increased used of higher-level questions with students and their growth in communicating with and working with students (Edwards and Newton 1994c).

In another study, high school sophomores viewed 10- to 15-minute videotapes of their teacher being Cognitively Coached twice a week for six weeks, and the control group of students viewed videotapes of the teacher reporting the results of the coaching conversation for the same period of time (Muchlinski 1995). Those who viewed the tapes of the Cognitive Coaching conversations increased in the use of verification behaviors in problem solving. The differences approached significance.

Students of teachers who were trained in Cognitive Coaching were assessed in their question-asking skills for quantity and quality (Ushijima 1996b). In one semester, 78 percent of the students increased the quantity of questions asked, 74 percent of the students improved the quality of the questions they asked, and 65 percent of the students reduced the number of irrelevant responses they gave. In one year, 85 percent of the students increased the quantity of questions asked, 91 percent of the students improved the quality of the questions they asked, and 46 percent of the students reduced the number of irrelevant responses they gave. Students of teachers who participated in more coaching cycles gained more than did students of teachers who participated in fewer coaching cycles.

In that same study, students of teachers who were trained in Cognitive Coaching were also assessed in their math problem-solving skills (Ushijima 1996b). In one year, 86 percent of the students showed gains in their math problem-solving skills. Students of teachers who participated in more coaching gained more than did students of teachers who participated in less coaching.

The use of Cognitive Coaching was associated with teacher use of more higher-order thinking skills in a study by Hull et al. (1998). Teachers trained in Cognitive Coaching grew significantly on the Encouragement of Higher Order Thinking Skills subscale of the Teacher Survey (McCombs 1995) over a matched control group. In another study (McLymont 2000; McLymont and da Costa 1998), teachers trained in Cognitive Coaching found themselves consciously creating an atmosphere of trust and nonjudgmentalness in their classrooms, seeking to help their students arrive at decisions on their own.

In a study by Eger (2006), veteran teachers at the high school level reported that "Cognitive coaching had a significant impact on transforming their teaching" (61). They said, "they learned new strategies that created a calmer classroom, more discussions and openness between teachers and students, and a greater sense of ownership in teachers and students solving their own problems" (61).

Irons (2014) explored the impact of training in asking mediative questions (Costa and Garmston 2002) and coaching with a trained Cognitive Coach for 10 weeks on the questions that one middle school teacher and five mentors asked their students. Sessions were recorded and transcribed for a total of 173 minutes. Participants filled out questionnaires at the beginning of the study and at the end. Irons found that the participants valued receiving coaching on how they were applying what they learned in the seminar; moved from asking closed-ended questions to asking open-ended questions; developed the language pattern of using the word *space*; talked about how they had learned and grown as a result of the seminar; and indicated that their students were learning in the same areas that they were learning. In addition, they had implemented the elements of mediative questions into their practice.

Other Benefits for Students

Cognitive Coaching has also been associated with other benefits for students. Cognitive Coaching was used as part of a program to restructure a Title I program by having teachers teach collaboratively rather than pulling the students out of class (Hagopian, Williams, Carrillo, and Hoover 1996). Teachers received training in how to manage instruction, Cognitive Coaching, pedagogy, and assessment. After two years, teachers showed significant improvements in their perceptions that the model impacted teaching and student learning.

In the study by Hull et al. (1998), teachers who used Cognitive Coaching decreased significantly in referring students to special education, compared with teachers in a matched control group. This finding was substantiated in the focus groups when teachers said, "In the past when I had a student who was having difficulty, I would have referred him or her to special education right away. Now I ask a colleague to come in and coach me on how to make the student successful. As a result, we are able to make him or her successful in the regular classroom."

Veteran high school teachers in Eger's (2006) study said that their participation in Cognitive Coaching as a method of supervision changed their teaching styles, which "had an impact on the classroom environment" (62). They perceived their classrooms as being "a lot friendlier" (62) and believed that "they . . . were less reactive to situations and student behaviors" (62). They felt that they were more open with students "during class discussions" (62). In their students, they saw "more thinking about answers and better connection of ideas and concepts" (62). They observed that, "students are opening up and sharing things in discussions that did not occur before" (62). In addition, teachers felt "a greater sense of calm working with colleagues and students" (64). Participants perceived that Cognitive Coaching "influenced student behaviors, their thinking, and the climate of the classroom thus impacting student achievement" (64).

Eger (2006) found that when second- and third-year teachers received Cognitive Coaching from their mentors, they "perceived that improving their instruction, providing a variety of lessons, and better classroom management resulted in greater student participation, more involved learning, . . . ultimately having a greater impact on student achievement" (92).

In Powell and Kusuma-Powell's (2007) study in which 14 students coached each other in their writing, the researchers perceived that the students increased in self-confidence in their writing and thinking abilities. In addition, the students in the study reported that they were applying the nonverbal and verbal skills that they learned in their personal lives.

Fifteen teachers who had taken instruction in Sheltered Instruction Observation Protocol (SIOP) received Cognitive Coaching as they were implementing what they had learned with English-language learners (Batt 2010). Before receiving coaching, 53 percent of the teachers had implemented the model "to a great extent," while after coaching, 100 percent had implemented it "to a great extent" (1003). "Virtually all teachers reported positive effects on student learning as a result of SIOP implementation after coaching. . . . The self-reports corroborated with data obtained from observation ratings on the valid and highly reliable SIOP instrument (Echevarría et al. 2008, 1003). "This study suggests that coaching has a direct and significant effect on teachers' instruction. The cognitive coaching phase heightened implementation substantially, even after teachers had already had an extended period of long-term ongoing training and peer collaboration in each school" (1004). In addition, "all teachers reported gains in students' achievement as a result of heightened SIOP implementation following the coaching phrase, which both rewarded teachers for employing the framework and further motivated them" (1005).

THE IMPACT OF COGNITIVE COACHING ON
TEACHERS AND PRINCIPALS PROFESSIONALLY

Researchers have found that Cognitive Coaching affects individual teachers professionally. In a peer coaching program for professors in which they received 42 hours of training in Cognitive Coaching, the professors showed a maximum improvement in their ability to analyze and evaluate themselves (Garmston and Hyerle 1988). They also grew in self-perception and the ability to autonomously perform cognitive activities. In addition, they developed increased confidence in themselves and greater enthusiasm for teaching.

In a study by Edwards and Newton (1994a), teachers who had been trained in Cognitive Coaching said that some of the reasons they liked it were that it was respectful of teachers, it was nonjudgmental, and it made sense. In other studies, teachers who engaged in Cognitive Coaching were significantly more satisfied with teaching as a profession than were teachers in a matched control group who had not taken Cognitive Coaching training (Edwards and Newton 1994a, 1994b, 1994c).

Teachers were more positive in their comments about teaching and expressed more positive feelings about all aspects of their experiences as teachers than did teachers who did not take Cognitive Coaching training (Edwards and Newton 1995). Those trained in Cognitive Coaching listed 16 sources of dissatisfaction with their positions, and those who had not taken the training listed 57 sources of dissatisfaction.

Clinard et al. (1995) found that teachers who used Cognitive Coaching in working with student teachers reported "renewed enjoyment and enthusiasm about teaching in the classroom" (21). In addition, they became more motivated to remain in the field of education. They became more aware of areas in which they needed to grow professionally, improved in their ability to communicate with others, felt more confident in facilitating the thinking of others, and became more respectful of others. In a study by Clinard et al. (1997), supervising teachers indicated that by supervising their student teachers using Cognitive Coaching, they increased in their sense of professionalism.

In another study, second-, third-, and fourth-year elementary school teachers grew significantly in seven months over a matched control group in expressions of how they had developed professionally (Krpan 1997). Those who participated in Cognitive Coaching indicated that they had grown in awareness of their teaching practices as a result of coaching. They indicated that they had numerous opportunities to grow and change professionally.

Teachers grew significantly on Saphier's (1989) Teacher Professionalism and Goal-Setting subscale of the School Culture Survey over a matched control group after using Cognitive Coaching for a year (Edwards et al. 1998). Teachers who had higher scores on Teacher Professionalism and Goal Setting built rapport with others more frequently and participated in more coaching cycles. In addition, they increased in satisfaction with their positions and with their choice of teaching as a profession, compared to teachers in a matched control group (Edwards et al. 1998).

Edwards (2004) conducted interviews with teachers who were passionate about Cognitive Coaching. She found that those teachers were interested in "Becoming" in their lives. She identified five steps in the process: "Beginning the Journey, Learning for Becoming, Gathering Colleagues on the Path, Re-Identifying, and Continuing the Journey" (71).

In a study by Reed (2007), teachers experienced "a renewed sense of professionalism" (231) as a result of being coached by an instructional coach, which brought about "a more professional attitude," as well as "teachers' willingness to change their educational practice" (231).

Knaebel (2008) observed a literacy coach who was trained in Cognitive Coaching in a Reading First School in Indiana three to four times per week for four months in a qualitative single-case study. She also conducted informal interviews with her. Knaebel found that the coach possessed the following characteristics: credentials, knowledge of the research base, knowledge of what she was responsible to do, an in-depth understanding of the context in which she worked, and the ability to access the resources in her teachers. The competencies that she possessed included the ability to build trust and rapport, collaborate with her teachers, serve as a model, coach her teachers, conduct regular classroom visits, provide professional development to her teachers, and be organized.

Skytt et al. (2014) examined the self-reported competencies of 15 experienced principals who coached 23 beginning principals for two years. Both the coaches and the beginning principals were more competent and had acquired the behaviors fostering the competencies on the seven indicators on the Alberta Professional Practice Competencies for School Leaders (Government of Alberta 2012) by the end of the project.

Wooten Burnett (2015) studied seven physical education teacher candidates in a master's program who received three cycles of Cognitive Coaching (three planning conversations and three reflecting conversations) over a period of six weeks. They indicated that Cognitive Coaching helped them to grow professionally.

THE IMPACT OF COGNITIVE COACHING ON STAFF AS A TEAM

According to several studies, as faculty implemented Cognitive Coaching in their schools, the teamwork of the staff was enhanced and school cultures became more collaborative. In one study, teachers reported that, as a result of Cognitive Coaching training and implementation, they noticed more of a sense of community in the school, they talked more with each other about teaching, and the atmosphere in the school was more positive (Edwards and Newton 1994a). Teachers also indicated that they had more rapport with each other, they seemed more open to growth and new ideas, and they tended to evaluate themselves more frequently. They reported that positive influences on their attitudes toward Cognitive Coaching included the fact that it positively influenced their school's culture and that they were able to work together in coaching triads to support each other.

Ushijima (1996b) conducted a yearlong study of teachers who were using Cognitive Coaching. She found that team teaching improved, and teachers had more of a sense of community. Teachers trusted each other more, had greater resiliency, shared ideas more, felt more comfortable with taking risks, solved problems together, were more accepting of differences because of mutual respect, communicated more across grade levels, and gave and received more support and feedback in their work with children. In a study by Clinard et al. (1997), teachers who supervised student teachers using Cognitive Coaching indicated that they benefited because they were able to network with other educators.

Edwards et al. (1998) used the School Culture Survey (Saphier 1989) in a three-year study. Teachers trained in Cognitive Coaching grew significantly on the Administrator Professional Treatment of Teachers subscale over a matched control group. Teachers who felt that their administrators treated them more professionally also participated in more coaching cycles. In addition, teachers grew significantly on Saphier's Collaboration subscale of the School Culture Survey over a matched control group. Teachers who had higher scores in collaboration built rapport more frequently, participated in more coaching cycles, and believed that they had changed more as a result of their use of Cognitive Coaching. In another study (McLymont 2000; McLymont and da Costa 1998), teachers using Cognitive Coaching to improve their teaching of mathematics developed a collaborative coaching community as they worked together to discover new insights about their teaching.

Teachers who had taken training in Cognitive Coaching reported that the training reduced their sense of isolation and helped them grow in trust (Dougherty 2000). They also felt more of a sense of collegiality with other teachers in their school.

When school district leaders wove Cognitive Coaching into all levels of the district, veteran teachers pointed out "the improvement and incorporation of cognitive coaching in the culture and climate at various levels of the institution, whether it is the overarching district goals, changes in the classroom environments, or relationships among peers and students" (Eger 2006, 57). The teachers "used terms like 'it is the way we do business here,' 'it is a culture of continuous improvement,' and 'the district is a more thoughtful district, more reflective'" (57).

Skytt et al. (2014) used the School Climate Index for principals (DiPaola and Tschannen-Moran 2005) to measure the perceptions of school climate of 15 experienced principals who coached 23 beginning principals for two years. Both coaches and principals increased on all four subscales. The coaches increased more on the Academic Press subscale than the principals. School climate increased in both coaches' and principals' schools. Teacher Professionalism increased more in the principals' schools, followed by Collegial Leadership, Community Engagement, and Academic Press.

COGNITIVE COACHING AND SUPERVISORY RELATIONSHIPS

Researchers have also explored the use of Cognitive Coaching in supervisory relationships. First-year teachers who received Cognitive Coaching from trained coaches expressed more satisfaction with the supervision they received than did first-year teachers who received traditional supervision (Edwards 1993). In another study, Mackie (1998) compared teachers who used the Cognitive Coaching format with teachers who received traditional supervision. Those who used Cognitive Coaching rated the overall quality of the observation process significantly higher. They also indicated that the coaching process had more impact on their teaching practices, as well as on their attitudes toward teaching.

Tarnasky (2000) found that community college department chairs tended to use elements of Cognitive Coaching and laissez-faire supervision when they supervised both full-time and part-time faculty because they tended to relate in more of a collegial manner with those whom they supervised, although they were not aware of the formal strategies.

Hauserman, Edwards, and Mastel (2013) conducted four focus groups with 19 teachers at a school in which the teachers were asked to compare their impressions of walk-throughs with Cognitive Coaching. The teachers had experienced walk-throughs by their administrators for a year, followed by Cognitive Coaching from their administrators for a year. The first theme of the study was "Reflection as a Catalyst for Change" (14). The teachers indicated that the walk-throughs did not assist them in growing professionally, while Cognitive Coaching allowed them to reflect on their practice and gain insights. The second theme was "Support and Trust" (16). The teachers felt that their administrators supported them in the Cognitive Coaching process, and they trusted their administrators. The third theme was "Openness and Collaboration" (17). The teachers felt less isolated with Cognitive Coaching, and they were more willing to share their ideas with administrators and with other teachers. The teachers felt encouraged to grow professionally, and they increased in their abilities to reflect. In short, the teachers preferred Cognitive Coaching over walk-throughs.

COGNITIVE COACHING IN MENTORING
AND TEACHER PREPARATION PROGRAMS

Cognitive Coaching has been examined in mentoring and teacher preparation programs. In one study, teachers in a graduate program of educational administration and leadership focusing on collaboration were coached each semester on their formative portfolios (Geltner 1994). In addition, they wrote reflectively on each document or artifact. They reported that Cognitive Coaching was one of the most powerful aspects of the program. They said that when they were in difficult situations later in their careers, they found themselves thinking about their coaching experiences and realizing that they had the resources they needed within themselves to be successful.

Student teachers who had received 10 hours of training in Cognitive Coaching along with their supervising teachers reported that Cognitive Coaching provided them with a greater understanding of why teaching occurs the way it does, facilitated trust with their cooperating teacher, caused them to think deeply as they planned lessons, provided a common language for them to share with their cooperating teachers, and helped them anticipate the lesson in the planning conversation and bring closure to the lesson in the reflecting conversation (Townsend 1995). In addition, Cognitive Coaching provided a structure that gave them time to think about their teaching, required the supervising teacher to use the recommended coaching practices of nonjudgmental responses, helped them to own the lesson and to feel a sense of power, helped them when they received support, and worked best when the supervising teacher had confidence in Cognitive Coaching. Cooperating teachers reported that Cognitive Coaching helped them to become more conscious about their teaching, validated their beliefs about teaching, worked well for both progressive and traditional teachers, encouraged them to have collegial relationships with other coaches, enhanced the reflection of teachers who were already reflective, and helped them to realize the value of listening by using paraphrasing and probing. When student teachers who had used Cognitive Coaching in their student-teaching experiences were no longer encouraged or supported in using their skills, however, they tended to not take the time to engage in coaching (Townsend 1995).

Beginning teachers in the Student-Teacher Expanded Program (STEP), an alternative teacher-training program, participated in one-year internships and received mentoring from teachers who had been trained in Cognitive Coaching (Burk, Ford, Guffy, and Mann 1996). They were compared with teachers in a traditional 12-week student teaching program. Those who participated in the STEP program increased significantly over the control group on ratings by self, supervising teacher, and university supervisor on all subscales of the Proficiencies for Teachers Survey, including Learner-Centered Knowledge, Learner-Centered Instruction, Equity in Excellence, Learner-Centered Communications, and Learner-Centered Professional Development. The areas of Learner-Centered Knowledge and Equity in Excellence were the most robust.

Clinard et al. (1997) conducted a study in which supervising teacher mentors used Cognitive Coaching with their student teachers. They found that the training was an excellent way to prepare mentor-teachers. They concluded that Cognitive Coaching impacted mentors in all areas, including their work with student teachers, their own students, and other teachers. The supervising teachers indicated that by supervising student teachers using Cognitive Coaching, they increased their sense of professionalism and were able to network with other educators.

When master teachers used Cognitive Coaching with their student teachers, the student teachers reflected on their practice, as evidenced by data from both the master teachers and the student teachers (Brooks 2000a, 2000b). The student teachers reported that the coaching they received from the master teachers helped them to reflect and grow. In addition, the student teachers noticed a difference between master teachers who had been trained in Cognitive Coaching and master teachers who had not received the training.

Coy (2004) investigated the yearlong Transition to Teaching Program, of which Cognitive Coaching was a key component. She found that during the year, mentors moved from a "sense of failure and frustration as Mentors" to a "sense of personal growth and empowerment" (142). In addition, they moved from not "understand[ing] their roles as mentors" to having "rejuvenated energy levels" (142). They went from having "limited success attributed to 'Random Acts of Collaboration' between mentor and protégé" to "recogniz[ing] sound mentoring practices" and "recogniz[ing the] need for protégés to reflect on their practices" (142). Coy also found that the mentors were able to establish trust and rapport, and they moved from telling the protégé what to do to asking questions and paraphrasing in order "to assist the protégé to be more reflective and self-directed" (146). She discovered that at the end of the project, "the focus of the mentors . . . was the development of the protégé, not the 'cloning' of the mentor" (140).

Eger (2006) found that the more commonalities the mentor and second- or third-year teacher had (grade level, department, subject taught, etc.), "the greater impact mentoring appeared to have when it came to the four phrases of their teacher thinking and behaviors" (99).

In a study of a mentoring program by Robinson (2010), both novice teachers and mentor-teachers mentioned Cognitive Coaching as being an important part of the program. In another study, McCloy (2011) worked with a student teacher and a mentor-teacher, providing five preobservation protocols based on the National Board for Professional Teaching Standards propositions, as well as postobservation protocols based on Cognitive Coaching. Rather than the focus being on the student teacher, the focus was on the mentor-teacher. The student teacher asked the questions of the mentor-teacher. McCloy found that the mentor-teacher benefited professionally from the experience. The mentor-teacher reported that the quality of the relationship with her student teacher facilitated her learning, she learned from her student teacher, she learned to examine her own lesson plans more critically, she noticed the difference between the naïveté of her student teacher and her own experience and felt affirmed, and she felt a sense of accountability for providing accurate information to her student teacher.

Jaede, Brosnan, Leigh, and Stroot (2014) examined the influence of Cognitive Coaching on 28 middle school and high school mentor-teachers in an urban setting. They found that the mentor-teachers' use of Cognitive Coaching enabled them to become better mentors by focusing on the thinking and learning of their interns. They were able to assist their interns in becoming more autonomous, "help the interns develop their own perspective about teaching, and create their own identity as a teacher in an urban context" (22). In addition, they found that the mentors changed their view of the role they played. They came to see themselves "as teacher educators, mediators of intern learning" (23). They were focused on helping their interns grow each day, and they created "a collaborative partnership with the shared goal of identifying evidence of student learning" (23). They also came to see themselves as teacher-leaders with "collegial collaboration with colleagues, students and parents" (25).

Wooten Burnett (2015) studied seven physical education teacher candidates in a master's program who received three cycles of Cognitive Coaching (three planning conversations and three reflecting conversations) over a period of six weeks. According to the teacher candidates, Cognitive Coaching "impacted their overall teaching experience" (109).

THE IMPACT OF COGNITIVE COACHING ON TEACHERS' PERSONAL LIVES

Teachers reported that Cognitive Coaching not only impacted them professionally, it impacted them personally. In an early study, teachers listed usability of skills in all areas of their lives as the number one source of satisfaction with Cognitive Coaching. They listed self-growth as the second source of satisfaction (Edwards and Newton 1994a).

Awakuni (1995) conducted a study in which teachers participated in Cognitive Coaching for a year. Teachers reported having increased confidence in themselves and a greater sense of self. In a study by Clinard et al. (1995), teachers who used Cognitive Coaching with student teachers grew in their positive feelings about themselves, and they altered their attitudes with people in their families.

In Schlosser's (1998) study, teachers indicated that they first used Cognitive Coaching in their personal lives before using it in their professional lives because trust had already been established in those relationships. In another study, teachers who had taken eight days of Cognitive Coaching training indicated that they had developed personally as a result of the training (Beltman 2009).

Wooten Burnett (2015) studied seven physical education teacher candidates in a master's program who received three cycles of Cognitive Coaching (three planning conversations and three reflecting conversations) over a period of six weeks. Seven teachers participated in the control group. She also collected data from semistructured interviews, an open-ended survey, and the three cycles of planning and reflecting conversations. The participants indicated that the coaching they received "helped them become more aware of their self-development" (97) and that it had helped them grow personally.

RECOMMENDATIONS FOR IMPLEMENTING COGNITIVE COACHING

Researchers have also investigated best practices for implementing Cognitive Coaching. Their findings can be categorized into 12 recommendations.

Recommendation #1: Establish Long-Term, District-Level Support to Provide Training and to Support Teachers as They Are Implementing Cognitive Coaching

Numerous researchers have found the importance of school districts providing long-term, district-level support to implement Cognitive Coaching. District-level administrators and site-level administrators need to communicate about their expectations, value Cognitive Coaching, and make it a priority, according to Weatherford and Weatherford (1991). Johnson (1997) found that it is important for school districts to make a long-term commitment to Cognitive Coaching. If not, an elite group of those who have been trained is created. In addition, teachers may learn that innovations in teaching are "here today, gone tomorrow." Without long-term support, teachers are expected to use their planning time and lunch periods for coaching.

Teachers will benefit from practicing Cognitive Coaching in meetings and other district events so that they can see the importance and value of coaching and internalize the skills (Johnson 1997). Teachers will benefit from having other teachers in their school who are trained in Cognitive Coaching so that they will have coaching partners (Johnson 1997).

Aldrich (2005) and Dougherty (2000) also observed that, for Cognitive Coaching to become embedded in school cultures, school districts and organizations need to support teachers over time. According to Eger (2006), when school district leaders wove Cognitive Coaching into all levels of the district, veteran teachers "used terms like 'it is the way we do business here,' 'it is a culture of continuous improvement,' and 'the district is a more thoughtful district, more reflective'" (57).

Reed (2007) discovered the importance of allowing three to five years to implement Cognitive Coaching in a school or district. In her study, "the teachers involved . . . did not have great success in the very first year and even faced some difficulty during the second year. However, they recognized in years three to five, they had developed strong educational techniques which accounted for their students' success" (236). "Implementation of any innovation or reform is a process. It takes time to build capacity and understanding within a faculty/staff" (236). According to Reed (2007), it is important for school districts to involve the teachers' union early in the process of deciding to use instructional coaches in order to make sure that they understand and support their use.

Recommendation #2: Enlist Principals' Support and Modeling of Cognitive Coaching

As with any innovation, principal support and modeling of Cognitive Coaching are key to helping teachers use it. Administrators need to be in favor of the process and participate in the training along with the teachers when Cognitive Coaching is implemented for maximum success, according to Weatherford and Weatherford (1991). In addition, they need to model the process, value it, and make it a priority. In a study by Edwards and Green (1999b), teachers who were in schools with principal support for Cognitive Coaching were more likely to persist in a three-year project in which Cognitive Coaching was used to help teachers implement standards-based education than those in schools without principal support. Teachers tended to stay in the project when their principals participated in Cognitive Coaching training, supported the use of Cognitive Coaching, and modeled coaching behaviors. McLymont and da Costa (1998) also found that principal support of Cognitive Coaching was a key determinant of its success.

When principals have received training in Cognitive Coaching and supported the instructional coach, teachers are generally more willing to work with the instructional coach and value the coaching process (Reed 2007). In addition, the principal needs to hold teachers accountable for participating in coaching and understand that the instructional coach is supposed to serve as a coach to the teachers and not as an assistant principal. "Principals and the campus leadership must not only be a part of the planning process but also understand what is expected of them during the implementation process and what they must establish as expectations for their campus" (234).

Reed (2007) also found that three types of resistance to using instructional coaches included "resistance as a pervasive attitude by individual teachers, . . . resistance of the union, . . . [and] resistance to change" (176). "There appeared to be a synergy between the themes of campus leadership and teachers' willingness and openness to learn. As resistance increased, the opportunity for implementation of the Cognitive Coaching^{SM} process decreased" (226).

Recommendation #3: Be Aware of Implementation Concerns and Use Tools Such as the Concerns-Based Adoption Model (CBAM) States of Concern and Levels of Use When Implementing Cognitive Coaching

Administrators should be aware that teachers will have different stages of concern and implement Cognitive Coaching at different rates. In Donnelly's (1988) study, administrators who were implementing Cognitive Coaching went through stages of concern and levels of use that were detailed in the Concerns-Based Adoption Model (CBAM) (Hall and Hord 2001). They tended to implement aspects of Cognitive Coaching that were similar to old methods of supervision first, and those that were unique to Cognitive Coaching were implemented later. In Donnelly's study, new administrators displayed early stages of concern, and experienced administrators displayed Stage Four concerns.

Schlosser (1998) also found that both teachers and administrators went through the levels of the Concerns-Based Adoption Model (Hall and Hord 2001). Teachers in his study reported that the use of nonjudgmental behaviors, although powerful ways of mediating the thinking of others, seemed unnatural to them at first until they became accustomed to using them.

Reed (2007) discovered that teachers' concerns about the implementation of Cognitive Coaching aligned with the stages of concern in the Concerns-Based Adoption Model (CBAM), developed by Hall and Hord (2001). She suggested that when instructional coaches use Cognitive Coaching with teachers, they need to explain the Cognitive Coaching process, including the different types of conversations, the various responses that they make, and so forth so that the teachers will be able to understand what the coaches are doing. When teachers understand the process, they will become more accepting of it. "While teachers may not need the eight day Cognitive Coaching[SM] seminar training model, they do need an awareness of the process and terminology used by an instructional coach, principal, or assistant principal. They need to understand how the instructional coach uses the Cognitive Coaching[SM] process as a tool" (234).

Recommendation #4: Recognize That All Teachers Can Benefit from Being Involved in Cognitive Coaching

Researchers have compared teachers by demographic areas such as gender and number of years of teaching experience and found that all teachers can benefit from being involved in Cognitive Coaching. Foster (1989) found no significant differences in teacher responses on the Teacher Thought Processes questionnaire, a measure of teacher reflective thought, for number of years of teaching experience in the areas of teaching, planning, analyzing and evaluating, and applying. She also found no differences between elementary and secondary teachers in their perceptions of the impact that Cognitive Coaching had on their thought processes.

No differences were found according to years of experience in how teachers reported that Cognitive Coaching impacted their instructional process in the areas of planning, teaching, analyzing, applying, or overall instructional process (Alseike 1997). No significant differences were found between males and females who received Cognitive Coaching in their receptivity to coaching; their report of its impact on their planning, analysis, and applying; nor on their states of mind of efficacy, flexibility, consciousness, interdependence, and overall holonomy (Alseike 1997). Females scored higher than males on teaching, and males scored higher than females on craftsmanship.

In the same study, no differences were reported by number of years of teaching for the states of mind of flexibility, craftsmanship, consciousness, and interdependence (Alseike 1997). Teachers with 16 to 20 years of teaching experience had higher overall holonomy, and teachers with six to 20 years of experience had higher efficacy than did teachers with zero to five years of experience or teachers with over 21 years of experience. In Eger's (2006) study, no differences were found in the effect of Cognitive Coaching on the four phases of teacher thinking by "grades taught, number of years teaching, number of years in the district and number of years using cognitive coaching" (54).

Recommendation #5: Create Norms of Collaboration

It is essential that administrators create norms of collaboration in buildings to encourage teachers to use their Cognitive Coaching skills. In Johnson's (1997) study, teachers indicated that norms of "leave me alone" and "you-never-want-to-shine" in the teaching culture worked against their using Cognitive Coaching. If they rose above the rest, they tended to be asked to take on additional responsibilities, adding to their already full workload. In addition, teacher schedules and the structure of the buildings in which they worked tended to keep teachers away from their colleagues and contribute to feelings of isolation, making coaching and building relationships with other teachers a challenge (Johnson 1997). While Johnson's study was done in the 1990s when school norms were different from what they are in 2014, administrators will still benefit from creating norms of collaboration.

Recommendation #6: Invite Voluntary Participation

As with any innovation, it is important for teachers to be able to choose whether to participate or not (Krpan 1997; Smith 1997; Weatherford and Weatherford 1991). Coaching needs to be voluntary rather than prescribed and mandated.

Recommendation #7: Establish a Trusting Environment

Tschannen-Moran (2014) and Bryk and Schneider (2002) wrote about the importance of having trusting environments in schools. Trust between the members of the coaching team is critical to success (Krpan 1997; Weatherford and Weatherford 1991). Johnson found that it is important for teachers to not only establish trusting relationships with their coaching partners but to establish trusting relationships in the larger school organization as well. In another study, the researchers found that nonjudgmental behaviors contributed to the development of trust in coaching relationships (McLymont and da Costa 1998).

Aldrich (2005) conducted a study on online coaching and discovered that in order for Cognitive Coaching to succeed online, participants must have trust and rapport. They must also use the skills of acknowledging, paraphrasing, and asking mediational questions. In addition, participants must engage with each other and with the content.

Reed (2007) found that instructional coaches will benefit from taking time to build relationships of trust with the teachers whom they are coaching. They need to be "approachable" (197), "respectful" (197) of the teachers, and "knowledgeable" (201) about instruction. Teachers need to see them as being able to provide resources, and they need to be "nonjudgmental" (202). They need to provide "open communication" (195) so that the teachers do not feel like the coaches are above them. According to Reed, "Teachers worked closely with the instructional coach when there was a high level of trust" (243).

Recommendation #8: Involve Teachers Right Away in Using Their Coaching Skills

It is important for administrators to involve teachers right after they have attended Cognitive Coaching training in using their coaching skills. In Foster's (1989) study, when teachers participated in seven or more Cognitive Coaching conferences, they perceived that the conferences had a high impact on their thought processes in planning, teaching, analyzing and evaluating, and applying. Teachers who participated in four to six coaching conferences reported that Cognitive Coaching had an average impact on planning, teaching, and applying, and a high impact on analyzing and evaluating. Teachers who participated in one to three coaching conferences reported that Cognitive Coaching had an average impact on their thought processes in the four areas, and teachers who participated in no coaching conferences reported that Cognitive Coaching had a low impact on their thought processes in the four areas. In another study, teachers who used Cognitive Coaching more frequently obtained significantly higher scores on the Self-Reflection Survey: Cognitive Coaching Rating Scale (Schuman 1991) than did those who used it less (Edwards and Newton 1994b). Subscales included Planning, Teaching, Analyzing and Evaluating, and Applying.

Alseike (1997) reported that the more coaching the teachers received, the more they indicated that Cognitive Coaching impacted their planning, applying, and overall teaching, as well as their interdependence. Teachers who participated in seven or more Cognitive Coaching training sessions scored significantly higher than those who had not attended any sessions, or had only attended one or two sessions, on teaching, applying, and overall instructional processes (Alseike 1997). In addition, teachers who had attended seven or more trainings scored significantly higher in the states of mind of efficacy and consciousness than did teachers who had not attended any trainings or had attended one or two trainings, even though they had received coaching from experienced coaches. Alseike also found that teachers who received Cognitive Coaching from experienced coaches, yet had never coached another teacher, scored significantly lower in teaching, applying, and the state of mind of efficacy than did teachers who had coached another teacher seven or more times.

Edwards and Green (1999a) discovered that practice in using Cognitive Coaching skills is essential in order to bring about the growth that is possible with Cognitive Coaching. When teachers became involved immediately after receiving their training in implementing coaching skills and participating in coaching, they tended to persist in a three-year project using Cognitive Coaching to implement standards-based education (Edwards and Green 1999b).

Recommendation #9: Structure Time for Cognitive Coaching

Participants talked about the importance of having time to use their Cognitive Coaching skills. Fortunately, many schools provide coaches, so coaching is built into the school day. This was not the case when many of the studies were done. Still, administrators and teachers need to be creative in finding time and money to enable teachers to practice Cognitive Coaching, according to Weatherford and Weatherford (1991).

Lack of time to do coaching was listed as the number-one source of dissatisfaction with Cognitive Coaching in one study (Edwards and Newton 1994a). Cognitive Coaching takes time, and teachers already have many demands on their time (Beltman 2009; Townsend 1995). Teachers who engaged in Cognitive Coaching who had taken seven to 12 days of training and had taught eight years or more indicated that shortage of time was a major hindrance to coaching and engaging in reflection about their teaching (Johnson 1997).

In Schlosser's (1998) study, teachers reported a tension between appreciating having the opportunity to work and talk with their colleagues and the lack of time in the school day. Coy (2004) discussed the importance of administrators providing time for Cognitive Coaching interactions between mentors and their protégés.

Five participants who were involved in an online coaching study said that lack of time to do online coaching and to attend training sessions inhibited the effectiveness of their practice (Aldrich 2005). They also listed not being able to communicate in person and problems with using technology as inhibitors.

While "teachers perceived time as an obstacle that inhibited the implementation and use of the Cognitive Coaching[SM] process, . . . teachers . . . identified time as a major factor in why they built such a strong relationship" (Reed 2007, 203). "Leadership at both the campus and district level has to recognize the importance of building relationships and commit the time needed for development of those relationships. This is one of the most critical components, yet it is the one often overlooked" (235). In Batt's (2010) study, in order to save time and decrease the amount of time that students needed to spend with a substitute teacher, the teachers who were being coached employed phone conversations, e-mail discussions, and sticky notes on manuals with their coaches.

Recommendation #10: Recognize That Teachers Tend to Use Cognitive Coaching Skills on an Informal Basis More Frequently Than They Use the Formal Planning Conversation, Observation, and Reflecting Conversation

Teachers using Cognitive Coaching during the course of a year reported that they used Cognitive Coaching skills informally more frequently than they used them on a formal basis, including a planning conversation, and observation, and a reflecting conversation (Awakuni 1995). Still, whether teachers were coached formally, including the planning conversation, classroom observation, and reflecting conversation, or informally, they identified Cognitive Coaching as having a positive impact on their teaching (Alseike 1997).

Recommendation #11: Invite Teachers to Use Their Coaching Skills in Many Contexts

Teachers can use their coaching skills in a variety of contexts, including online exchanges.

Foster (1989) reported no differences in perceptions of impact on thought processes between teachers who were coached by an administrator and teachers who were coached by another teacher. In a study by Alseike (1997), no differences were noted in the impact of Cognitive Coaching on the instructional processes of planning, teaching, analyzing, and applying, nor on the states of mind between teachers who had been coached by the principal, a building resource teacher, another teacher, or a combination of the three.

Participants in an online coaching study said that text-based nonverbal cues, the sharing of asynchronous journal reflections, the ability to send private messages, and "the invisibility of the use of the maps and sentence stems" (Aldrich 2005, 204) were unique features that they enjoyed as they conducted coaching conversations on the Internet. They also enjoyed the chat capability, the whiteboards, the training they received, the availability of the moderator, and the fact that their coaching sessions were archived. The more people participated in online coaching conversations, including the more messages that they posted and the more sessions they conducted, the more they met the conditions for effective online coaching.

Recommendation #12: Distinguish between Coaching and Evaluation

Distinctions need to be made between coaching and evaluation if the principal is doing the coaching so that teachers know when they are being coached and when they are being evaluated (Weatherford and Weatherford 1991). Teachers who had taken seven to 12 days of training in Cognitive Coaching indicated that when principals focused on evaluating teachers rather than on supervising them, they created a climate that did not encourage them to develop professionally (Johnson 1997).

QUESTIONS FOR ADDITIONAL RESEARCH

Many areas have been investigated in Cognitive Coaching research, and many areas remain to be investigated. The following sections include ideas for additional studies to investigate its impacts on self-directedness, the five states of mind, students, the professional lives of teachers, school staff, principals, supervisory relationships, mentoring and teacher preparation programs, the personal lives of teachers, and the larger community.

Questions about the Impact of Cognitive Coaching on Self-directedness

Researchers have gathered evidence about the effects of Cognitive Coaching on the self-directedness of educators using various types of data. What additional types of data might be available that would indicate self-directedness? How might educators grow on instruments of self-directedness as a result of participating in Cognitive Coaching? Do any differences exist in growth in self-directedness between when educators coach others and when they receive coaching? Which aspects of Cognitive Coaching contribute the most to helping teachers grow in self-directedness?

Questions about the Impact of Cognitive Coaching on the Five States of Mind

Numerous researchers have also studied the effects of Cognitive Coaching training and use on the five states of mind, both as a whole and individually (i.e., efficacy, flexibility, consciousness, craftsmanship, and interdependence). What additional instruments might be developed to measure the five states of mind in various contexts? Researchers could also use instruments that have been developed to measure each of the states of mind. In addition, researchers could correlate teacher growth in the five states of mind with student learning and achievement.

Questions about the Impact of Cognitive Coaching on Students

So far, researchers have primarily studied the outcomes of teachers coaching each other, or the outcomes of teachers being coached by administrators, university professors, or full-time coaches. A major area for future research involves studying what might happen when teachers coach students. In addition, what might be some of the outcomes if teachers taught coaching skills to students so that they could coach each other? In what ways might that impact student growth and thinking? How might student learning be accelerated through the process? How much might student test scores increase? What might be the impact on students' self-directedness? How might student increases in self-directedness influence their behaviors in the classroom, their study habits, their relationships with other students, and their learning?

Some possible areas to explore could be student time on task, self-directedness of students, students' views of themselves as self-directed learners, student identity, student interactions with one another in general, student ability to solve problems, frequency of students asking questions, number of discipline referrals, student grades, student development of higher-order thinking skills, and student growth in the five states of mind. Other possible areas to explore might be how student relationships with each other would change, how student relationships with the teacher could change, how student relationships with their parents might change, and how students resolve difficulties. What might happen at the different grade levels? Would similar outcomes occur at all grade levels, or would students in some grade levels change more quickly?

Questions about the Impact of Cognitive Coaching on Teachers' Professional Lives

In what additional ways might Cognitive Coaching impact teacher professionalism? What might be some of the aspects of Cognitive Coaching that provide teachers with more enjoyment of teaching as a profession? What changes could occur in the ways that teachers viewed themselves professionally? In what ways might Cognitive Coaching impact teachers' desire to learn and try new skills, get higher degrees, and take more classes? How might Cognitive Coaching impact teachers' emphasis with students on the five states of mind? What impact would Cognitive Coaching have on teachers' sense of feeling part of the school community? Still other researchers could investigate the progression that teachers make as they learn to coach and as they implement and internalize Cognitive Coaching values, beliefs, maps, and skills. What progression do teachers go through as they internalize the identity of a mediator of thinking? In what ways would teacher–parent conferences be different in schools in which teachers used Cognitive Coaching on a regular basis? How might Cognitive Coaching impact the number of conflicts that teachers had with parents or that teachers had with children?

Questions about the Impact of Cognitive Coaching on School Staff

More research could be done to investigate the effects of Cognitive Coaching on entire schools. Researchers might examine principal–teacher interactions in the school. For example, what changes might occur in the ways in which the principal interacts with individual staff members? What changes could occur in the ways that the principal interacts with the staff as a group? What might be some changes in the ways that the staff members interact with the principal? What could be some changes in the ways that staff members interact with each other? What changes might occur in weekly staff meetings?

How might Cognitive Coaching help teachers grow in collective efficacy? What might be the results of higher levels of collective efficacy on students and their ability to achieve? Other possibilities could be to investigate how personal relationships between teachers in the school could change. What changes might occur in how teams of teachers interacted with other teams? What could be some of the changes in how teams interacted within their teams? What changes might teachers notice in their friendships with other teachers in the school? Would fewer arguments and disagreements occur?

Further possibilities could include investigations into the impact of Cognitive Coaching on schoolwide problems. When problems occur, what are some of the strategies that teachers and administrators use to resolve them? Might fewer difficulties and issues arise as a result of all staff using Cognitive Coaching? Fewer major problems and conflicts should occur because Cognitive Coaching training and use could provide teachers with the skills to work together and coach each other in an atmosphere of trust and rapport in order to solve problems while they are still small. Researchers could also examine how Cognitive Coaching might contribute to schools becoming learning organizations (Senge, Cambron-McCabe, Lucas, Smith, Dutton, and Kleiner 2012). What might happen to teachers' desire to learn? How could that desire be passed on to students?

Questions about the Impact of Cognitive Coaching on Principals

Researchers could investigate the outcomes from principals receiving coaching and coaching each other. How might coaching influence what they do on a daily basis? In what ways might they grow from coaching and being coached? How might the way in which they interact with teachers, parents, and students change? How might parent involvement change? Might student attendance increase as a result of principals being coached and coaching students? What might be some of the effects of principals being coached on teacher collective efficacy, school climate, school culture, and organizational climate? How might teacher implementation of initiatives change?

Questions about the Impact of Cognitive Coaching on Supervisory Relationships

Researchers could investigate the use of Cognitive Coaching in supervisory relationships. When principals use Cognitive Coaching in supervisory relationships, how might teachers grow in their proficiency compared with traditional supervision? What might be the outcomes of principals using the calibrating conversation with teachers? How might teacher evaluations differ as a result of principals coaching teachers? What might be some of the effects on school culture, school climate, faculty collegiality, and faculty satisfaction with position and with teaching as a profession?

What percentage of time do principals use the four support functions of coaching, collaborating, consulting, and evaluating? How might principal use of the four support functions influence teacher thinking, teacher self-directedness, and teacher growth in the five states of mind? How might that affect teacher assessment of school culture, school climate, and organizational climate?

Questions about the Impact of Cognitive Coaching on Mentoring and Teacher Preparation Programs

Researchers could also conduct additional studies on the use of Cognitive Coaching in mentoring and teacher preparation programs. If teacher candidates received Cognitive Coaching training in their teacher training program, how might that enhance their student teaching experience? How might that impact the length of time they remain in the teaching profession? If time were provided for student teachers to coach each other, how might their practice improve? How might the success of student teachers who had Cognitive Coaching training and subsequent coaching compare with the success of student teachers who did not receive training in Cognitive Coaching?

Questions about the Impact of Cognitive Coaching on Teachers' Personal Lives

Researchers could also conduct studies on the impact of Cognitive Coaching on teachers' personal lives. How might relationships with spouses, partners, or significant others change? In what ways could teachers' relationships with their children change? Might relationships change in different ways depending on the ages of the teachers' children? What could be some trends in how teachers' friendships outside school might change? Which of the Cognitive Coaching maps and tools do teachers use most frequently in their personal lives?

Other researchers might examine the teacher as a person. What impact could Cognitive Coaching have on teacher emotions, such as happiness, sadness, enthusiasm, or anger? How might Cognitive Coaching impact teacher curiosity and creativity? What might be some changes in teacher self-concept? What changes could occur in teacher identity and how teachers see themselves? What might be some possible changes in teachers' abilities, desires, and goal setting?

Questions about Cognitive Coaching and the Larger Community

An even larger systemic inquiry might involve examining the impact of a school's use of Cognitive Coaching on the surrounding community. How might schools change in their interactions and relationships with the local community? What changes might begin to occur in families as a result of students experiencing coaching at school?

Researchers could investigate what happens in other schools in a school district in which one school begins using Cognitive Coaching. What impact might one school's use of Cognitive Coaching have on teachers in other schools? When teachers at a school that uses Cognitive Coaching interact with teachers in other schools at distric-twide in-services and at other meetings, what might be some of the effects?

Questions about Implementing Cognitive Coaching

Researchers could also conduct studies to determine the most effective strategies for implementing Cognitive Coaching. What processes are in place in school districts that implement Cognitive Coaching at the highest levels? In addition to the 12 recommendations that researchers have already identified, what else might be important for districts to consider in order to get the greatest return on their investment?

Possible Avenues for Collecting Data

A range of possibilities exists for collecting data on the research questions above. One possible avenue for collecting data might be in-depth interviews with teachers, students, parents, and members of the community. Other possible sources of data might be videos or audiotapes of coaching conversations, teacher journals, teacher testimonies, lesson plans, teacher language, teacher questions, commercially prepared instruments, observations, and instruments in educational journals or papers. Other possibilities include measuring levels of use and stages of concern from the Concerns-Based Adoption Model (CBAM) (Hall and Hord 2001) and observing interactions in schools and classrooms. Researchers could administer a variety of instruments before, during, and after Cognitive Coaching has been implemented.

CONCLUSION

This chapter included studies on the impact of Cognitive Coaching on self-directed learning, the five states of mind, and the states of mind of efficacy, flexibility, consciousness, craftsmanship, and interdependence. In addition, the chapter included studies on the impact of Cognitive Coaching on students, the professional lives of educators, supervisory relationships, the staff as a team, mentoring and teacher preparation programs, and teachers' personal lives. Research on the 12 recommendations for implementing Cognitive Coaching was also included. The chapter concluded with suggestions for further research.

Much research has been done to investigate the impacts of Cognitive Coaching, and much remains to be done. As school districts involve teachers in investigating the outcomes of Cognitive Coaching training and use with students, teachers, the school, and the community, the field will advance even further.

NOTE

We are extremely grateful to Jenny Edwards for developing this chapter. In addition to her own research, Dr. Edwards has maintained a clearinghouse for studies about Cognitive Coaching and has been an important contributor to our learning.—A. L. Costa and R. J. Garmston

REFERENCES

Aldrich, R. S. 2005. *Cognitive Coaching practice in online environments.* PhD diss., retrieved from ProQuest Dissertations and Theses. (UMI No. 3197394)

Alseike, B. U. 1997. *Cognitive Coaching: Its influences on teachers.* PhD diss., retrieved from ProQuest Dissertations and Theses. (UMI No. 9804083)

Ashton, P., R, Webb, and C. Doda. 1983. *A study of teachers' sense of efficacy.* Gainesville: University of Florida.

Avant, R. C. 2012. *Instructional coaching and emotional intelligence.* PhD diss., retrieved from Pro-Quest Dissertations and Theses. (UMI No. 3509025)

Awakuni, G. H. 1995. *The impact of Cognitive Coaching as perceived by the Kalani High School core team.* PhD diss., retrieved from ProQuest Dissertations and Theses. (UMI No. 9613169)

Baker, K. L. 2008. *The effects of Cognitive Coaching on initially licensed teachers.* PhD diss., retrieved from http://libres.uncg.edu/ir/uncw/f/bakerk2008-1.pdf.

Bal, A. P, and Ö. Demir. 2011. Cognitive Coaching approach in view of pre-service teachers. *Ahi Evran Üniversitesi Kirsehir Egitim Fakültesi Dergisi (KEFAD)* 12(4): 325–40.

Bandura, A. 2006. Guide to the construction of self-efficacy scales. In F. Pajares and T. Urdan (eds.), *Self-efficacy beliefs of adolescents,* vol. 5, 307–37. Greenwich, CT: Information Age Publishing.

Batt, E. G. 2010. Cognitive Coaching: A critical phase in professional development to implement sheltered instruction. *Teaching and Teacher Education* 26:997–1005. doi.org/10.1016/j.tate.2009.10.042

Beltman, S. 2009. Educators' motivation for continuing professional learning. *Issues in Educational Research* 19(3): 193–211.

Betoret, F. D. 2006. Stressors, self-efficacy, coping resources, and burnout among secondary school teachers in Spain. *Educational Psychology: An International Journal of Experimental Educational Psychology* 26(4): 519–39.

Bjerken, K. S. 2013. *Building self-directed teachers: A case study of teachers' perspectives of the effects of Cognitive Coaching on professional practices.* PhD diss., retrieved from ProQuest Dissertations and Theses. (UMI No. 3564120)

Brooks, G. R. 2000a. *Cognitive Coaching training for master teachers and its effects on student teachers' ability to reflect on practice.* PhD diss., retrieved from ProQuest Dissertations and Theses. (UMI No. 3054851)

———. 2000b. Cognitive Coaching for master teachers and its effect on student teachers' ability to reflect on practice. *Delta Kappa Gamma Bulletin* 67(1): 46–50.

Bryk, A., and B. Schneider. 2002. *Trust in schools: A core resource for improvement.* New York: Russell Sage Foundation.

Burk, J., M. B. Ford, T. Guffy, and G. Mann. 1996, February. *Reconceptualizing student teaching: A STEP forward.* Paper presented at the annual meeting of the American Association of Colleges for Teacher Education, Chicago, Ill.

Caprara, G. V., C. Barbaranelli, P. Steca, and P. S. Malone. 2006. Teachers' self-efficacy beliefs as determinants of job satisfaction and students' academic achievement: A study at the school level. *Journal of School Psychology* 44:473–90. doi:10.1016/j.jsp.2006.09.001

Chang, D., C.-D. Lee, and S.-C. Wang. 2014. The influence of Cognitive Coaching on teaching reflection and teaching effectiveness: Taking teachers participating in formative teacher evaluation in elementary and secondary schools as examples. *Journal of University of Taipei* 45(1), 61–80. doi:10.6336/JUT.4501.004

Clay, M. 1993. *An observation survey of early literacy achievement.* Portsmouth, N.H: Heinemann.

Clinard, L. M., T. Ariav, R. Beeson, L. Minor, and M. Dwyer. 1995. *Cooperating teachers reflect upon the impact of coaching on their own teaching and professional life.* Paper presented at the annual meeting of the American Educational Research Association, San Francisco, Calif.

Clinard, L. M, L. Mirón, T. Ariav, I. Botzer, J. Conroy, K. Laycock, and K. Yule. 1997, March. *A cross-cultural perspective of teachers' perceptions: What contributions are exchanged between cooperating teachers and student teachers?* Paper presented at the annual meeting of the American Educational Research Association, Chicago, Ill.

Costa, A. L., and R. J. Garmston. 2002. *Cognitive Coaching: A foundation for renaissance schools.* Norwood, MA: Christopher-Gordon.

Coy, L. J. 2004. *A case study of a professional development initiative focused on novice teacher mentoring* PhD diss., retrieved from ProQuest Dissertations and Theses. (UMI No. 3155974)

da Costa, J. L., and G. Riordan. 1996. *Teacher efficacy and the capacity to trust.* Paper presented at the annual meeting of the American Educational Research Association, New York.

Diaz, K. A. 2013. *Employing National Board Certification practices with all teachers: The potential of Cognitive Coaching and mentoring.* PhD diss., retrieved from ProQuest Dissertations and Theses. (UMI No. 3557981)

DiPaola, M. F., and M. Tschannen-Moran. 2005. *School Climate Index.* Retrieved from https://webmail.fielding.edu/owa/redir.aspx?C=otg62NIW20uysJikgeEQWmHuBywI7NEIiHoY1TLlGwcTvH4rgokEKYCEbs8EMA34CV-eyHhHzjI.&URL=http%3a%2f%2fmxtsch.people.wm.edu%2fResearchTools%2fSCI_OMR.pdf.

Donahue-Barrett, K. 2014. *The influence of literacy coaching on teacher knowledge and practice within the writing workshop model.* PhD diss., retrieved from ProQuest Dissertations and Theses. (UMI No. 3624785)

Donnelly, L. 1988. *The Cognitive Coaching model of supervision: A study of its implementation.* Unpublished master's thesis, California State University, Sacramento.

Dougherty, P. A. 2000. *The effects of Cognitive Coaching training as it pertains to: Trust building and the development of a learning community for veteran teachers in a rural elementary school.* PhD diss., retrieved from ProQuest Dissertations and Theses. (UMI No. 3054864)

Dutton, M. M. 1990. *Learning and teacher job satisfaction.* PhD diss., retrieved from ProQuest Dissertations and Theses. (UMI No. 90-26940)

Dweck, C. 1976. *Implicit theories of intelligence scale for children—self form.* Retrieved from http://growthmindseteaz.org/caroldweck.html.

Echevarría, J., D. J. Short, and M. E. Vogt. 2008. *Implementing the SIOP model through effective professional development and coaching.* Boston, MA: Pearson/Allyn & Bacon.

Edwards, J. L. 1993. *The effect of Cognitive Coaching on the conceptual development and reflective thinking of first year teachers.* PhD diss., retrieved from ProQuest Dissertations and Theses. (UMI No. 9320751)

———. 2004. The process of becoming and helping others to become: A grounded theory study. In I. F. Stein, F. Campone, and L. J. Page (Eds.), *Proceedings of the second ICF coaching research symposium,* 69–78. Washington, DC: International Coach Federation.

Edwards, J. L., and K. Green. 1997. *The effects of Cognitive Coaching on teacher efficacy and empowerment.* Research Rep. No. 1997-1. Evergreen, CO: Author.

———. 1999a, April. *Growth in coaching skills over a three-year period: Progress toward mastery.* Paper presented at the annual meeting of the American Educational Research Association, Montreal, Canada.

———. 1999b, April. *Persisters versus nonpersisters: Characteristics of teachers who stay in a professional development program.* Paper presented at the annual meeting of the American Educational Research Association, Montreal, Canada.

Edwards, J. L., K. Green, C. A. Lyons, M. S. Rogers, and M. Swords. 1998, April. *The effects of Cognitive Coaching and nonverbal classroom management on teacher efficacy and perceptions of school culture.* Paper presented at the annual meeting of the American Educational Research Association, San Diego, Calif.

Edwards, J. L., and R. R. Newton. 1994a, February. *The effects of Cognitive Coaching on teacher efficacy and empowerment.* Research Rep. No. 1994-1. Evergreen, CO: Author.

———. 1994b, July. *The effects of Cognitive Coaching on teacher efficacy and thinking about teaching.* Research Rep. No. 1994-2. Evergreen, CO: Author.

———. 1994c, October. *Qualitative assessment of the effects of Cognitive Coaching training as evidenced through teacher portfolios and journals.* Research Rep. No. 1994-3. Evergreen, CO: Author.

———. 1995, April. *The effects of Cognitive Coaching on teacher efficacy and empowerment.* Paper presented at the annual meeting of the American Educational Research Association, San Francisco, Calif.

Eger, K. A. 2006. *Teachers' perception of the impact of Cognitive Coaching on their teacher thinking and behaviors.* PhD diss., retrieved from ProQuest Dissertations and Theses. (UMI No. 3223584)

Ellison, J. 2003. Coaching principals for increased resourcefulness. In J. Ellison and C. Hayes (eds.), *Cognitive Coaching^SM: Weaving threads of learning and change into the culture of an organization,* 13–25. Norwood, MA: Christopher-Gordon.

Ellison, J., and C. Hayes. 1998. *Energy sources team self-assessment survey.* Highlands Ranch, CO: Kaleidoscope Associates.

Evans, R. E., Jr. 2005. *Utilizing Cognitive Coaching to enhance the implementation of recommended middle school instructional strategies.* PhD diss., retrieved from ProQuest Dissertations and Theses. (UMI No. 3189304)

Fine, J. C., and S. W. Kossack. 2002. The effect of using rubric-embedded Cognitive Coaching strategies to initiate learning conversations. *Journal of Reading Education* 27(2): 31–37.

Foster, N. J. 1989. *The impact of Cognitive Coaching on teachers' thought processes as perceived by cognitively coached teachers in the Plymouth-Canton Community School District.* PhD diss., retrieved from ProQuest Dissertations and Theses. (UMI No. 8923848)

Garmston, R., and D. Hyerle. 1988, August. *Professor's peer coaching program: Report on a 1987-88 pilot project to develop and test a staff development model for improving instruction at California State University.* Sacramento, CA: Author.

Garmston, R., C. Linder, and J. Whitaker. 1993, October. Reflections on Cognitive Coaching. *Educational Leadership* 51(2): 57–61.

Geltner, B. B. 1993. *Integrating formative portfolio assessment, reflective practice, and Cognitive Coaching into preservice preparation.* Paper presented at the annual convention of the University Council for Educational Administration, Houston, Texas.

———. 1994. *The power of structural and symbolic redesign: Creating a collaborative learning community in higher education.* Descriptive Report #141. (ERIC Document Reproduction Service no. ED374757)

Gibson, S., and M. H. Dembo. 1984. Teacher efficacy: A construct validation. *Journal of Educational Psychology* 36(4): 569–82. doi.org/10.1037/0022-0663.76.4.569.

Glenn, R. A. 1993. *Teacher attribution: Affect linkages as a function of student academic and behavior failure and teacher efficacy (academic failure).* PhD diss., Memphis State University. (Dissertation Abstracts International, 54/12-A, AAD94014958)

González, L. J. 2009. *Los cinco estados de la mente en el counseling espiritual.* PhD diss., Universidad Iberoamericana, México, D. F., retrieved from http://www.uia.mx/.

Government of Alberta. 2012. *The Alberta School Leadership Framework: Building leadership capacity in Alberta's education system.* Edmonton, Canada: Government of Alberta.

Grinder, M. 1996. *ENVoY: A personal guide to classroom management.* 3rd ed. Battle Ground, WA: Michael Grinder and Associates.

Guskey, T. R., and P. Passaro. 1993. *Teacher efficacy: A study of construct dimensions.* Lexington: College of Education, University of Kentucky. (ERIC Document Reproduction Service no. ED 359202)

Hagopian, G., H. B. Williams, M. Carrillo, and C. C. Hoover. 1996, April. *The 2-5 collaborative in-class model: A restructuring of the Title I program.* Paper presented at the annual meeting of the American Educational Research Association, New York.

Hall, G. E., and S. M. Hord. 2001. *Implementing change: Patterns, principles, and potholes.* Boston, MA: Allyn and Bacon.

Hauserman, C., C. Edwards, and N. Mastel. 2013. *The role of Cognitive Coaching in enhancing instruction.* Manuscript submitted for publication.

Henry, A. G. 2012. *Cognitive Coaching: An examination of the reflective journaling of teacher candidates.* PhD diss., retrieved from ProQuest Dissertations and Theses. (UMI No. 3510381)

Hoover-Dempsey, K. V., O. C. Bassler, and J. S. Brissie. 1987, Fall. Parent involvement: Contributions of teacher efficacy, school socioeconomic status, and other school characteristics. *American Educational Research Journal* 24(3): 417–35.

Hull, J., J. L. Edwards, M. S. Rogers, and M. E. Swords. 1998. *The Pleasant View experience.* Golden, CO: Jefferson County Schools.

Humphries, C. A., E. Hebert, K. Daigle, and J. Martin. 2012. Development of a physical education teaching efficacy scale. *Measurement in Physical Education and Exercise Science* 16: 284–99.

Irons, N. A. 2014. *Coaching for questioning: A study on the impact of questioning.* Unpublished Capstone Action Research Project. Fielding Graduate University, Santa Barbara, CA.

Jaede, M., P. Brosnan, K. Leigh, and S. Stroot. 2014, April. *Teaching to transgress: How Cognitive Coaching^{SM} influences the apprenticeship model in pre-service urban teacher education.* Paper presented at the annual meeting of the American Educational Research Association, Philadelphia, PA.

Johnson, J. B. 1997. *An exploratory study of teachers' efforts to implement Cognitive Coaching as a form of professional development: Waiting for Godot.* PhD diss., retrieved from ProQuest Dissertations and Theses. (UMI No. 9729048)

Kirkpatrick, D. 1998. *Evaluating training programs: The four levels.* San Francisco, CA: Berrett-Koehler.

Knaebel, D. R. 2008. *Exploring the experiences of a literacy coach in a Reading First school in Indiana.* PhD diss., retrieved from ProQuest Dissertations and Theses. (UMI No. 3318963)

Krpan, M. M. 1997. *Cognitive Coaching and efficacy, growth, and change for second-, third-, and fourth-year elementary school educators.* Master's thesis, retrieved from ProQuest Dissertations and Theses. (UMI No. 1384152)

Liebmann, R. 1993. *Perceptions of human resource developers from product and service organizations as to the current and desired states of holonomy of managerial and manual employers.* PhD diss., retrieved from ProQuest Dissertations and Theses. (UMI No. 9327374)

Lin, C.-J. 2012. *The influence of Cognitive Coaching on the planning and use of instructional conversations, with a focus on mathematics instruction.* PhD diss., retrieved from ProQuest Dissertations and Theses. (UMI No. 3569062)

Lipton, L. 1993. *Transforming information into knowledge: Structured reflection in administrative practice.* Research/Technical Paper #143.

Loeschen, S. 2012. *Generating reflection and improving teacher pedagogy through the use of Cognitive Coaching in a mentor/beginning teacher relationship.* PhD diss., retrieved from ProQuest Dissertations and Theses. (UMI No. 3513140)

Los Angeles Unified School District. n.d. *Mathematics Instructional Guides Grades K–5.* Retrieved from http://notebook.lausd.net/portal/pageid=33,813644&_dad=pt.

Mackie, D. J. 1998. *Collegial observation: An alternative teacher evaluation strategy using Cognitive Coaching to promote professional growth and development.* PhD diss., retrieved from ProQuest Dissertations and Theses. (UMI No. 9826689)

Maginnis, J. L. 2009. *The relationship clinical faculty training has to student teacher self-efficacy.* PhD diss., retrieved from ProQuest Dissertations and Theses. (UMI No. 3387682)

McCloy, D. 2011. *Learning teaching reciprocal learning.* PhD diss., retrieved from ProQuest Dissertations and Theses. (UMI No. 3453503)

McCombs, B. 1995. *Teacher survey.* Aurora, CO: Mid-Continent Regional Educational Laboratory.

McDonough, S. 1991. *The supervision of principals: A comparison of existing and desired supervisory practices as perceived by principals trained in Cognitive Coaching and those without Cognitive Coaching training.* Unpublished master's thesis, California State University, Sacramento.

———. 1992, Spring. How principals want to be supervised. *Visions* 9(3): 4–5, 7b.

McLymont, E. F. 2000. *Mediated learning through the coaching approach facilitated by Cognitive Coaching.* PhD diss., retrieved from ProQuest Dissertations and Theses. (UMI No. NQ59634)

McLymont, E. F., and J. L. da Costa. 1998, April. *Cognitive Coaching: The vehicle for professional development and teacher collaboration.* Paper presented at the annual meeting of the American Educational Research Association, San Diego, CA.

McMahon, P. J. 1997. *Cognitive Coaching and special education advanced practicum interns: A study in peer coaching.* Master's thesis, retrieved from ProQuest Dissertations and Theses. (UMI No. 1385959)

Moche, R. 1999. *Cognitive Coaching and reflective thinking of Jewish day school teachers.* PhD diss., retrieved from ProQuest Dissertations and Theses. (UMI No. 9919383)

———. 2000 Fall/2001 Winter. Coaching teachers' thinking. *Journal of Jewish Education* 66(3): 19–29. doi.org/10.1080/0021624000660304.

Muchlinski, T. E. 1995. *Using Cognitive Coaching to model metacognition during instruction.* PhD diss., retrieved from ProQuest Dissertations and Theses. (UMI No. 9538459)

Newman, E. J. 1993. *The effect of teacher efficacy, locus-of-control, and teacher enthusiasm on student on-task behavior and achievement.* PhD diss., retrieved from ProQuest Dissertations and Theses. (UMI No. 9334264)

Podell, D. M., and L. C. Soodak. 1993, March/April. Teacher efficacy and bias in special education referrals. *Journal of Educational Research* 86(4): 247–53.

Poole, M. G. 1987. *Implementing change: The effects of teacher efficacy and interactions among educators.* PhD diss., retrieved from ProQuest Dissertations and Theses. (UMI No. 9121469)

Poole, M. G., and K. R. Okeafor. 1989, Winter. The effects of teacher efficacy and interactions among educators on curriculum implementation. Journal of Curriculum and Supervision 4(2): 146-161.

Powell, W., and O. Kusuma-Powell. 2007, Summer. Coaching students to new heights in writing. *Educational Leadership: Engaging the Whole Child* (online only) 64.

Reed, L. A. 2007. *Case study of the implementation of Cognitive Coaching by an instructional coach in a Title I elementary school.* PhD diss., retrieved from ProQuest Dissertations and Theses. (UMI No. 3270804)

Rennick, L. W. 2002. *The relationship between staff development in balanced literacy instruction for kindergarten teachers and student literacy achievement.* PhD diss., retrieved from ProQuest Dissertations and Theses. (UMI No. 3051831)

Rich, P. C. 2013. *Perceptions of Cognitive Coaching of alternatively certified reading teachers situated in two high poverty urban schools: A case study.* PhD diss., retrieved from ProQuest Dissertations and Theses. (UMI No. 3605228)

Rinaldi, L. 2013. *The effects of learning about the five states of mind on elementary children in grades 3, 4, and 5.* PhD diss., retrieved from ProQuest Dissertations and Theses. (UMI No. 3591907)

Robinson, B. K. 2010. *Building a pathway to support through professional development and induction: A case study examining an induction program for novice educators.* PhD diss., retrieved from ProQuest Dissertations and Theses. (UMI No. 3426698)

Robinson, J. M. 2011. *Supporting National Board candidates via Cognitive Coaching conversations and communities of practice.* PhD diss., retrieved from ProQuest Dissertations and Theses. (UMI No. 3449849)

Saphier, J. 1989. *The school culture survey.* Acton, MA: Research for Better Teaching.

Schlosser, J. L. 1998. *The impact of Cognitive Coaching on the thinking processes of elementary school teachers.* PhD diss., retrieved from ProQuest Dissertations and Theses. (UMI No. 9821080)

Schuman, S. 1991. *Self-reflection survey: Cognitive Coaching rating scale.* Federal Way, WA: Federal Way Public Schools.

Senge, P., N. Cambron-McCabe, T. Lucas, B. Smith, J. Dutton, and A. Kleiner. 2012. *Schools that learn: A fifth discipline fieldbook for educators, parents, and everyone who cares about education.* Rev. updated ed. New York: Crown Business.

Simmons, J., G. M. Sparks, A. Starko, M. Pasch, A. Colton, and J. Grinberg. 1989. *Exploring the structure of reflective pedagogical thinking in novice and expert teachers: The birth of a developmental taxonomy.* Paper presented at the annual meeting of the American Educational Research Association, San Francisco, CA.

Skytt, J. K., C. P. Hauserman, W. T. Rogers, and J. B. Johnson. 2014. *Cognitive Coaching: Building school leadership capacity in Alberta's education system Leader2 Leader Project (L2L).* Program report. Edmonton, Canada: The Alberta Teachers' Association.

Slinger, J. L. 2004. *Cognitive Coaching: Impact on students and influence on teachers.* PhD diss., retrieved from ProQuest Dissertations and Theses. (UMI No. 3138974)

Smith, M. C. 1997. *Self-reflection as a means of increasing teacher efficacy through Cognitive Coaching.* Master's thesis, retrieved from ProQuest Dissertations and Theses. (UMI No. 1384304)

Sommers, W., and A. Costa. 1993. Bo Peep was wrong. *NASSP Bulletin* 77(557): 110–13. doi.org/10.1177/019263659307755719.

Tarnasky, R. F. 2000. *Instructional supervision in selected Colorado community colleges: The role of the department chair in working with part-time and full-time faculty.* PhD diss., retrieved from ProQuest Dissertations and Theses. (UMI No. 9973923)

Townsend, S. 1995. *Understanding the effects of Cognitive Coaching on student teachers and cooperating teachers.* PhD diss., retrieved from ProQuest Dissertations and Theses. (UMI No. 9544000)

Tracz, S. M., and S. Gibson. 1986, November. *Effects of efficacy on academic achievement.* Paper presented at the annual meeting of the California Educational Research Association, Marina del Rey, CA.

Tschannen-Moran, M. 2014. *Trust matters: Leadership for successful schools.* San Francisco, CA: Jossey-Bass.

Tschannen-Moran, M., and C. R. Gareis. 2004. *Principal Sense of Self-Efficacy Scale.* Retrieved from http://mxtsch.people.wm.edu/ResearchTools/PSE_OMR.pdf.

Tschannen-Moran, M., and A. Woolfolk Hoy. 2001. Teacher efficacy: Capturing an elusive construct. *Teaching and Teacher Education* 17(7): 783–805.

Ushijima, T. M. 1996a. *Five states of mind scale for Cognitive Coaching: A measurement study.* PhD diss., retrieved from ProQuest Dissertations and Theses. (UMI No. 9720306)

———. 1996b. *The impact of Cognitive Coaching as a staff development process on student question asking and math problem solving skills.* Research Rep. No. 1996-1. Honolulu, HI: Author.

Uzat, S. L. 1999. *The relationship of Cognitive Coaching to years of teaching experience and to teacher reflective thought.* PhD diss., retrieved from ProQuest Dissertations and Theses. (UMI No. 9947709)

Vincenz, L. 1990. *Development of the Vincenz empowerment scale.* PhD diss., retrieved from ProQuest Dissertations and Theses. (UMI No. 9031010)

Weatherford, D., and N. Weatherford. 1991. *Professional growth through peer coaching: A handbook for implementation.* Unpublished master's thesis, California State University, Sacramento.

Wooten Burnett, S. 2015, in press. *Cognitive Coaching^SM: The impact on teacher candidates' teacher efficacy.* PhD diss., ProQuest Dissertations and Theses.

15

How Leaders Support Learning in an Agile Organization

It is not the strongest that survive environmental changes, it is the most flexible—the most intelligently agile.

Megginson (1963), writing about a statement
often attributed to Charles Darwin

Early in the development of our Cognitive Coaching training experiences, we conducted a workshop for principals in a Southern California school district. The superintendent requested he be given some time at the opening of the workshop to make some announcements to the district administrators. He announced that due to budgetary cutbacks, some administrative positions were to be eliminated. "Some of you will not be rehired for the next school year," he stated. "I will examine your performance over the past year and announce my decision in the next few weeks about who will be rehired and who will be released. . . . And now I'll turn it over to Art and Bob."

Needless to say, Cognitive Coaching did not prosper in that school district. And since that time we have become acutely aware of the culture of organizations and the attributes of leaders who create fertile ground in which the seeds of Cognitive Coaching will germinate, take root, and flourish: a culture of curiosity, thinking, and learning. This final chapter describes the attributes of an organization—an agile organization—and the dispositions and actions of the leaders of those organizations that are congruent with the principles, vision, and mission of Cognitive Coaching.

COGNITIVE COACHING: A HIGHER PURPOSE

What is our higher purpose? It is our belief in schools that are communities in which self-directedness is the dominant value; that learning, not teaching, is the central aim;

that developing the cognitive capacity of each inhabitant is a realized vision; that leadership is the responsibility of all the members of the learning community; that the states of mind inform the life, direction, and values of the school; and that teachers find schools as communities of support, inspiration, and individual and collective inquiry.

Most conditions of schooling hardly support these aims. According to a recent study published by the Alliance for Excellent Education, 13 percent of the nation's 3.4 million teachers move schools or leave the profession every year, costing states up to $2 billion. Researchers estimate that over 1 million teachers move in and out of schools annually, and between 40 and 50 percent quit within five years (Neason 2014).

The primary reason teachers leave is because they're dissatisfied. They leave because of isolated working conditions and inadequate support. These losses disproportionately affect high-poverty, urban, and rural schools, where teaching staffs often lack experience, thus deepening the divide between poor and wealthy students, with the most experienced teachers teaching the more privileged populations. Many teachers, though working hard and achieving success for students, become frustrated, exhausted, and have declining morale as they operate in an atmosphere of tensions due to:

- External mandates from upper echelons with insufficient time, training, and materials to teach them and the inability to influence the system
- Evaluations of teachers based upon one year's test scores irrespective of achievements in previous years
- Mismatches between test content and curriculum for which teachers had little involvement but are being held accountable
- Political wrangling about state and national programs and unfunded mandates
- Insufficient and nonuniform application and implementation of technology
- Unions, administrators, and boards adopting adversarial relations rather than working collaboratively toward collective solutions and common goals
- Teacher isolation when schools are organized around teacher separation rather than interdependence
- Fiscal woes and uncertainty of present and future financial conditions of their school district
- Extreme and challenging variation in students' abilities, developmental levels, learning styles, economic backgrounds, and cultural differences
- Inappropriate and inaccurate comparisons of other countries' educational systems regardless of teacher working conditions, culture, or political support

The effects of excessive and sustained stress on cognition, creativity, and social interaction are well documented. In barren, intellectually "polluted" school climates, some teachers understandably grow depressed. Teachers' vivid imagination, altruism, creativity, and intellectual prowess soon succumb to the humdrum daily routines of unruly or apathetic students, irrelevant curriculum, bureaucratic mandates, impersonal surroundings, and equally disinterested co-workers. In such an environment, the likelihood that teachers will value the development of students'—and their own—intellect and imagination is marginal.

At no time has the higher purpose we state been more important. Both the world and the schools in which students seek to learn are bewilderingly multidimensional. As reported in *Forbes* (Rogowsky 2014), data are doubling every 12 to 18 months and the number of smartphone users will reach 6.4 billion users in a world population of about 7 billion. Cloud computing, a $41-billion industry in 2011, is projected to be a $241-billion business by 2020. Eight out of 10 children report life is too complicated. A third of working professionals experience health issues brought on by workplace stress. Political biases are becoming stronger than racial biases (Brooks 2014). In 1960, 5 percent of Republicans and Democrats said they would be displeased if their child married someone from the other party. By 2010, 49 percent of Republicans and 33 percent of Democrats said they would object. As early as 1998, Kegan wrote that the mental demands of modern life were too complex for the majority of adults. By the time this book is being read, advances in learning theory, schooling, political discourse, technology, social media, and international relations will have left even more adults behind. What is required for schools to become agile is the emergence of a new kind of leadership: leaders who passionately communicate and reinforce high aspirations for students, who ignite and focus teacher energy, who develop the leadership capacities of teachers, who distribute informed decision making within their schools, and who, recognizing that complex (natural) systems organize themselves, foster the development of work cultures whose dominant features are self-directedness, collaboration, and inquiry. The principal in an agile school—in a simpler time, called an instructional leader—is now more than that. She/he creates the conditions in which teachers, given their daily work and learning, become the leaders of instruction. The principal carries the vision, sensitivities, and skills sets of a cultural change agent and developer of adults.

SCHOOLS ARE MESSY PLACES

Our finest moments are most likely to occur when we are feeling deeply uncomfortable, unhappy or unfulfilled. For it is only in such moments, propelled by our discomfort, that we are likely to step out of our ruts and start searching for different ways or truer answers.

—M. Scott Peck, American psychologist

Leaders have long recognized that schools are messy environments in which linear reasoning with command-and-control policies do not produce desired effects. Agile organizations dare to provoke the status quo, abandoning the principle of a single locus of control, adapting programs and processes to meet changing needs of students in a changing environment, using co-evolution, self-organizing thinking, and valuing of "messiness" as leadership practices that are consistent with complex adaptive systems. That educators have clung to old models of governance may in part be reason for current dysfunctions in institutions of learning. Schools should not be the old-style military organizations in which leaders decide and subordinates implement. Even militaries are pivoting away from this orientation. Atkinson and Moffat (2007, xix) write, "Agility is the gold standard for Information Age militaries. Facing uncertain futures and new sets of threats in a complex, dynamic, and challenging security environment, militaries around the world are transforming themselves, becoming more information-enabled and network-centric." They go on to say, "Forces must continually co-evolve in a changing environment" (40). In schools, these forces are teaching and learning networks.

Vulnerability does not sound like a leadership trait in the military, yet the stigmatization of depression and suicide and the actual suicide rate at Fort Carson dramatically plunged when the commanding general cried in front of his top officers, recounting the pain of his son's suicide. This was unusual, "nonmilitary" behavior, yet it launched a series of practical steps, including the creation of an embedded Behavioral Health Team that assigned a clinical physician to each brigade. This person would deploy to Afghanistan or Iraq and return with the troops when the brigade rotated home. General Mark Grahm's surprising intervention led to benefits throughout the army (Dreazen 2014). Agile leaders dare to work beyond the box of traditional leadership models.

The new identity of the "instructional leader" in schools is to design, construct, and maintain the conditions for teachers to become highly capable in organizing, activating, and assessing learning in the classroom (Leithwood and Seashore-Lewis 2011). The mediational leader allocates resources to the cognitive development of the community—the students, teachers, administrators, parents, and classified personnel.

The 1930s conception of schools as factories in which inputs resulted in predictable outcomes has been challenged by many writers. Frymeir, as long ago as 1987, noted that what teachers do is determined more by the structure of the workplace than by experience, degrees, or training. Fullan (1993, 2001) wrote that all the good work done in schools is merely tinkering unless teachers connect with their moral purpose and develop change agentry, which requires inquiry and collaboration. Wheatley (2006), writing about organizations as systems, reminds us that it is the relationship between things that is important. Cuban (2010) commented on the tendency to apply linear logic to the work of schools and offers a useful distinction between complex and complicated systems. Complicated systems, he says, are managed by linear thinking. Even activities as intricate as brain surgery or placing a human on the moon depend on mathematical algorithms, precise maps, and dedication to detail. They require command over many moving parts, each of which responds to stimulus in predictable ways. Complex systems, however, are more like raising a child. Endless variables emerge, each interacting with and influencing the others. Put simply, such complex systems cannot be controlled, and to attempt to do so would be to deny organizations their potential fidelity, agility, and unified focus as institutions of learning.

So it is with schools and small groups within schools. Tiny events make major disturbances. A person who consistently paraphrases in a group evokes more careful speaking and listening from others. High ratios of positivity in teams evoke more inquiry than in teams with less positively oriented statements (Losada 1999). Terms from meteorology are also used to describe complicated systems as dynamic (linear—think *automobile engine*) and dynamical as a synonym for complex (non-linear—think *hurricane*, which is caused by less than a one-half-degree Fahrenheit increase in temperature at the surface of the ocean) (Garmston and Wellman 2013). In *Leading in a Culture of Change*, Fullan (2001, 8) says "in a nonlinear . . . world, everything exists only in relationship to everything else, and the interactions among agents in the system lead to complex, unpredictable outcomes. In this world, interactions, or relationships, among its agents are the organizing principle."

Not only are schools in their entirety an anathema to linear management, in the past two decades or so the study of small groups has found them also to be complex and adaptive systems. They are "driven by interactions both among and between group members and between the group and its embedding contexts" (Arrow, McGrath, and Berdahl 2000, 3). In a word, they are dynamical, with nonlinear interactions where everything affects everything else and in which energy affects operations as much, if not more, than material things like curriculum guides or schedules in the life and learning of the group. From this perspective, groups are understood as interactions between individual and collective purposes. Groups in schools in which listening is prized and practiced—schools in which planning, reflecting, and problem resolving are routinely done with the skills sets and aims of Cognitive Coaching—are those that will experience success in adapting to continuing environmental changes while maintaining a sense of personal and collective efficacy.

The interdependence groomed within such schools is a leading resource for their responsiveness to change. As presented in chapter 2, social identities form by recognizing the interconnections among individuals, schools, communities, or cultures. Interdependence includes a sense of kinship that comes from a unity of being, a sense of sharing a common habitat (a class, school, or neighborhood), and a mutual bonding for common goals, shared values, and shared conceptions of being. Sergiovanni (1994) tells us that the German sociologist Ferdinand Tonnies called this way of being *gemeinschaft*. *Gemeinschaft* is a community of mental life. In some respects, Sergiovanni says, these values are still central to indigenous people in Canada and the United States. We have witnessed the peaceful coexistence within such comunities in Mennonite groups in British Columbia and indiginous alliances in Bali. *Gemeinschaft* contrasts with *gesellschaft*, in which community values have been replaced by contractual ones. This is the case in most modern organizations. Interdependent people's sense of self is enlarged from a conception of *me* to a sense of *us*. They understand that individuality is not lost as they connect with the group; only egocentricity is set aside.

Yet another set of terms has been used to describe these distinctions as *technical* and *adaptive* (Heifetz, Linsky, and Grashow 2009). Technical change requires people to apply existing skills in new ways; adaptive change calls for acquiring new tools and ways of thinking outside the current paradigm.

Table 15.1.   Linear and Nonlinear Systems Compared

Authors and Emphasis	Linear	Nonlinear
Garmston and Wellman	Dynamic: focused on parts and things	Dynamical: focused on whole and energy
Heifetz	Technical	Adaptive
Cuban	Complicated	Complex

Cognitive Coaching supports adaptive change in which leaders pay as much attention to energy as things. Margaret Wheatley (2006) reminds us that it is the relationship between things that is critical. She states, "Systems influence individuals, and individuals call forth systems" (27). Garmston and Von Frank (2012) assert that it is the quality of human relationships that is the energy source that produces work in organizations. Garmston and Wellman (2013) claim that shifts in identity—to being collaborators and inquirers—is essential for schools to become adaptive. In the professional climate in which agile schools exist, teachers and others feel satisfaction by mediating the growth of others. Rather than entering such hierarchical relationships of expert and novice, their commitment is to fostering self-directed learning in students, themselves, and colleagues.

Schools and groups within schools are nonlinear systems despite a façade of command-and-control mechanisms and policy manuals for subordinates to follow. Departments, grade-level groups, committees and administrative teams are nonlinear feedback networks that are continuously involved in ongoing processes of positive and negative feedback (Stacey 1996). Paradoxically, cognitive conflict is necessary for sound decision making (Garmston and Wellman 2013) yet ratios of nearly 6 to 1 positive to negative interactions within teams distinguish high-performing groups from mid- to-low performing groups (Losada 1999). These groups also balance inquiry with advocacy. Agile schools know that despite the organizing principles from which schools were formed, schools are indeed complex and, as such, wisely attend to energy, human relationships, and the dedication to fostering self-directedness in teachers and staff.

Leaders of agile organizations boldly work within uncertain environments, unpredictable demands, and the staff they inherit by initiating programs that develop faculty capacity. They may work from intuition and gut, as much as science, in understanding what their organizations can become. Most productively, the agile leader makes efforts toward changes in work culture because, as a second order/adaptive change, it has the greatest potential for significant and long-lasting results. They follow this with multiyear support and resource allocation. The Education Division of the Orange County Department of Education, a service center in Southern California, provides one example. In 1995 the California Department of Education launched a commission to develop a curriculum and assessment model to raise student learning, especially for students who normally fell below grade-level expectations. After exploring various models of professional improvement, Assistant Superintendent MacDonnell organized a retreat for the division's leaders to explain her vision of how Cognitive Coaching and Adaptive Schools could capacitate division members, making them even more effective in their interactions with school districts. She felt that she had been unable to communicate her personal passion for the vision in a way that staff could internalize so agreed to be coached publicly by one of the retreat trainers.

> During the coaching session, she revealed her vision, passion, and concerns about serving historically underserved students. She voiced her desire for professional efficacy for all involved in creating change in how all students must be well-served, especially those identified as underserved. The audience was spellbound as they listened to MacDonnell reveal her thinking about the well-trained staff and her vision for this work. She shared that her deepest passion was to focus the division's work on student achievement. Hearing her story, many in the room experienced a breakthrough in their own thinking about the power of this vision and its potential. Many staff members had tears in their eyes by the end of the coaching session. (MacDonnell and Lindsey 2011, 35)

With her leadership, the division maintained a 20-year focus on three primary foci identified in 1995: Cognitive Coaching, Cultural Proficiency, and Adaptive Schools. MacDonnell and Lindsey write, "this organization valued its human resources with the financial commitment to long-term, meaningful professional development. A large part of the investment in human capital was to develop our own experts. We now have cognitive coaching agency trainers and adaptive schools associate trainers" (2011, 38). We visited the division several years after MacDonnell's retirement and discovered that norms, skills sets, and values initiated under her leadership are intact.

In the agile schools we've witnessed, similar achievements in transformed work cultures occur. Cognitive Coaching is introduced not as another professional development activity but becomes a sustaining part of the organization. Regular refreshment and extended learning sessions are provided. Orientation is planned for new faculty members. Internal leaders and directors of ongoing professional development are established. We have seen this evolve in Jefferson County Schools in Colorado, Fairfax County Schools in Virginia, the Dar es Salam School in Tanzania, and the international school in Kuala Lumpur. Agile schools often integrate programs that support group development such as Adaptive Schools (Garmston and Wellman 2013) and Habits of Mind for students (Costa and Kallick 2008). Teachers teach students the skills sets of Cognitive Coaching and apply them to student projects like senior mentoring projects for incoming freshman.

It has been noted that schools are often fragmented with too many initiatives bearing little relationship to one another. In agile schools, leaders help teachers relate common principles across existing programs and use tenets of Cognitive Coaching as one set of criteria for new adoptions. Many of the Cognitive Coaching skills sets used in coaching are also applicable to teaching. Teachers employ mediative questioning in classrooms, paraphrasing with groups and individual conferences with student and parents. Learned elements of rapport are essential to maximize the parent–teacher relationship, and in many schools teachers introduce students to paraphrasing, posing questions and concepts of efficacy and craftsmanship. Overheard in a kindergarten class was a student saying, "Please, teacher, I didn't understand. Can you paraphrase that?"

In Hayward, California, ninth-grade students learn how to peer coach other students using skills sets from Cognitive Coaching. This program equips students with metacognitive tools to support their growth in their academic and personal lives. In an anonymous survey, 88 percent of students agreed that through peer coaching they "have learned skills that will help me be successful" and that "peer coaching has improved my ability to communicate with others" (Barbosa-Gonzales 2009).

In schools that are also agile organizations, collective identities have morphed into being teaching organizations in which development of each individual within the organization is the primary aim. Like the bricklayer when asked to explain his job reported that he was building a cathedral, teachers in agile organizations see their mission as the full development of human potential in the students with whom they work. Like a mission to end human suffering, such grand aspirations inspire, focus, and energize educators. Such passion stands up against restrictive district policies.

In East Los Angles, a principal who was told she was referring too many children for special services responded, "So your directions to me are to abandon the welfare of a percentage of our kids . . . do I have that right?" The directive withered in the face of this stance. Mission, we are told by Robert Dilts, is even a greater determinant of what people do than identity, when that mission is held as sacred and almost a spiritual aim (Dilts 2014).

One Michigan school principal stated:

Every building has had coaching conversations in front of the school board. Our board presentations used to be a dog-and-pony show where the buildings were showing off the latest and greatest. Now it is a conversation reflecting on our practice. In fact, it has been so powerful, one teacher did a coaching conversation involving representatives from the district and also the school board members about the superintendent's first year. Once you get into it, everybody sees the power in it. We have a superintendent who trusts us and developed that trust across the district to make it work. He knew it would take time and built in the time. It's not a matter that we just said "Cognitive Coaching is going to be embedded in our culture"; we made the commitment and we had a leader who was willing to say this is what it's going to take.

LEADING THE AGILE ORGANIZATION

Gone are the days of the heroic leader who bears the responsibility for solving the problems and directing the instruction in the daily life of the classroom. Top-down directives reinforce an environment of fear, distrust, and internal competitiveness that reduces collaboration and cooperation. They foster compliance instead of commitment, yet only genuine commitment can bring about the courage, imagination, patience, and persistence necessary in the agile organization. For those reasons, leadership must be distributed among diverse individuals and teams who share responsibility for creating the organization's future. The new work of instructional leaders in agile organizations is to design, construct, and maintain *the conditions* for teachers to become highly capable in organizing, activating, and assessing learning in the classroom. This requires finding and seeking time for teachers to network and to meet together to plan, to share information about individual students, and to learn with and from each other. Agile administrators create and protect reasonable workloads and provide opportunities and time for teacher growth and learning.

Following are several suggested actions that agile leaders can take to create an agile organization:

- *Develop a vision with your school staff and community—a vision that inspires the staff toward an important aim.* The full attainment of an aim, like going east or becoming a culture of inquiry, is illusive. There is always more distance to travel. For example, the vision of a company called Menlo Innovations (Sheridan 2013) is "Putting an end to suffering in the world." The vision of the Institute for Habits of Mind is "Educating for a more thought-full world." Kiva, a nonprofit lending company, has as its mission statement: "To connect people through lending for the sake of alleviating poverty." The school's vision is the future into which teachers, students, and parents gaze.

- *Hire great team members.* They should *not* be a collection of individuals, hired only because of their knowledge of their content or discipline. Rather, include those who surprise you with their divergent thinking, who understand how to come together and function as a team, who are passionate about teaching and learning, who are great problem solvers, who know how to think creatively and flexibly, who are open to continuous learning, and who have a sense of humor (Sheridan 2013).
- *Encourage autonomy, self-direction, and risk-taking.* Then stand back and let the staff do their jobs. Give teachers more leeway to do things on their own—designing curriculum, gathering assessment data, resolving parent complaints, and handling student discipline.

Autonomy makes such a difference because micromanagement, the opposite of autonomy, puts people in a threatened state. A reduction in autonomy is experienced by the brain in much the same way as a physical attack. This fight-or-flight reaction is triggered when a perceived threat activates the amygdala, the release of hormones, and other autonomic activity that makes people reactive. They are now primed to respond quickly and emotionally. When this fight-or-flight reaction kicks in, even if there is no visible response, productivity falls and the quality of decisions is diminished. When the neural circuits for being reactive drive behavior, some other neural circuits associated with executive thinking, (e.g., controlling oneself, paying attention, innovating, planning, and problem solving) become less active.

By giving staff members genuine autonomy, and groups collective autonomy, a school can reduce the frequency, duration, and intensity of this threat state. The perception of increased choice in itself activates reward-related circuits in the brain, making people feel more at ease.

Over time, sustained lack of autonomy is an ongoing source of stress, which by itself can habitually lead the brain to be more reactive than reflective. Sustained stress can also decrease the performance of important learning and memory brain circuits, as well as the performance of the prefrontal cortex, which is so important for reflection.

If school leaders try to enforce strict rules that make teachers feel micromanaged, the physiological state associated with the fight-or-flight reaction would probably lead to the opposite outcome (Newton and Davis 2014).

- *Expect the best from your faculty.* Expectations, as shown in the 1992 Pygmalion study (Rosenthal and Jacobson), influence performance. Dopamine may have been at work, report Bill and Ochan Powell (2014), as the teacher's pleasure with a student's performance inspires even more effort and attainment. What we expect, we get, becoming a form of prediction influencing the efforts of others.
- *Ask the "why" of education* (Sinek 2011). Teachers' goals are tightly connected with their perception of the purpose of education. When teachers are evaluated and schools are judged based on student achievement test scores, it sets up a signal about what is valued as the main purpose: Today, in most schools, outcomes of education test scores are the most important outcome of a child's education. From a neuroscience perspective, the system couldn't be better designed to bring out the worst in everybody.

The practice of focusing on test scores reinforces an external locus of control. Teachers are inclined to blame parents, students, the administration, inadequate, materials, or the miserable system, so it wasn't their fault that their scores were low. If the goal is to raise test scores, the teacher might try to convince parents and administrators that he/she is already doing everything possible. This fight-or-flight reaction leads to low levels of effectiveness and high turnover rates.

However, when school policies demonstrate and assess why this school exists, that what is important to be achieved are the larger, long-range goals (e.g., maximizing students' potentials, developing such lifelong dispositions as loving learning, persisting, curiosity, thinking interdependently, and resilience), teachers feel inspired, challenged, and dedicated to the larger outcomes of schooling.

David Chojnacki, executive director of the Near East South Asia Schools, initiated a book study based on Costa and Kallick's (2014) book using social networking among leaders of several schools in the area. The group began exploring the development of mechanisms to measure the attainment of school missions, thus lifting the message of what is important above test scores to the harder to measure, and more important, aspects of schooling (Chojnacki 2014).

Helping teachers and parents understand the larger "why" of education builds altruism; by helping students learn, teachers' intrinsic motivation rapidly expands. "Our goals become my goals" (Lipton and Wellman 2013a, 63).

- *Enlist the principal's support and modeling of Cognitive Coaching.* Modeling behaviors has a transfer effect, especially when leaders do the modeling and leaders take time to describe their intentions and how they are using certain behaviors to support that intention. See Recommendation #2 in chapter 14.

- *Create a climate of trust and pride* (see chapter 6). Trust and pride spiral throughout the ecosystem of the agile school. That's because humans in social settings tend to pick up the mood and attitudes of others nearby. Hatfield, Cacioppo, and Rapson (1994) refer to this phenomenon as "emotional contagion." One person's emotions can rapidly influence those of a group. When people see others act in a certain way, their mirror neurons are activated as if they had taken the actions themselves, even if they don't directly imitate that behavior.

  Watching someone else in a situation can have an impact on the brain similar to that of experiencing it directly. The staff room, department meetings, and faculty meetings are perfect locations for viral behavior, transmitted through observation. As long as people see the difference it makes, a change in a few individuals' neural patterns can move rapidly through the school culture.

- *Deliberately structure the organization for collaboration and interdependence.* When the board of education of Crystal Lake school district in Illinois decided to build a new high school, then-superintendent Joe Saban gathered the architects and staff together to dialogue about plans. They wanted to "build in" the principles of Cognitive Coaching. Interdependence was a high priority. As a result, the teachers' and departments' separate offices were abandoned. Instead, one large room containing all the teachers' desks was designed. Teachers from all departments could observe, work, hear, plan with each other, and think transdisciplinary. Lipton and Wellman (2013b) state, "the most interesting mysteries lie at the intersection of minds" (62). For example, at Menlo Innovations, computers are purposely shared by two employees, causing them to collaborate and build capacity of flexibility. Furthermore, computers are not standardized. Rather there is intentionally a mix of operating systems, speed, memory configuration, and so on. Furthermore, each week the pairs switch (Sheridan 2013)!

- *Support the use of Cognitive Coaching skills in informal settings, parent conferences, with students, and in meetings.* This is in line with Recommendation #10 in chapter 14 and is a significant indication of culture change.
- *Develop an atmosphere of continuous learning and knowledge growth.* Bob Garmston, in "Developing Smart Groups" (Costa and O'Leary 2013, 78–80), identifies five actions leaders can take to develop smarter groups:

    1. Construct diverse groups. When members of a group have diverse sets of mental tools, the process of group decision making is less likely to get stuck on less-than-optimal solutions and is more likely to result in superior ways of doing things.
    2. Include members with higher social sensitivity. Groups are more productive when they include members who not only listen well but also are alert and respond to other members' body language, facial expressions, and voice intonations.
    3. Develop skills in the ways groups talk. The group's ability to create a connectedness among its members and be open to creativity, to respect and honor the individual parts while expecting a higher outcome from the whole, is essential to the ability of the group to work well. Groups whose members align their efforts rather than pursue their own agendas for individual status are able to create higher collective intelligence and produce better outcomes.
    4. Promote positivity in the workplace. In groups, the amount of positivity influences the amount of inquiry and the ratio at which people talk about others rather than themselves. The more positivity in a group, the higher the achievement on measurable goals.
    5. Reflect on experience. Smart groups see encounters as learning opportunities. They use both formal and informal means of assessing what is working and what needs refinement. Reflection is the key to growth. Effective teams are conscious of what makes them effective. They possess knowledge about being a productive group or the persons in charge of districts, schools, grade-level teams, or departments are well advised to invest in the kinds of group development activities in which groups are encouraged to self-reflect, learn

from their experiences, and modify their practices. Reflection, not repetition, is the key to growth.

Schools are intended to produce learning—that goes not only for students but also for teachers and administrators. Leaders, therefore, are intent on developing teachers' craftsmanship—deepening teacher knowledge not only about the disciplines being taught but also developing expertise in their work, using knowledge about effective teaching, learning about how students learn, differences in learning styles, and so on (see chapters 8 and 9).

Much is being written about learning for careers, for college, and for success in life (Wagner 2008; Conley 2010; U. S. Department of Education, Office of Educational Technology 2013; Mansilla and Jackson 2011; Costa and Kallick 2015). These authors identify needed skills, knowledge, and dispositions for the uncertain future that students face today. Such learnings include:

Problem solving and critical thinking
Collaboration across networks and leading by influence
Agility and adaptability
Initiative and entrepreneurship
Effective written and oral communication
Accessing and analyzing information
Curiosity and imagination
Open-mindedness
Inquisitiveness
Analyzing the credibility and relevance of sources
Reasoning, argumentation, and explaining proof and point of view
Comparing, contrasting ideas, analyzing, and interpreting competing or conflicting evidence
Knowing how to arrive at an accurate answer
Finding many ways to solve problems
Persistence

Cultivating these dispositions may require reframing teaching and learning (Costa and Kallick 2014). Teachers and leaders, therefore, must find ways of organizing the curriculum, assessment, and the workplace to better understand and incorporate these new educational outcomes for tomorrow's citizens. This will require deepening teacher knowledge, developing expertise to do the work, using knowledge about effective teaching, teacher thinking, and school organization to make high-quality decisions.

In the typical high school, there are towers of knowledge. These silos create territories, deter transdisciplinary learning, and isolate teachers. Certain disciplines are perceived to have more worth than others. Through credit requirements; time allotments; allocation of resources; national, state, and local mandates; standards; testing; and so forth, schools send covert messages to students and the community concerning which subjects are of greater worth. This fractionalization across departments results in incongruent goals among the different people involved. We need, therefore, to put together teams of teachers who have been artificially separated by departments and to redefine their task from teaching their isolated content to instead develop multiple intellectual capacities of students. Peter Senge (2006) contends that we are all natural systems thinkers and the findings in cognitive research are compatible and supportive of the need to move from individual to collective intelligence, from disciplines to themes, from independence to relationships.

One Michigan school superintendent stated:

I wouldn't jump into this if I didn't what to change in our district. Don't take coaching as your model if you're not willing to change, because it is going to happen. I think that an important piece too is that we had PLCs [professional learning communities]—some were functioning better than others, and some people had more training than others, but that's where we started. We had that in place, and we had Habits of Mind being implemented in our classrooms. Then we started the Adaptive Schools work. These are all the concepts, skills, and strategies we need to run the PLCs that we started with. It's very much a cycle with so many connections.

CONCLUSION

Chapter 1 opened with the quote from Brian Stetler (2014, 52), senior media correspondent for CNN and the host of *Reliable Sources*. He stated: "Hypercomplexity—our world is changing at an unprecedented rate, becoming more complex and globally interactive. Clear-cut, unambiguous understandings are no longer an option." Thus, we believe the capacity for self-directed and continuous learning has become the most important capacity needed for survival in the future. Fortunately, the brain's adaptability supports our ability to change, to grow, and to continuously develop our intellectual capacities throughout a lifetime. And that's what Cognitive Coaching is all about: producing self-directed learners and leaders with the disposition for continuous, lifelong learning.

Technological/information age organizations are searching for ways to deal with the complexities, uncertainties, and rapidity of change of the 21st and 22nd centuries. Modern educational organizations are learning the virtues of agility. Agile organizations are inclusive, responsive, robust, innovative, flexible, and adaptive. Their leaders realize that these virtues are invested in the intellectual prowess of their employees. Therefore, they devote their human, material, and financial resources to protecting, enhancing, and liberating the intellectual capacities of their employees:

creativity, imagination, collaboration, entrepreneurship, innovation, networking, and learning how to learn (Atkinson and Moffat 2007; Costa, Garmston, and Zimmerman 2014; Hargreaves and Fullan 2012).

The relationship presumed by Cognitive Coaching is that teaching is a professional act and that coaches support teachers in becoming more resourceful, informed, and skillful professionals. Cognitive Coaches do not work to change overt behaviors but rather attend to the internal thought processes of teaching as a way to improve instruction. Behaviors change as a result of refined perceptions and cognitive processes.

Agile organizations view:

- The human mind as having no limits except those in which we choose to believe
- Humans as makers of meaning, and knowledge as constructed, both consciously and unconsciously, from experience
- All people at all ages as having the capacity to continue to develop intellectually
- All members of the school community as continual and active learners
- Leadership as the mediation of both the individual's and the organization's capacity for self-renewal
- Organizations, treated as complex adaptive systems, as capable of organic growth

Agile schools acknowledge interdependent communities of autonomous human beings bound by core values, common goals, caring, respect for diversity, and the ability to struggle together. Its members are reflective, examining their products and processes in a continuing climate of self-renewal. Agile schools encourage the development and contribution of each person's unique personal and professional identities.

This requires an adaptive, rather than a technical, shift in mind-set, a global view of mission driven by uncompromising values and beliefs about a better world.

APPENDICES

A

How to Make Student Learning Central to School Conversations

Descriptions of Practice

Drawn from the California State Department of Education (1997)

The following pages include some suggestions for achieving the goal of making student learning central to school conversations.

Student learning is, after all, the end goal of any set of teaching standards. Professional communities focus on student learning as the end and teaching as the means. As teachers from various grade levels and disciplines examine external standards for student learning, invite them to seek agreement on what constitutes good student work in their school in the subject matter they are addressing. If necessary, modify the external standards so that a common version for "our students" exists. Next, facilitate processes through which teachers can agree on developmental schemes and feedback systems that are coherent and consistent across classrooms. As teachers periodically examine actual student work from various classrooms, they construct practical knowledge about their students related to teaching standards. Three questions drive these collective inquiries: What are our standards for student work? Does this work meet our expectations? If not, what do we need to do to see that students are successful?

RELATE STUDENT STANDARDS TO TEACHING STANDARDS

Relate a set of teaching standards based on common agreement about learning and curriculum. Significant research has been invested in developing standards for teaching, and practicing teachers have offered many refinements. Standards for teaching will always address the six domains of inquiry (though often under different organizational schemes; see chapter 9).

LINK RESEARCH-BASED TEACHING STANDARDS
TO TEACHERS' DAILY PRACTICE

The several frameworks listed in the references describe the attributes of effective teaching. They describe a continuum of increasingly more sophisticated competencies, from beginner or novice through intermediate stages to levels of greater experience and mastery. Because these standards are descriptive and nonevaluative, they provide objective, research-based, and theoretically sound descriptors of what constitutes effective teaching.

Such documents suggest processes that are clearly formative in nature. Faculties can use such standards to guide inquiry into local definitions of effective teaching. These local elaborations can be used by teachers to prompt reflection about their own teaching and connections to student learning; to develop professional goals; and to guide, monitor, and self-assess progress of professional growth.

DISCUSS THE STANDARDS IN SMALL GROUPS,
GRADE LEVELS, OR DEPARTMENTS

Talk about what it means to align one's teaching practices with what is known about effective instruction in subject matter for your students. Invite groups to select one standard to work on. Define the standard operationally: What would it look and sound like? For example, what would the standard look and sound like in a third-grade class of mainly bilingual students while teaching reading? How would a standard look and sound in an 11th-grade honors physics class?

LEGITIMIZE SELF-MODIFICATION THROUGH REFLECTIVE PRACTICE

Teaching frameworks are useful resources for helping teachers reflect on their own practices. After they describe how a standard might sound or look in their classrooms, invite teachers to set goals related to where their teaching fits on the continuum of development. Ask them to consider: What might I do to achieve the next higher level of competency? Expert teachers are never complacent about their performance; they always have more to learn.

TALK ABOUT REAL STUDENTS AND REAL STUDENT WORK

Frequently examine student work together. Compare it to expectations. Is the work good enough? What can be done to "reinvent instruction" so that all students succeed? Relate improvement efforts to teaching standards.

COUCH THE STANDARDS IN PROJECTS OF ACTION RESEARCH

The most convincing data are not reports of distant and historical research but teachers' own results from using new processes and approaches in their classrooms with their own students and subject matter. Help teachers to establish and employ techniques of data collection to assess the results on student behavior and performance when they use new processes or standards.

PROMOTE IN-CLASSROOM OBSERVATION
OF TEACHING AND LEARNING

Provide release time for all teachers to observe one another as they apply teaching standards in their classrooms. After a conversation in which the standards are selected and operationalized, the observing teachers can collect evidence of indicators of the performance of such standards. During a reflecting conversation, share the data, allow the teacher to self-assess, and plan for achieving the next higher level of performance on the continuum.

B

Inner Coaching

Following are three sample internal dialogues to illustrate what it might sound like if you were talking to yourself about an upcoming walk-through and follow-up conversation:

Table Appendix B.   Inner Coaching

1. Jim is the teacher and will be observed my Kathleen, a teacher colleague.

Internal dialogue

Questions I am asking myself:	Possible responses:
What are my feelings toward this teacher? What are his/her strengths and talents? How well is he/she fitted for this assignment?	I'm eager to observe Jim. He's a great addition to the primary grade staff. As an ex-marine back from the Middle East, he has captured the respect of the parents and staff. His second-graders adore him. He can lift up two of the second-graders at a time with only one arm! They climb on his lap when he reads to them. He's a great father figure and role model for many of the kids who come from single-mom homes. He's already started a soccer team of primary grade students. While I don't care much for his tattoos, the kids seem to accept them.
What are my intentions for this observation and subsequent conversations? What are my long-range and immediate goals and outcomes for this teacher?	My long-range goal for Jim is to keep him in the profession. So many first-year teachers get discouraged and quit. But I think Jim has a bright future and I want him to feel efficacious and satisfied that he has made the right choice. My more immediate goals are to build trust and to have him see me as a resource to him
How will I know/by what indicators might I know that these desired goals have been achieved?	Does he share some feelings/satisfactions/confusions, problems with me? Does he invite me back? Does he seem relaxed, humorous, interested, and engaged in the conversation? Is he "kid focused"?

(continued)

Table Appendix B. *Continued*

Questions I am asking myself:	Possible responses:
What identity do I want to project about myself? What role am I to play?	I don't want to come across as "motherly" even though I'm much older and have had many more years of teaching experience than he. I want to come across as a colleague of equal status. He has a lot of new and refreshing ideas. I could learn a lot from him as well. We could learn from each other.
What behaviors will I be aware of in myself that are congruent with that role/identity?	I want to establish a rapport between us. I need to inquire with humility into his goals, aspirations, values, and attitudes toward his students. I'd like to know more about his feelings of acceptance by the primary grade staff. I need to listen both verbally and nonverbally. I must paraphrase, clarify, and remain nonjudgmental.
How will I gather feedback about the effectiveness of my skills in conducing the observation and conversation?	I must remember to invite him to reflect on this conversation and what it meant to him. I want him to give me some feedback as to what I did or didn't do that was meaningful to him. I'll want to know what I should do differently next time.

2. Maryann is a high school teacher of Spanish and will be observed by Chuck, a mentor who has been assigned to coach her.

Internal dialogue

Questions and interrogative self-talk:	Possible responses:
What are my feelings toward this teacher? what are his/her strengths and talents? How well is he/she fitted for this assignment?	Maryann is popular and a hard worker. She's been teaching here for 26 years. The students respond positively to her and she has good rapport with parents. I like working with her, though don't know much about her personally. She does not, however, contribute to PLC committee work. I think she feels somewhat isolated from the rest of the staff because she teaches in a specialty field that is difficult to integrate with other subjects. Furthermore, her classroom is in a portable at the far end of our campus
What are my intentions for this observation and subsequent conversations? What are my long-range and immediate goals and outcomes for this teacher?	I'd like to encourage her participation in committee work and in the conversation following this observation help her see the valued role she could play in professional group work. She knows the school having taught here so long and is knowledgeable about the large Spanish-speaking community. She has even taught the parents of some of our students and is well remembered.
What will be the focus of this observation? What is its relationship to our school plan and curriculum documents?	I will be looking for examples of contributions she can make to others as well as areas of uncertainty is which she might request support.
How will I know/by what indicators might I know that these desired goals have been achieved?	Ideally, Maryann will talk candidly with me about her work, areas where she sees growth potential and areas in which she is very proud.
What identity do I want to project about myself? What role am I to play?	I've not taught as many years as Maryann has so I intend to be seen as a reaffirming yet neutral mediator. I want to challenge Maryann to remain a continuous learner even though she is nearing retirement. She has much to offer.

Questions and interrogative self-talk:	Possible responses:
What behaviors will I be aware of in myself that are congruent with that role/identity?	Reporting data without judgment, empathically listening, and broaching the topic of collaborative work with other staff members.
How will I gather feedback about the effectiveness of my skills in conducting the observation and conversation?	If, by the end of our conversation, Maryann makes a commitment to contribute in some way or collaborating in even a small way to others on the staff.

3. Alicia, the principal of the school, is preparing her walk through the classroom of Michael, a middle-school social studies teacher.

Internal dialogue

Interrogative self-talk:	Possible responses:
What are my feelings toward this teacher? What are his/her strengths and talents? How well is he/she fitted for this assignment?	Michael is a recent hire. Although he's successfully taught sixth grade in an elementary school for several years, this is his first year at a middle school. He changed because he wanted to try teaching in his subject area (social studies) rather than being responsible for all the subjects at the elementary level. He is enthusiastic and learner centered. It is yet to be seen if this is the best grade-level fit for him.
What are my intentions for this observation and subsequent conversations? What are my long-range and immediate goals and outcomes for this teacher?	I want to establish that my visits are routine and not to be seen as threatening. My long-range goals are to support Michael in surviving and prospering during the first few months of instruction and supporting him in becoming a self-confident teacher of social studies for these sometimes baffling students during their "transition years" to adolescence.
What will be the focus of this observation? What is its relationship to our school plan and curriculum documents?	The middle-school teachers are working to help students develop strategies for listening to each other. I am interested in learning the degree of supporting. I want to learn how he is accommodating this learning through the use of class meetings, discussion groups, and inquiry sessions.
How will I know/by what indicators might I know that these desired goals have been achieved?	I'm hoping Michael pays little attention to me during the visit and that I am free to wander and interview kids about what they are doing and learning. I predict that Michael will ask me to judge him and I will resist. I want him to be self-evaluative.
What identity do I want to project about myself? What role am I to play?	Nonjudgmental observer, data gatherer. Because my role as an administrator is also to evaluate Michael, I must be clear about my intentions with him at this time. This is *not* an evaluative observation; it is a growth and learning opportunity. I'm not sure he distinguishes between the two yet so I must be sure to let him know the difference and that his evaluation will be at a later date. Come to think of it, I should not even use the word "evaluation" in our conversation today.

(continued)

Table Appendix B.   *Continued*

Interrogative self-talk:	Possible responses:
What behaviors will I be aware of in myself that are congruent with that role/ identity?	I will silently slip into the room and initially take a place at the rear of the room where I can observe. When students are at their work tasks, I will wander and speak with some of them. During our reflection conversation, I will avoid making value judgments and avoid giving advice. I will ask Michael how he thinks the lesson went and what evidence he has to support his judgments.
How will I gather feedback about the effectiveness of my skills in conducing the observation and conversation?	I will ask Michael to reflect on our conversation and to give me some feedback as to what I did or didn't do to make him feel comfortable. I will inquire as to what he is carrying forth from this conversation.

C

Mediative Questioning

Effective Cognitive Coaches are aware of the direct correlation between the structure of their questions and the production of thought processes. As such, they deliberately use language and syntax to construct questions in ways that mediate desired mental processes in the other person's mind.

The following sample questions come from planning and reflective conversations. The questions are designed to mediate specific mental processes. These questions are only examples, however, and they are not meant to be prescriptive.

The coach's intent with these questions is to engage, mediate, and thereby enhance another's cognitive functions. The questions are adroitly designed and posed to deliberately engage the intellectual functions of teaching. The long-range intent, of course, is self-coaching: to help the other person reach a point where he or she habitually and autonomously initiates such questions without the coach's intervention.

The left-hand column in each chart contains desired cognitive thoughts and processes. The right-hand column contains sample questions. Specific syntax cues in the questions are in boldface type. As you analyze these examples, you may wish to refer to chapter 3 to review the criteria for constructing powerful, mediative questions.

Table Appendix C.   Planning and Reflecting Conversations

Planning Conversation	
If the desired cognitive thought or process is to:	**Then the coach might ask:**
Describe (State the purpose of the lesson.)	**What** outcomes do you have in mind for your lesson today?
Envision (Translate the lesson purposes into descriptions of desirable, observable student behaviors.)	As you see this lesson unfolding, **what will** students be doing?

(*continued*)

Table Appendix C. *Continued*

If the desired cognitive thought or process is to:	Then the coach might ask:
Predict (Envision teaching strategies and behaviors to facilitate students' performance of desired behaviors.)	As you envision this lesson, **what do you see** yourself doing to produce those student outcomes?
Sequence (Describe the sequence in which the lesson will occur.)	**What will** you be doing **first? Next? Last?** How will you close the lesson?
Estimate (Anticipate the duration of activities.)	As you consider the opening of the lesson, **how long** do you anticipate that will take?
Define (Formulate procedures for assessing outcomes by envisioning, defining, and setting success indicators.)	**What will you see** students doing or hear them saying **that will indicate** to you that your lesson is successful?
Metacogitate (Monitor his or her own behavior during the lesson.)	**What will you be aware of** in students' reaction **to know if** your directions are understood?
Self-Assess (Identify a process for personal learning.)	As a professional, what are **you hoping to learn more about your own** practices as a result of this lesson?
Describe (Depict the data-collecting role of the observer.)	**What will you want me to look** for and give you feedback about while I am in your classroom?

Reflecting Conversation

If the desired cognitive thought or process is to:	Then the coach might ask:
Assess (Express feelings about the lesson.)	As you reflect on your lesson, **how do you feel** it went?
Recall and Relate (Recollect student behaviors observed during the lesson to support those feelings.)	**What did** you see students doing (or hear them saying) **that made you feel** that way?
Recall (Recollect their own behavior during the lesson.)	**What do you recall** about your own behavior during the lesson?
Compare (Draw a comparison between student behavior performed with student behavior desired.)	**How did** what you observe **compare** with what you planned?
Infer (Abstract meaning from data.)	Given this information, **what do you make of it**?
Draw Conclusions (Assess the achievement of the lesson purposes.)	As you reflect on the goals for this lesson, **what can you say about** your students' achievement of them?
Metacogitate (Become aware of and monitor their own thinking during the lesson.)	**What were you thinking when** you decided to change the design of the lesson? OR **What were you aware of** that students were doing **that signaled you** to change the format of the lesson?
Infer from Data (Draw hypotheses and explanations from the data provided.)	What **inferences** might you **draw** from these **data**?

If the desired cognitive thought or process is to:	Then the coach might ask:
Analyze (Examine why the student behaviors were or were not achieved.)	What hunches do you have to **explain why** some students performed as you had hoped while others did not?
Describe Cause-Effect (Draw causal relationships.)	**What did you do** (or not do) to **produce the results** you obtained?
Self-Assess (Construct personal learnings.)	What personal learnings did you gain from this experience?
Apply (Prescribe alternative teaching strategies, behaviors, or conditions.)	As you plan future lessons, **what insights** have you developed that **might be carried forth** to the next lesson or other lessons?
Evaluate (Give feedback about the effects of this coaching session and the coach's conferencing skills.)	As you think back over our conversation, **what has this** coaching session **done for you? What** is it that I did (or didn't) do that was **of benefit** to you? **What assisted** you? **What** could I **do differently** in future coaching sessions?

D

Calibrating Conversation Script

Table Appendix D. Calibrating Conversation Script

Conversation		Regions of the Map
Coach	**Coachee**	
You have the self-assessment instrument for presenters, and you've had a chance to look at it and rank yourself as to where you are. There are several different areas there— which one would you like to focus on?	Well, under communication, I really wanted to focus on verbal, and if we have any additional time I'd like to look at nonverbal. So my first choice is verbal communication while presenting.	Select a focus.
So when you look at this, and where you placed yourself, what are some of the data you have with both frequency and proficiency?	What stands out for me is paraphrasing, in terms of I marked myself highly skilled in paraphrasing and high frequency because that's something I do deliberately, and yet I'm at a level of unconscious competence—it's just who I am. So that stood out for me. Choosing appropriate voice was another one I marked highly skilled and that I use quite frequently. I'm really conscious when I'm presenting to groups of using the approachable voice intentionally and using the credible voice intentionally with groups and also the pitch and modulation at other times.	Identify existing level of performance or placement on a rubric and give supporting evidence.
So a couple of different things that stand out for you that you're already aware of. You have high skill and high frequency in these, so that's going really well for you.	Yes. Then I looked at some stretches for me where I'd really like to grow. I looked at being pretty proficient in probing for specificity; however, the frequency that I do that is another	

(*continued*)

Table Appendix D. *Continued*

Conversation		Regions of the Map
Coach	**Coachee**	
	thing. I know how to do it, but I'm forgetting to do it or not doing it as frequently as I could—as the opportunities present themselves.	
So you know you're capable of probing, but you're not doing it as often as you think you might be able to.	Absolutely. And something that surprised me was I'm not confident in redirecting resistance. And so as I reflected on that, I thought at first, I'm pretty proficient, then I changed that to partially proficient. The reason I think I'm partially proficient is I don't have any fist-fights during my presentations (laughs) and people feel heard, yet I want to become more conscious and skillful in redirecting resistance. So I want to notice it and have skills to navigate it.	
So two things: being more intentional about probing and redirecting resistance.	Yes.	Specify desired placement and explore values, beliefs, and identity congruent with desired placement.
At what level of competence would you like to be?	Oh, highly skilled, definitely.	
So your goal is to advance to a higher level of this rubric, and you're considering things that you hold dear to you. What causes you to want to move forward in those two areas?	I hold dear the needs of participants being met, and I want to move forward in those areas because I think I can serve better. When participants are given the opportunity to become more specific in their thinking, to hone and refine their thinking and their practices, that's going to serve them better. And with redirecting resistance, I hope I'm not missing anything. I want to serve participants better by noticing tensions that live in the room within individuals and the group.	
So really on both counts it comes back to the participants and having them get the most out of your presentation.	Yes.	
When you think about experiences that you've had with presenters who were really powerful in helping you get clarity so you could walk away with deeper understanding or	Well, let's take the redirecting resistance or tension. I notice Bob Garmston with redirecting resistance and what I notice is that he slows the pace. With a question	Establish behavioral indicators for new placement on rubric or level of performance.

Conversation		Regions of the Map
Coach	**Coachee**	
presenters who were highly conscious of resistance, what are some of the behaviors you noticed about them?	that's expressing some tension, he breathes. I notice that he shows thoughtfulness, breathes, and paraphrases that person. I know that he's thinking and what he's doing is intentional. Another thing I've noticed from presenters who I think are acknowledging resistance is using the feel, felt, found pattern, and I'd like to do that more readily. In fact the last time I had resistance in a workshop I didn't think about it until afterward. I realized I could have used the pattern of feel, felt, found with that tension. Another thing I've seen presenters do, and I'd want to be more mindful of, is offering a polarity paraphrase. It helps relieve the tensions and helps people expand their perspective. So I would love to have those things come up right in the moment when I'm feeling that tension—being so skillful that they come to mind and I go right into them.	
So you want it to come very naturally like the paraphrasing does. So for you a real deeply held belief is that it's the presenter's responsibility to be aware of those things and do something about it.	Definitely, absolutely. It's the presenter who is there to support learning and help people to know that tensions are a part of learning and any those tensions are part of teaching so being more adept at that is something I really seek.	
So an ideal place for you is being fully present, fully conscious, not only with content but also with presentation skills.	Absolutely. You bring up the presentation processes and the content and I think keeping an eye on both in terms of navigating tensions is important. Thank you for framing it in that way.	
What might be some people or strategies that would help you get from awareness to the unconscious competence level? What might be some supports that you will draw upon to move to that level in this area?	I definitely want to read more deeply in the *Presenter's Fieldbook* and add those strategies. What I do is put them on 3×5 cards so they're living in the room with me. And I'll scan for opportunities to actually practice using them. That's one thing that I can do. Another is to actually observe other presenters	Describe support needed to get to new placement and commit to action.

(continued)

Table Appendix D. *Continued*

Conversation		Regions of the Map
Coach	**Coachee**	
	I know who are proficient or highly skilled at acknowledging or navigating resistance. That's another thing I'd love to do. Also, acknowledging what I'm already doing around that so that I have baseline data. I know that I get through presentations and people generally have good feelings about it. And I know that I'm a good listener, so there are some things that I'm doing that I'm probably not really aware of, and so really paying attention next time I present in terms of what is it that I already do to help people navigate tensions.	
So really three big things: using a strategy that's worked for you, which is to have some 3x5 cards as reminders of what you're working on; to observe others with that particular lens of how they manage that; and then to rely upon the things you know you're doing well and incorporating those into this resistance piece.	Yes.	
So we've had an opportunity to chat about this. When you think about where you were when you marked this and where you are now, what are some things that have happened for you as a result of our time together?	Actually I've really appreciated the experience of thinking about this in terms of what does it look like when it's being done. So having an opportunity to say there are some behaviors that match these labels and to become more specific in terms of what will I do and what will I be looking for. Also to say aloud what you've been thinking internally is just great. I feel supported in this conversation by just being able to speak it, and to talk about it openly also helps me to commit personally to doing something about these areas of growth for myself.	Reflect on the coaching process, explore refinements, and explore ways of using this process on your own.
So a combination of the reflection, getting more specific about what it looks like, and just having a chance to make your thinking more visible.	It's also experiencing what's possible in a presentation for me. Working with this self-assessment has brought back to mind all of	

Conversation		Regions of the Map
Coach	**Coachee**	
	the nuances that go into doing a presentation that really supports learners. So having this talk has really reminded me and expanded the possibility of growth for me and growth for my participants.	
Thank you. What hunches do you have about how you might use this same type of process without me in the future?	I will hear myself asking some of the same types of questions that you asked me.	

E

Sources of Standards for
Teachers' and Leaders' Performance

There are numerous sources of standards of teacher and leader performance. The following are but a few.

Council of Chief State School Officers. n.d. "Interstate School Leaders Licensure Consortium (ISLLC) Standards have recently been developed by the Council of Chief State School Officers (CSSO) in collaboration with the National Policy Board on Educational Administration (NPBEA)." www.ccsso.org.

Danielson, C. 2007. *Enhancing professional practice. A framework for teaching.* Alexandria, VA: Association for Supervision and Curriculum Development.

Hattie, J. 2003, October. *Teachers make a difference: What is the research evidence?* Auckland: University of Auckland, Australian Council for Educational Research.

Interstate Teacher Assessment and Support Consortium Standards Model: Core teaching standards: A resource for state dialogue. 2011, April. Washington, DC: Council of Chief State School Officers.

Marzano, R. J. 2007. *The art and science of teaching.* Alexandria, VA: Association for Supervision and Curriculum Development.

Marzano, R., T. Frontier, and D. Livingston. 2011. *Effective supervision: Supporting the art and science of teaching.* Alexandria, VA: Association for Supervision and Curriculum Development.

Marzano, R., D. Pickering, and J. Pollock. 2001. *Classroom instruction that works: Research-based strategies for increasing student achievement.* Alexandria, VA: Association for Supervision and Curriculum Development.

National Board for Teacher Certification Standards. Available from the National Board for Teacher Certification Standards, 1525 Wilson Blvd., Suite 700, Arlington, VA 22209. www.nbpts.org.

Office for Standards in Education of the UK. 2004. *Why colleges succeed.* Crown Copyright: UK Government. Retrieved 1 August 2013 from http://dera.ioe.ac.uk/id/eprint/5184.

Pathwise Assessment Criteria, Framework induction Program. 2005. Available from Educational Testing Service (ETS). www.ets.com.

Saphier, J., M. A. Haley-Spica, and R. Gower. 2008. *The skillful teacher: Building your teaching skills.* Acton, MA: Research for Better Teaching.

Silver, H., et al. 2014. *The theory of classroom effectiveness framework.* Franklin Lakes, NJ: Thoughtful Education Press, Silver Strong. http://www.thoughtfulclassroom.com/.

Sternberg, R., and J. Horvath. 1995, August–September. A prototype view of expert teaching. *Educational Researcher* 24(6): 9–17.

Tsui, A. 2003. *Understanding expertise in teaching: Case studies in ESL teaching*. Cambridge: Cambridge University Press. 978-0-521-63569-1.

Tucker, P. D., and J. H. Stronge. 2005. *Linking teacher evaluation and student learning*. Alexandria, VA: Association for Supervision and Curriculum Development.

F

Other Resources about Self-Directedness

Although the chapters in this book reflect developments in many areas, companion books provide additional resources for developing self-directedness in students, teachers, and leaders. They include:

Costa, A., R. Garmston, and D. Zimmerman. 2014. *Cognitive Capital: Investing in teacher quality*. Teachers College Press.
   Explores the application of Cognitive Coaching to teacher groups becoming partners with administrators in setting and monitoring school improvement initiatives.
Costa, A., and B. Kallick. 2004. *Assessment strategies for self-directed learning*. Thousand Oaks, CA: Corwin Press.
———. 2009. *Learning and leading with habits of mind: 16 characteristics for success*. Alexandria, VA: Association for Supervision and Curriculum Development.
———. 2014. *Dispositions: Reframing teaching and learning*. Thousand Oaks, CA: Corwin Press
Dweck, C. 2000. *Self-theories: Their role in motivation, personality, and development*. Philadelphia, PA: Psychology Press.
Edwards, J. 2001. *Inviting students to learn: 100 tips for talking effectively with your students*. Alexandria, VA: Association for Supervision and Curriculum Development.
Ellison, J., and C. Hayes. 2003. *Cognitive Coaching: Weaving threads of learning and change into the culture of an organization*. New York: Roman and Littlefield.
   Explores how the principles of Cognitive Coaching transform a school system's identity by applying congruent practices.
———. 2006. *Effective school leadership: Developing principals through Cognitive Coaching*. New York: Roman and Littlefield.
   Grounded in research, this book offers practical examples of coaching conversations for principals illuminating a coach's thinking and decision making.
Feuerstein, R., L. Falik, and R. Feuerstein. 2015 *Changing minds and brains: Insights on 60 years of mediated learning experience: The last work of Reuven Feuerstein*. New York: Teachers College Press.
Garmston, R. 2011. *I don't do that anymore: A memoir of awakening and resilience*. Charleston, SC: CreateSpace.
   From high school dropout to advanced degrees and from worldwide contributions to professional development, Garmston's story illustrates self-determination and perseverance through numerous challenges.

Garmston, R., and B. Wellman. 2009. *The Adaptive School: A sourcebook for developing collaborative groups.* Norwood, MA: Christopher Gordon.

It is in the conceptual realm that this book makes a vital contribution. *The Adaptive School* addresses the essential processes of schooling (culture, community, communications) that are too often overlooked in the frenzy of responding to continuing demands for change. Included are practical structures, strategies, and tools.

Garmston, R., and D. Zimmerman. 2013. *Lemons to lemonade: Resolving problems in meetings, workshops, and PLCs.* Thousand Oaks, CA: Corwin.

Collaborative work requires informed attention in which all participants share responsibility for effective interactions. This how to book reveals mental approaches and leadership tools for resolving a host of common, and not so common, problems that surface when groups meet.

Grift, G. 2014 *Transformative talk Cognitive Coaches share their stories.* New York: Rowman and Littlefield.

Becoming a Cognitive Coach requires a commitment time and mental energy resources, and an openness to transforming your identity as a helping person. Is it worth it? You bet! This book contains stories of the learning members individually and collectively, on school culture, students, and, of course, themselves.

Jackson, Y. 2011. *The pedagogy of confidence: Inspiring high intellectual performance in urban schools.* New York: Teachers College Press

Pink, D. 2009. *Drive: The surprising truth about what motivates us.* New York: Penguin Books.

The author argues that human motivation is largely intrinsic, and that the aspects of this motivation can be divided into autonomy, mastery, and purpose. He argues against old models of motivation driven by rewards and fear of punishment, dominated by extrinsic factors such as money. Traditional rewards are not as effective as we think.

G

List of Videos, Topics, and Length of Time in Minutes

Following is a list of videos that accompany the text. Videos may be viewed by logging on to the HTML as noted. Each video is coded with a number. The first number identifies the chapter and if the chapter contains more than one video, the second number indicates sequence of videos. For example, Video 2-1 would be chapter 2, video number 1. In some cases there is more than one video segment. The third number indicates the segment: 3-1.2. Also included are some suggestions to direct your attention to the salient features of this video. So that you can anticipate the time needed for viewing, the duration of each video in minutes and seconds is also listed.

While each video is intended to focus on a certain coaching skill or map, the demonstration will also include multiple skills. For example, while viewing the video on the planning conversation, you will also be aware of the use of rapport, questioning, paraphrasing, pausing, and so on.

Appendix G. List of Videos, Topics, and Length of Time in Minutes

Video #	Access	Content	Focus	Time
Video 3-1 Three segments 3-1.1 3-1.2 3-1.3	http://www.thinkingcollaborative.com/supporting-audio-videos/3-1/	Rapport	There are two segments in which Jane and Monica demonstrate rapport. Observe the matching of their body posture. In the second video notice the mirror movement of the coach's head in response to Monica's. In a third segment with Carolee as coach, notice the body configurations matching in more casual postures.	6.00
Video 3-2	http://www.thinkingcollaborative.com/supporting-audio-videos/3-2/	Wait Time	The coach, Carolee, uses wait time after having asked Kris a question (wait time 1). Chris processes internally for over 7 seconds before responding the first time. In response to a second question, Chris reflects nearly as long before responding. These responses are consistent with Rowe's research showing that wait time of 3-5 seconds elicits deeper thinking.	1.5
Video 3-3 Two segments 3-3.1 3-3.2	http://www.thinkingcollaborative.com/supporting-audio-videos/3-3/	Paraphrasing Paraphrasing 2	Viewers will notice Ochan's appreciation of the abstracting paraphrase that led her to transfer the concept of deliberateness to another area of her work. Finally, Carolee summarizes Ochan's learning in this reflection conversation.	.38 .23
Video 3-4.1 Three segments	http://www.thinkingcollaborative.com/supporting-audio-videos/3-4/	Questioning 1 (How do you make decisions?) (What do you look for in the data?)	Jane, the coach, asks Shatta, a Planning Central Office director, two meditational questions: How do you make decisions? (What do you look for in the data?)	.55
Video 3-4.2	http://www.thinkingcollaborative.com/supporting-audio-videos/3-4.2/	Questioning 2 (a question about values)	In video 3-4.2, during a calibrating conversation. Sue, the coach, asks Carolyn a mediating question about values.	.16
Video 3-4.3	http://www.thinkingcollaborative.com/supporting-audio-videos/3-4.3/	Questioning 3 (a question relating fear to goals)	In Video 3-4,3, during a reflecting conversation, Carolee, the coach, poses a mediating question for Ochan, exploring Ochan's expression of fear (internal content) of goals. (external content)	.40

Video 5.1	http://www.thinkingcollaborative.com/supporting-audio-videos/5-1/	Shifting Support Functions (consulting on Habits of Mind)	Jane, the coach, switches the support function from coaching to consulting about Habits of Mind with a new teacher, Erin. Notice how Jane's body language breaks rapport to signal a change in support function.	2.42
Audio 10-1	http://www.thinkingcollaborative.com/audio-10-1/	I'm Articulate	In this audio we are reminded of how language can evoke stereotypes.	4.5
Video 10-1	http://www.thinkingcollaborative.com/supporting-audio-videos-10-1/	Eye Accessing Cues	Carolee asks Ochan a question, which initiates Ochan's internal search. Notice her eye movements. From what you have learned about eye movements and representational systems, what are some inferences you might make about Ochan's internal search?	.36
Video 11-1	http://www.thinkingcollaborative.com/supporting-audio-videos-11-1/	Planning Conversation	In this video James, the teacher, is being coached by Alejandro. In this conversation viewers are invited to focus on the steps in the planning conversation map. You may wish to use the planning map in figure 11.1 to observe the journey.	12.00
Video 11-2	http://www.thinkingcollaborative.com/supporting-audio-videos-11-2/	Reflecting Conversation	In this video Carolee coaches Ochan through a reflective conversation. You may wish to use the reflecting map in figure 11.2 to follow the journey.	12.00
Video 12-1 Two segments	http://www.thinkingcollaborative.com/supporting-audio-videos-12-1/	Cognitive Shift	In the first video, Jane, the coach, paraphrases Dana. Watch Dana's response noting ways her behaviors suggest what is happening internally. In the second video Carolee asks a question triggering an observable cognitive shift in Ochan.	1.02
Video 12-2	http://www.thinkingcollaborative.com/supporting-audio-videos-12-2/	Cognitive Shift	In the second video Carolee asks a question triggering a cognitive shift in Ochan.	1.05
Video 13-1	http://www.thinkingcollaborative.com/supporting-audio-videos-13-1/	Calibrating Conversation	In this video Sue coaches Carolyn through a calibrating conversation. Also note the use of a third point—the list of standards—in the conversation.	15 min

Glossary of Terms

Agile organization: In the context of this book, refers to the current revolution in organizations from a top-down, authoritarian, rule-bound structure to one of trust, collaboration, networking, and of relationships and the intellect, placing a premium on our greatest resource: human minds in relationship with one another.

Autobiographical listening: Occurs when the listener associates a colleague's story with his or her own experiences. This kind of listening includes making judgments and comparisons. (See also *Set-asides*.)

Calibrating conversation: Uses an externally generated, locally adopted third-point description of teaching excellence with the purpose of supporting the capacities and practice of self-directed learning and instructional improvement. It draws on a teacher's drive for craftsmanship and their disposition of continuous learning.

Capabilities: These are the metacognitive maps, models, and skills that individuals use to guide behavioral choices. For the coach, these metacognitive functions include knowing one's intentions and choosing congruent behaviors, setting aside unproductive patterns of listening and responding, adjusting personal style preferences, and navigating within and among various coaching maps and support functions.

Coaching cycle: Describes how a coach uses both the planning and reflective maps before and after an event in which the coach will be present as observer and data collector.

Cognitive Coaching: A nonjudgmental, interactive strategy focused on developing and utilizing cognitive processes, liberating internal resources, and accessing the five states of mind as a means of more effectively achieving goals while enhancing self-directed learning.

Collaborating: Providing support in a mentoring or collegial relationship in which coach and colleague share and test ideas, determine focus for inquiry, and gather and interpret data to inform collaborative practice.

347

Consulting: A support service in which the colleague develops teaching effectiveness by drawing on the more extensive experience of the consulting educator.

Consciousness: The human capacity to represent information about what is happening outside and inside the body in such a way that it can be evaluated and acted upon by the body. To be conscious is to be aware of one's thoughts, feelings, points of view, and behaviors and the effect they have on the self and others. (See also *States of mind.*)

Constructivism: A theory of learning based on a belief that human beings have an innate quest to make meaning from their experiences. When humans are perplexed by anomalies, discrepancies, or new information, they have a natural inclination to make sense of it. This theory is used to guide pedagogy, curriculum, and assessment of learning.

Craftsmanship: The human drive to hone, refine, and constantly work for improvement. Includes striving for precision, elegance, refinement, and fidelity. (See also *States of mind.*)

Cultural Proficiency: A model for individual transformation and organizational change. It is the *policies and practices* of an organization, and the *values and behaviors* of an individual, that enable that organization and person to interact effectively with one's clients, colleagues, and community in a culturally diverse environment (Lindsey, et al. 2009).

Efficacy: Engaging in cause-and-effect thinking, spending energy on tasks, setting challenging goals, persevering in the face of barriers and occasional failure, and forecasting future performances accurately. Efficacy is linked to a belief that one's work will make a difference and is related to being optimistic, confident, and knowledgeable. Efficacy is also linked to having an internal locus of control. (See also *States of mind.*)

Evaluation: Measurement of and judgment about performance, based on external criteria or standards.

Flexibility: The human capacity to perceive from multiple perspectives and to endeavor to change, adapt, and expand the repertoire of response patterns. Involves humor, creativity, risk taking, and adaptability. (See also *States of mind.*)

Holonomy: An individual's cognitive capacity to be autonomous and interdependent at the same time. The ability to function as a member of the whole while still maintaining separateness. A person's cognitive capacity to accept the concept that he or she is whole in terms of self and yet subordinate to a higher system. The idea of holonomy is based on Arthur Koestler's work.

Identity: A mental model we construct of ourselves as unique individuals constructed from the meanings we make of our interactions with others and how we perceive that others see us.

Interdependence: The human need for reciprocity, belonging, and connectedness; the inclination to become one with the larger system and community. (See also *States of mind.*)

Inquisitive listening: Occurs when the listener becomes curious about portions of a colleague's story that are not relevant to the problem at hand. Inquisitive listening sometimes involves mind reading and scrutinizing. (See also *Set-asides.*)

Leader: A role, not a position. In agile organizations, each person, regardless of position, provides leadership. Leaders advocate ongoing examination of decisions consistent with mission and vision. Teachers, principals, secretaries, and students contribute leadership. Asking the difficult question is an example of leadership. Modeling empathy, curiosity, craftsmanship, and consciousness are acts of leadership. Acts that focus collective energy on the important (not the urgent), on asking "Why are we doing things this way?" are examples of leadership. Engaged participants (Garmston and Wellman 2013) are group members who are metacognitively alert. All members of a school community engage in school reform (Lambert, Walker, Zimmerman, and Cooper 2002).

Limbic system: The limbic system comprises several brain structures associated with memory and emotion.

Mediate: To be the medium of bringing about a result. Literally, it means "middle."

Mediator: One who interposes himself or herself in the middle between a person and some event, problem, conflict, or other perplexing situation and intervenes in such a way as to enhance meaning and self-directed learning. Based on Feueuerstein's "Mediated Learning Experience" (MLE).

Mental map: A pathway held in the mind and used by a coach as a guide to interaction; a scaffold for a conversation. For example, Bloom's taxonomy is a mental map guiding a sequence of cognitive tasks.

Neurotransmitters: These are chemical messengers that carry, boost, or modulate signals between neurons and other cells in the body. They are related to mood, concentration, learning, capacity to attend, to think, and to have organs like heart and lungs functioning. Excitatory neurotransmitters stimulate the brain, inhibitory neurotransmitters calm the brain and balance mood. Over 100 have been discovered so far.

- **Dopamine** (both excitatory and inhibitory) has a number of important functions in the brain; this includes regulation of motor behavior, pleasures related to motivation, and also emotional arousal. It plays a critical role in the reward system.
- **Glutamate** (excitatory) is the body's most prominent neurotransmitter. It is important for neural communication, memory formation, learning, and regulation.
- **Serotonin** (inhibitory) is produced by and found in the intestine (approximately 90 percent) and the remainder in central nervous system neurons. It

functions to regulate appetite, sleep, memory and learning, temperature, mood, behavior, muscle contraction, and function of the cardiovascular system and endocrine system.

Pace and lead: A coaching map used to restore a colleague's resourcefulness. Pacing and leading validates perceptions of a problematic state and helps the person envision a desired state. (See also *Problem-resolving conversation.*)

Paralanguage: Describes what is communicated and understood by the qualities of voice, body, gestures, and other nonverbal behaviors that exist alongside the words we speak. (The prefix *para* means "alongside.")

Planning conversation: A structured interaction conducted before participating in an event, resolving a challenge, or attempting some task. The intent is to mediate the cognitive processes of planning. The coach may or may not be present during the event or available for a follow-up conversation. (See also *Coaching cycle.*)

Problem-resolving conversation: A conversation conducted when a colleague feels stuck, helpless, unclear, or lacking in resourcefulness; experiences a crisis; or requests assistance. The coach's aim is to have the colleague access the necessary internal resources for resolution. (See also *Pace and lead.*)

Reflecting conversation: A structured interaction conducted after participating in an event or completing a task, the intent of which is to mediate the cognitive processes of reflection. The coach may or may not have participated in or witnessed the conversation. (See also *Coaching cycle.*)

Resources: Internal elements used for decision making, problem solving, and effective action. Resources include electrochemical energy released by neurotransmitters and peptides related to accessing states of efficacy, flexibility, craftsmanship, consciousness, and interdependence. (See also *Problem-solving conversation* and *Pace and lead.*)

Self-directed learning: The capacity for self-managing, self-monitoring, and self-modifying.

Set-asides: Noticing and freeing oneself from several normal ways of listening and responding in order to be more effective as a Cognitive Coach. Three set-asides are described in this glossary: autobiographical listening, inquisitive listening, and solution listening. Three special set-asides are used during pacing and leading. They are:

- Closure is setting aside the need to understand the end of the story and instead looking for nonverbal cues indicating that a colleague has achieved the necessary internal resources.
- Comfort is setting aside the need for feeling at ease in pacing and leading conversations and being willing to embrace ambiguity as a resource for both parties.

- Comprehension is setting aside a coach's need to know the history, background, and context of a problem. Seeking this information is useful when consulting, but it may be counterproductive when pacing and leading.

Solution listening: Occurs when a listener tries to solve a problem for another. It may involve interpretation and rehearsing.

States of mind: Basic human forces that drive, influence, motivate, and inspire our intellectual capacities, emotional responsiveness, high performance, and productive human action. Specifically, this book describes five states of mind: efficacy, flexibility, craftsmanship, consciousness, and interdependence. (See also the separate entry for each in this glossary.) States of mind are transitory depending on a variety of personal and situational factors, but they can be mediated by helping a colleague become conscious of them and accessing them. They can also be strengthened and made more accessible to self-coaching over time.

Support functions: Three services coaches may select from to further the aim of self-directed learning: Cognitive Coaching, collaborating, or consulting. (See definitions in this glossary.)

References

Abramov, I., J. Gordon, O. Feldman, and A. Chavarga. 2012. Sex and vision 1: Spatio-temporal resolution. *Biology of Sex Differences* 3:20.

———. 2012. Sex and vision 2: Color appearance of monochromatic lights. *Biology of Sex Differences* 3:21.

Acherman, D. 1990. *A natural history of the senses*. New York: Vintage Books.

Aguilar, E. 2014, October 9. 10 ways for administrators to support coaches: The art of coaching. *Teachers Education Week*.

Anderson, J. R. 1990. *Cognitive psychology and its implications*, 3rd ed. New York: Freeman.

Anderson, R. H. 1993. Clinical supervision: Its history and current context. In *Clinical supervision: Coaching for higher performance*. Lancaster, PA: Technomic.

Arlin, P. 1990. Wisdom: The art of problem finding. In R. J. Sternberg (ed.), *Wisdom: Its nature, origins and development*, 230–43. New York: Cambridge University Press.

Arrien, A., and S. Covey. 2006. *The speed of trust: The one thing that changes everything*. New York: Simon and Shuster.

Arrow, H., J. McGrath, and J. Berdahl. 2000. *Small groups as complex systems: Formation, coordination, development and adaptation*. Thousand Oaks, CA: Sage.

Atkinson, S. A., and J. Moffat. 2007. *The agile organization: From informal networks to complex effects and agility*. http://www.au.af.mil/au/awc/awcgate/ccrp/atkinson_agile.pdf.

Azar, B. 2010. Your brain on culture. *Monitor on Psychology*. American Psychology Association. Vol. 41, no. 10. Nov. 2010, p. 44.

Babad, E., F. Bernieri, and R. Rosenthal. 1991. Students as judges of teachers' verbal and non-verbal behavior. *American Educational Research Journal* 28(1): 211–34.

Bandler, R., and J. Grinder. 1975. *The structure of magic*. Palo Alto, CA: Science and Behavior Books.

Bandura, A. 1982, February. Self-Efficacy mechanism in human agency. *American Psychologist* 37(2).

———. 1997. *Self-efficacy: The exercise of control*. New York: Freeman.

Barbosa, J., A. Costa, and R. Garmston. 2014, November. Learning to listen; listening to learn. *Educational Leadership*.

Barbosa-Gonzales, J. 2009. *Peer coaching: An instructional manual*. Hayward, CA: Hayward Unified School District.

Bargh, J. 2014, January. Our unconscious mind. *Scientific American* 310(1): 30–37.

Basow, S. A., and K. Rubenfeld. 2003. "Troubles talk": Effects of gender and gender typing. *Sex Roles* 48(3–4): 183–87.

Bateson, G. 1972. *Steps to an ecology of mind: Collected essays in anthropology, psychiatry, evolution, and epistemology*. Chicago: University of Chicago Press.

———. 1979. *Mind and nature: A necessary unity*. Cresskill, NJ: Hampton Press.

———. 2000. *Steps to an ecology of mind: Collected essays in anthropology, psychiatry, evolution and epistemology*, rev. ed. Chicago: University of Chicago Press.

Bennis, W., and B. Nanus. 1985. *Leaders: The strategies for taking charge*. New York: Harper & Row.

Berliner, D. 1988. *The development of expertise in pedagogy*. Paper presented at the meeting of the American Association of Colleges for Teacher Education, New Orleans.

———. 1994. Expertise: The wonders of exemplary performance. In John N. Mangieri and Cathy Collins Block (eds.), *Creating powerful thinking in teachers and students*, 141–86. Ft. Worth, TX: Holt, Rinehart, and Winston.

Berman, P., and J. W. McLaughlin. 1977. *Federal program supporting educational change: Factors affecting implementation and continuation*. Santa Monica, CA: Rand Corp.

———. 1978. *Federal Programs Supporting Educational Change*, vol. 8: *Implementing and sustaining innovation*. Prepared for the U.S. Office of Education, Department of Health, Education, and Welfare. R-1589/HEW May 1978.

Berry, D., D. Davis, R. Ford, C. Hirz, N. Holt, V. James, and C. Weinberger. 2003. *Instructional strategies that work: A tool kit for educators, based on the work of Robert Marzano and Debra Pickering*. Englewood, CO: S.T.A.R. Mentor Program, Cherry Creek School District.

Bettman, E. 1984. *Without bias: A guidebook for nondiscriminatory communication*, 2nd ed. New York: Anti-Defamation League.

Biddle, B. 2000. *Role transitions in organizational life: An identity-based perspective*. Mahway, NJ: Lawrence Erlbaum.

Blakelee, S. 2006, January 19. Cells that read minds. *New York Times*, C3.

Bloom, B. S., M. D. Engelhart, E. J. Furst, W. H. Hill, and D. R. Krathwohl (eds.). 1956. *Taxonomy of educational objectives: The classification of educational goals. Handbook L: Cognitive domain*. New York: David McKay.

Boase, M. Tools for communication. Lamda Mi. http://www.lamda-mi.com/tools-for-communication.

Bolte Taylor, J. 2008. *My stroke of insight: A brain scientist's personal journey*. New York: Viking.

Borg, S. 2006. *Teacher cognition and language education: Research and practice*. London: Continuum.

———. 2009. Introducing language teacher cognition. http://www.education.leeds.ac.uk/research/files/145.pdf.

Borko, H., R. Cone, D. Russo, and R. Shavelson. 1979. Teachers' decision making. In D. Peterson and H. Walberg (eds.), *Research on teaching*. Berkeley, CA: McCutchan.

Borko, H., and C. Livingston. 1989. Cognition and improvisation: differences in mathematics instruction by expert and novice teachers. *American Educational Research Journal* 26:473–98.

Boyatzis, R., and A. Jack. 2010. Coaching with compassion can 'light up' human thoughts. *Coaching with compassion: An fMRI study of coaching to the positive or negative emotional attractor*. Paper presented at the Academy of Management annual meeting in Montreal.

Boyatzis, R., M. Smith, and A. Beveridge. 2013, June. Coaching with compassion: Inspiring health, well-being, and development in organizations. *Journal of Applied Behavioral Science* 49:153–78. doi:10.1177/0021886312462236.

Bracey, H., J. Rosenblum, S. Aubrey, and R. Trueblood. 1990. *Managing from the heart*. New York: Delacorte Press.

Braisby, N., and A. Gellatly. 2012. *Cognitive psychology*. New York: Oxford University Press.

Bronson, P. 2007, August 3. How not to talk to your kids The inverse power of praise. *New York Magazine.*

Brooks, D. 2014, October 28. Why partyism is wrong. *New York Times*, A27.

Brooks, J. G., and M. Brooks. 2001. Becoming a constructivist teacher, in A. Costa (ed.), *Developing minds: A resource book for teaching thinking.* Alexandria, VA: Association for Supervision and Curriculum Development.

Bruner, J. 1960. *The process of education.* Cambridge, MA: Harvard University Press.

Bryk, A., and B. Schneider. 2002. *Trust in schools: A core resource for improvement.* New York: Russell Sage Foundation.

Buhrmann, D. 1984. *Living in two worlds: Communication between a white healer and her black counterparts.* Cape Town: Human & Rousseau.

Burgoon, J., D. Buller, and W. Woodall. 1996. *Nonverbal communication: The unspoken dialogue.* New York: McGraw-Hill.

Caine, R. N., and G. Caine. 2015. *Mind/brain learning principles in action: Teach for the development of higher order thinking and executive function.* 3rd Edition. Thousand Oaks, CA: Corwin Press.

Calderhead, J. 1996. Teachers: Beliefs and knowledge. In D. C. Berliner and R. C. Calfee (eds.), *Handbook of educational psychology*, 709–25. New York: Macmillan.

Calhoun, E. F. 1985, April. *Relationship of teachers' conceptual level to the utilization of supervisory services and to a description of the classroom instructional improvement.* Paper presented at the annual meeting of the American Educational Research Association. Chicago, Illinois.

California State Department of Education. 1997. *Descriptions of practice.* Sacramento: California Standards for the Teaching Profession.

Chojnacki, D. 2014. Personal correspondence.

Christov-Moore, L., E. A. Simpson, G. Coudé, K. Grigaityte, M. Iacoboni, and P. F. Ferrari. 2014, October. Empathy: Gender effects in brain and behavior. *Neuroscience and Biobehavioral Reviews* 46:604–627.

Clark, C., and P. Peterson. 1986. Teachers' thought processes. In M. C. Wittrock (ed.), *Handbook of research on teaching*, 255–96. New York: MacMillan.

Clark C., and R. Yinger. 1979. *Teachers' thinking: Research on thinking.* Berkeley, CA: McCutchan.

Cogan, M. 1973. *Clinical supervision.* Boston: Houghton Mifflin.

Coladarci, A. P. 1959, March. The teacher as hypothesis maker. *California Journal of Instructional Improvement* 2:3–6.

Condon, W. S. 1975. Multiple response to sound in dysfunctional children. *Journal of Autism and Childhood Schizophrenia* 5(1): 43.

Conley, D. T. 2010. *Redefining college readiness.* Eugene, OR: Education Policy Improvement Center.

Cook, N. 2002. *Tone of voice and mind: The connections between intonation, emotion, cognition and consciousness.* Amsterdam: John Benjamins.

Costa, A. 1991. *Developing minds: A resource book for teaching thinking.* Alexandria, VA: Association for Supervision and Curriculum Development.

Costa, A., and R. Garmston. 1985. Pacing and leading: A coaching strategy. In *Leadership training for Cognitive Coaching: A training syllabus.* Hayward, CA: Office of the Alameda County Superintendent of Schools.

———. 1991. Five human passions: The source of critical and creative thinking. In A. Costa (ed.), *Developing minds: A resource book for teaching thinking*, 18–22. Alexandria, VA: Association for Supervision and Curriculum Development.

———. 1998. Maturing outcomes. *Encounter: Educating for Meaning and Social Justice* 11(1): 10–18.

———. 2000. Peer assistance and review: Potentials and pitfalls. In G. Bloom and J. Geldstein (eds.), *The peer assistance and review reader*, 63–79. Santa Cruz: New Teacher Center, University of California, Santa Cruz.

———. 2002. *Cognitive coaching: A foundation for renaissance schools syllabus*. Norwood, MA: Christopher-Gordon.

Costa, A., R. Garmston, and D. Zimmerman. 2014. *Cognitive capital: Investing in teacher quality*. New York: Teachers College Press.

Costa, A., and B. Kallick. 2004. *Assessment strategies for self-directed learning*. Thousand Oaks, CA: Corwin.

———. 2008. *Learning and leading with habits of mind: 16 characteristics for success*. Alexandria, VA: Association for Supervision and Curriculum Development.

———. 2014. *Dispositions: Reframing teaching and learning*. Thousand Oaks, CA: Corwin

———. 2015. Dispositions: Critical pathways for deeper learning. In J. Bellanca (ed.), *Deeper learning: Beyond 21st-century skills*. Bloomigton, IN: Solution Tree.

Costa, A., L. Lipton, and B. Wellman. 1996. Shifting rules, shifting roles: Transforming the work evironment to support learning. In S. Caldwell (ed.), *New directions in staff development*, 92–115. Oxford, OH: National Staff Development Council.

Costa, A., and P. O'Leary. 2013. *The power of the social brain*. New York: Teacher College Press.

Covey, S. 1989. *The seven habits of highly effective people: Powerful lessons in personal change*. New York: Simon and Schuster.

Cozolino, L. 2006. The social brain. *Psychotherapy in Australia* 12:16–21.

Crick, F., and C. Koch. 1998.Consciousness and neuroscience. http://resolver.caltech.edu/Caltech-AUTHORS:20130816-103137714.

Csikszentmihalyi, M. 1990. *Flow: The psychology of optimal e xperience*. New York: Harper and Row.

Cuban, L. 2010. *The difference between "complicated" and "complex" matters*. https://larrycuban .wordpress.com/2010/06/08/the-difference-between-complicated-and-complex-matters/ (accessed 20 February 2015).

Dahllof, U., and U. P. Lundgren. 1970. *Macro- and micro-approaches combined for curriculum process analysis: A Swedish educational field project*. Goteborg, Sweden: University of Goteborg Institute of Education.

Damasio, A. 1994. *Descartes' error: Emotion, reason and the human brain*. New York: Avon Books.

———. 1999. *The feeling of what happens: Body and emotion in the making of consciousness*. New York: Harcourt Brace.

———. 2010. *Self comes to mind: Constructing the conscious brain*. New York: Pantheon Books.

Daniels, V. 2009. The analytical psychology of Carl Gustav Jung. *Matrix meditations: A 16-week program for developing the mind-heart connection*. Rochester, VT: Destiny Books.

Danielson, C. 2007. *Enhancing professional practice: A framework for teaching*. Alexandria, VA: Association for Supervision and Curriculum Development.

———. 2013. *Framework for teaching*. http://danielsongroup.org/framework/ (accessed 20 February 2015).

Danielson, C., and T. McGreal. 2000. *Teacher evaluation to enhance professional practice*. Alexandria, VA: Association for Supervision and Curriculum Development.

Darling-Hammond, L., and J. Bransford. 2005. *Preparing teachers for a changing world: What teachers should learn and be able to do*. San Francisco. Jossey-Bass.

Darling-Hammond, L., A. E. Wise, and S. P. Klein. 1995. *A license to teach: Building a profession for 21st-century schools*. Boulder, CO: Westview Press.

Deal, T. E., and A. A. Kennedy. 1982. *Corporate culture: The rites and rituals of corporate life*. Menlo Park, CA: Addison-Wesley.

Derber, C. 1979. *The pursuit of attention: Power and ego in everyday life*. New York: Oxford University Press.

Dewey, J. 1938. *Experience and education*. New York: Collier.

Dilts, R. 1994. *Effective presentation skills*. Capitola, CA: Meta.

———. 2003. *From coach to awakener*. Capitola, CA: Meta.

———. 2014. *Logical level alignment*. http://www.nlpu.com/Articles/LevelsAlignment.htm (accessed 20 February 2015).

Dilts, R., and T. Epstein. 1995. *Dynamic learning*. Capitola, CA: Meta.

Dilts, R., and R. McDonald. 1997. *Tools of the spirit*. Capitola, CA: Meta.

Dolcemascolo, M., D. Miori-Merola, and J. Ellison. 2014. Personal correspondence.

Drago-Severson, E. 1994. *Becoming adult learners: Principles and practices for effective development*. New York: Teachers College Press.

———. 2004. *Helping teachers learn: Principal leadership for adult growth and development*. Thousand Oaks, CA: Corwin Press.

———. 2011. School leadership for adult development: The dramatic difference it can make. *International Handbook of Leadership for Learning Springer International Handbooks of Education* 25:757–77. http://link.springer.com/bookseries/6189 (accessed 20 February 2015).

Dreazen, Y. 2014. *The invisible front: Love and loss in an era of endless war*. New York: Crown.

Du Bois, W. E. B. 1903. *The souls of black folks*. New York: Bantam Classics.

Du Four, R., and M. Fullan. 2013. *Cultures built to last: Systemic PLCs at work*. Bloomington IN: Solution Tree Press.

Dweck, C. 2006. *Mindset: The new psychology of success*. New York. Ballentine Books.

Ebmeier, H. 2003, Winter. How supervision influences teacher efficacy and commitment: An investigation of a path model. *Journal of Curriculum and supervision* 18(2): 110–41.

Edelman, G. 1987. *Neural Darwinism: The theory of neuronal group selection*. New York: Basic Books.

Edwards, J. 2015. *Cognitive Coaching: A synthesis of the research*. Denver, CO: Thinking Collaborative.

Edwards, J. L., and R. R. Newton. 1995. *The effects of cognitive coaching and empowerment*. Paper presented at the annual meeting of the American Educational Research Association, San Francisco.

Ehrlichman, H., and D. Micic. 2012, April. Why do people move their eyes when they think? *Current Directions in Psychological Science* 21(2): 96–100. http://cdp.sagepub.com/content/21/2/96.abstract.

Einspruch, E., and B. Forman. 1985. Observations concerning research literature on neuro-linguistic programming. *Journal of Counseling Psychology* 32(4): 589–69.

Eisner, E. 1994. *The educational imagination: On the design and evaluation of school programs*. Upper Saddle River, NJ: Prentice-Hall.

Eisner, E., and E. Valance. 1974. *Conflicting conceptions of the curriculum*. Berkeley, CA: McCutchan.

Elgin, S. 1980. *The gentle verbal art of self-defense*. New York: Dorset Press.

Ellemers, N., D. De Gilder, and S. Haslam. 2004. Motivating individuals and groups at work: a social identity perspective on leadership and group performance. *Academy of Management Review* 29(3): 459–78.

Ellison, J. 2010. Personal correspondence.

Ellison, J., and C. Hayes. 2012. Calibrating conversations to promote effective instructional practices and collaborative professional conversations. In *Cognitive coaching seminar advanced training packet*. Alaska Administrator Coaching Project. Highlands Ranch, CO: Center for Cogntive Coaching.

Elmore, R., and E. City. 2007, May/June. *Harvard Education Letter* 23(3).

Evered, R., and J. Selman. 1989, Autumn. Coaching and the art of management. *Organizational Dynamics* 18:116–32.

Featherstone, E. 1994. *Skin deep: Women writing on color, culture, and identity*. Freedom, CA: Crossing Press.

Feuerstein, R. 1999. Mediated learning experience. In A. Costa (ed.), *Teaching for intelligence II: A collection of articles*. Arlington Heights, IL: Skylights.

Feuerstein, R., L. Falik, and R. S. Feuerstein. 2014. *Changing minds and brains: Higher thinking and cognition through mediated learning*. New York. Teachers College, Columbia University.

Feuerstein, R., and R. S. Feuerstein. 1991. Mediated learning experience: A theoretical review. In R. Feuerstein, P. Kelin, and A. Tannenbaum (eds.), *Mediated learning experience (MLE): Theoretical, psychosocial and learning implications*. London: Freund.

Feuerstein, R., R. S. Feuerstein, and L. Falik. 2010. *Beyond smarter: Mediated learning and the brain's capacity for change*. New York: Teachers College Press.

Firth, G. 1998. Governance of school supervision. In G. Firth and E. Pajak (eds.), *Handbook of research on school supervision*, 928–31. New York: Macmillan.

Fredrickson, B. L. 2001. The role of positive emotions in positive psychology: The broaden-and-build theory of positive emotions. *American Psychologist* 56:218–26.

Freiberg, K., and J. Freiberg. 1996. *Nuts! Southwest Airlines' crazy recipe for business and personal success*. New York: Broadway Books.

Fromm, E. 1956. *The art of loving*. New York: Harper & Row.

Frymeir, J. 1987. Bureaucracy and the neutering of teachers. *PhiDeltaKapan*, 69(1): 9–16.

Fullan, M. 1982. *The meaning of educational change*. New York: Teachers College Press.

———. 1993. Why teachers must become change agents: The professional teacher. *Educational Leadership* 50:6, 12–17.

———. 2001. *Leading in a culture of change*. San Francisco: Jossey-Bass.

———. 2014. *The principal: Three keys for maximizing impact*. San Francisco: Jossey Bass.

Fuller, B., K. Wood, T. Rapoport, and S. Dornbusch. 1982, Spring. The organizational context of individual efficacy. *Review of Educational Press* 52:7–30.

Galin, D., and R. Ornstein. 1973. Individual differences in cognitive style: Reflective eye movements. *Neuropsychologia* 12:367–76.

Garfield, C. 1986. *Peak performers: The new heroes of American business*. New York: William Morrow.

Garmston, R. 1987, February. How administrators support peer coaching. *Educational Leadership* 44(5): 18–26.

———. 1988. *A guide to neuro-linguistic programming and counseling in education*. Course syllabus. Sacramento: School of Education, California State University.

———. 1989. Cognitive coaching and professors' instructional thought. *Human Intelligence Newsletter* 10(2): 3–4.

———. 2005. The presenters fieldbook: A practical guide. Norwood, MA: Christopher Gordon.

———. 2013. *Developing smart groups. In A. Costa and P. O'Leary (eds.), The power of the social brain: Teaching, learning and interdependent thinking, 75–83. New York: Teachers College Press*.

Garmston, R., and D. Hyerle. 1988, August. Professor's peer coaching program: Report on a pilot project to develop and test a staff development model improving instruction at California State University, Sacramento. Sacramento: California State University.

———. 1998. *Professors' peer coaching program: Report on a 1987–1988 pilot program to develop and test a staff development model for improving instruction at California State University*. Sacramento: California State University.

Garmston R., C. Linder, and J. Whitaker. 1993, October. Reflections on Cognitive Coaching. *New Roles, New Relationships* 51(2): 57–61.

Garmston, R., L. Lipton, and K. Kaiser. 1998. The psychology of supervision: From behaviorism to constructivism. In G. Firth and E. Pajak (eds.), *Handbook of research on school supervision*, 242–86. New York: Macmillan.

Garmston, R., and V. Von Frank. 2012. *Unlocking group potential to improve schools*. Thousand Oaks, CA. Corwin Press.

Garmston, R., and B. Wellman. 2013. *The Adaptive School. A sourcebook for developing collaborative Groups*. 2nd ed., rev. New York. Rowman & Littlefield.

Gergen, K. 2009. *Relational being: Beyond self and community*. Oxford. Oxford University Press.

Gerhart, D. 2014. Personal correspondence.

Given, B. 2002. *Teaching to the brain's natural learning systems*. Alexandria, VA: Association for Supervision and Curriculum Development.

Glaser, J. 2014. *Conversational intelligence: How great leaders build trust and get extraordinary results*. Brookline, MA: Bibliomotion.

Glickman, C. 1985. *Supervision of instruction: A developmental approach*. Boston: Allyn and Bacon.

———. 2001. Holding sacred ground: The impact of standardization. *Educational Leadership* 58(4): 46–51.

Goldberg, H. 1991. *What men really want*. New York: Penguin Books.

Goldhammer, R. 1969. *Clinical supervision: Special method for the supervision of teachers*. New York: Holt, Rinehart, and Winston.

Goleman, D. 1997. *Emotional intelligence: Why it can matter more than I.Q.* New York: Bantam.

———. 2006. *Social intelligence: The revolutionary new science of human relationships*. New York. Bantam.

Good, T. L., and J. Brophy. 1973. *Looking in classrooms*. New York: Harper & Row.

Gorbachev, M., and A. Arrien. 1995. *Working together: Producing synergy by honoring diversity*. Pleasanton, CA: New Leaders Press.

Gorman, C. 1992, January 20. Sizing up the sexes. *Time*, 42–48.

Gotterson, K. 2014. Personal correspondence.

Gray, J. 1992. *Men are from Mars, women are from Venus: A practical guide for improving communication and getting what you want in a relationship*. New York: Harper Collins.

Grimm, E., T. Kaufman, and D. Doty. 2014, May. Rethinking classroom observation. *Educational Leadership* 71(8): 24–29.

Grimmet, P. P., and A. M. MacKinnon. 1992. Craft knowledge and the education of teachers. *Review of Research in Education* 18:387.

Grinder, M. 1991. *Righting the educational conveyor belt*. Portland, OR: Metamorphous Press.

———. 1993. *Envoy: Your personal guide to classroom management*. Battle Ground, WA: Michael Grinder.

———. 1997. *The science of nonverbal communication*. Battleground, WA: Michael Grinder.

———. 2001. Personal communication.

Guild, P., and S. Garger. 1985. *Marching to different drummers*. Alexandria, VA: Association for Supervision and Curriculum Development.

Gwet, K. 2012. *Handbook of inter-rater reliability: The definitive guide to measuring the extent of agreement among multiple raters*, 3rd ed. Gaithersberg, MD: Advanced Analytics.

Hahn, S., and A. Litwin. 1995. Women and men. In R. A. Ritvo, A. Litwin, and L. Butler (eds.), *Managing in the age of change: Essential skills to manage today's workforce*. Burr Ridge, IL: Irwin Professional.

Hargreaves, A. 1994. *Changing teachers, changing times: Teachers' work and culture in the postmodern age*. New York: Teachers College Press.

Hargreaves, A., A. Boyle, and A. Harris. 2014. *Uplifting leadership: How organizations, teams, and communities raise performance*. San Francisco: Jossey Bass.

Hargreaves, A., and M. Fullan. 2012. *Professional capital: Transforming teaching in every school*. New York: Teachers College Press.

Harris P., and R. Moran. 2000. *Managing cultural differences*. Houston, TX: Gulf.

Hartoonian, B., and G. Yarger. 1981. *Teachers' conceptions of their own success*. Washington, DC: ERIC Clearinghouse on Teacher Education, no. S017372.

Harvey, O. J. 1966. System structure, flexibility, and creativity. In *Experience, structure, and adaptability*, 39–65. New York: Springer.

———. 1967. *Conceptual systems and attitude change: Attitude, ego involvement, and change*. New York: Wiley.

Hasson, U., A. A. Ghazanfar, B. Galantucci, S. Garrod, and C. Keysers. 2012. Brain-to-brain coupling: Mechanism for creating and sharing a social world. *Trends in Cognitive Science* 16(2): 114–21.

Hatfield, E., J. Cacioppo, and R. L. Rapson. 1994. *Emotional contagion*. New York: Cambridge University Press.

Hattie, J. 2003, October. *Teachers make a difference: What is the research evidence?* University of Auckland: Australian Council for Educational Research, https://cdn.auckland.ac.nz/assets/education/hattie/docs/teachers-make-a-difference-ACER-(2003).pdf.

———. 2012. Visible learning for teachers: Maximizing impact on learning. New York: Routledge.

Heifetz, R., M. Linsky, and A. Grashow. 2009. *The practice of adaptive leadership: Tools and tactics for changing your organization and the world*. Cambridge, MA. Cambridge Leadership Associates.

Ho, A., and T. Kane. 2014. *The reliability of classroom observations by school personnel*. Cambridge, MA: Harvard Graduate School of Education, Bill and Melinda Gates Foundation. www.metproject.org.

Hofstadter, D. R. 2000. Analogy as the core of cognition. In J. Gleick (ed.), *The best American science writing 2000*. New York: Ecco Press.

Hoy, W. K., S. R. Sweetland, and P. A. Smith. 2002. Toward an organizational model of achievement in high schools: The significance of collective efficacy. *Educational Administration Quarterly* 38:77–93.

Hoy, W. K., C. J. Tarter, and A. Woolfolk Hoy. 2006. Academic optimism of schools: A force for student achievement. *American Educational Research Journal* 43(3): 425–46.

Hunt, D. E. 1977–1978. Conceptual level theory and research as guides to educational practice. *Interchange* 8:78–80.

———. 1980. *Teachers' adaptation: Reading and flexing to students' flexibility in teaching*. New York: Longman Press.

Huppke, R. 2014, June 2. Conformity vs. authenticity in the office: You can and should be professional at work and still be true to yourself. *Chicago Tribune*. http://articles.chicagotribune.com/2014-06-02/business/chi-biz-0602-work-advice-huppke-20140602_1_powerpoint-sylvia-ann-hewlett-confidence (accessed 20 February 2015).

Iacoboni, M. 2008. *Mirroring people*. New York: Farrar, Strauss, and Giraux.

Instructional Coaching Model. 2010. District Publication, Spokane Public Schools. http://www.spokaneschools.org/cms/lib/WA01000970/Centricity/Domain/629/Instructional%20Coaching%20Model%20FINAL%20-%206.10.pdf (accessed 20 February 2015).

Jackson, P. W. 1968. *Life in classrooms*. New York: Holt, Rinehart, and Winston.

Jackson, Y. 2001. Reversing underachievement in culturally different urban students: Pedagogy of confidence. In A. Costa (ed.), *Developing minds: A resource book for teaching thinking*, 222–28. Alexandria, VA: Association for Supervision and Curriculum Development.

Jansson, L. 1983. Mental training: Thinking rehearsal and its use. In W. Maxwell (ed.), *Thinking: The expanding frontier*. Philadelphia: Franklin Institute.

Jensen, E. 1996. *Brain-based learning*. Del Mar, CA: Turning Points.

———. 1998. *Teaching with the brain in mind*. Alexandria, VA: Association for Supervision and Curriculum Development.

———. 2014, October 14. Make it a habit to connect recent brain research with practical classroom strategies. *Eric Jensen's Brighter Brain Bulletin*. October 14, 2014.

Joyce, B., and B. Showers. 1980. Improving inservice training: The messages of research. *Educational Leadership* 37(5): 379–85.

———. 2002. S*tudent achievement through staff development*, 3rd ed. Alexandria VA; Association for Supervision and Curriculum Development.

Jung, C 1933. *Modern man in search of a soul*. Orlando, FL: Harcourt.

Kahn, W. A. 1992. To be fully there: Psychological presence at work. *Human Relations* 45(4): 321–49.

Keene, E., and S. Zimmermann. 1997. *Mosaic of thought*. Portsmouth, NH: Heinemann.

Kegan, R. 1994. *In over our heads: The mental demands of modern life*. Cambridge, MA: Harvard University Press.

———. 2000. What "form" transforms? A constructive-developmental approach to transformative learning. In J. Mezirow (ed.), *Learning as transformation: Critical perspectives on a theory in progress*, 35–70. San Francisco: Jossey-Bass.

Kegan, R., and L. Lahey. 2001. *How the way we talk can change the way we work: Seven languages for transformation*. San Francisco: Jossey-Bass.

———. 2009. *Immunity to change: How to overcome it and unlock the potential in yourself and your organization*. Boston, MA: Harvard University Press.

King, P., and K. Kitchener. 1994. *Developing reflective judgement: Understanding and promoting intellectual growth and critical thinking in adolescents and adults*. San Francisco: Jossey-Bass.

Kinsbourne, M. 1972. Eye and head turning indicates cerebral lateralization. *Science* 58:539–41.

———. 1973. The control of attention by interaction between the cerebral hemispheres. In S. Kornblum (ed.), *Attention and Performance*, vol. 4: 239–56. New York: Academic Press.

Knight, J. 2007. *Instructonal coaching: A patnership approach to improving instruction*. Thousand Oaks, CA. Corwin.

———. 2014, May. What you learn when you see yourself teach. *Educational Leadership* 71(8): 18–23.

Knowles, M. S., E. F. Holton III, and R. A. Swanson. 2005. *The adult learner*, 6th ed. Burlington, MA: Elsevier Butterworth Heinemann.

Kocel, K., D. Galin, R. Ornstein, and E. Merrin. 1972. Lateral eye movement and cognitive mode. *Psychonomic Science* 27(4): 223–24.

Kochman, T. 1983. *Black and white styles in conflict*. Chicago: University of Chicago Press.

Koestler, A. 1972. *The roots of coincidence*. New York: Vintage Books.

Kotulak, R. 1997. *Inside the brain*. Kansas City, MO: Andrews McMeel.

Kozhevnikov, L. 2007. Cognitive styles in the context of modern psychology: Toward an integrated framework of cognitive style. *Psychological Bulletin* 133(3): 464–81.

Kuhn, Thomas S. 1962. *The structure of scientific revolutions*. Chicago: University of Chicago Press.

Kupersmith, W., and W. Hoy. 1989. *The concept of trust: An empirical assessment*. Presentation at the American Educational Research Association annual meeting, New Orleans.

Laborde G., and C. Saunders. 1986. *Communication trainings: Are they cost effective?* Mountain View, CA: International Dialogue Education Associates.

Ladd, E. 2014. Personal correspondence.

Lambert, L., D. Walker, D. Zimmerman, and J. Cooper. 2002. The constructivist leader, 2nd ed. New York: Teachers College Press.

Land, G., and B. Jarman. 1992. *Break-point and beyond*. New York: Harcourt Brace Jovanovich.

Landers, D., W. Maxwell, J. Butler, and L. Fagen. 2001. Developing thinking skills through physical education. In A. Costa (ed.), *Developing minds: A resource book for teaching thinking*. Alexandria, VA: Association for Supervision and Curriculum Development.

Lankton, S. 1980. *Practical magic: A translation of basic neurolinguistic programming into clinical psychotherapy*. Cupertino, CA: Meta.

Lankton, S., and C. Lankton. 1983. *The answer within: A clinical framework for Ericksonian hypnotherapy*. New York: Brunner/Mazel.

Leary, M. R., and J. P. Tangney. 2003. *Handbook of self and identity*. New York: Guilford Press.

Leat, D. 1995. The costs of reflection in initial teacher education. *Cambridge Journal of Education* 25(2): 161–74.

Ledoux, J. 1996. *The emotional brain: The mysterious underpinnings of life*. New York: Simon and Schuster.

Leithwood, K., and K. Seashore-Lewis. 2011. *Linking leadership to student learning*. San Francisco: Jossey Bass.

Leithwood., K.Seashore-Louis, S. Anderson, and K. Wahlstrom. 2004. *How leadership influences student learning*. New York: Wallace Foundation.

Lewis, T., F. Amini, and R. Lannon. 2000. *A general theory of love*. New York: Random House.

Lieberman M. 2013. *Social: Why our brains are wired to connect*. New York: Crown.

Likert, R. 1967. T*he human organization: Its management and value*. New York: McGraw-Hill.

Lindsey, D., L. Jungwirth, P. Jarvis, and R. Lindsey. 2009. *Culturally proficiently learning communities: Confronting inequities through collaborative curiosity*. Thousand Oaks CA. Corwin Press.

Lindsey, D., R. Martinez, and R.Lindsey. 2007. *Culturally proficient coaching: Supporting educators to create equitable schools*. Thousand Oaks CA. Corwin Press.

Lindsey, R. B., K. Nuri Robins, and R. D. Terrell. 2009. *Cultural proficiency: A manual for school leaders*. Thousand Oaks, CA: Corwin Press.

Lipton, L., and B. Wellman. 2011. *Leading groups: Effective strategies for building professional community*. Charlotte, VT: MiraVia.

———. 2012. *Learning focused supervision: Assessing and developing professional practice using the framework for teaching*. Sherman, CT: MiraVia.

———. 2013a. Creating communities of thought. In A. Costa and P. O'Leary (eds.), *The power of the social brain: Teaching learning and interdependent thinking*. New York: Teachers College Press.

———. 2013b. *Learning-focused supervision: Developing professional expertise in standards-derived systems*. Charlotte, VT: MiraVia LLC.

Lipton, L., B. Wellman, and C. Humbard. 2001. *Mentoring matters: A practical guide to learning-focused relationships*. Sherman, CT: MiraVia.

Lohmar, D. 2006. Mirror neurons and the phenomenology of intersubjectivity. *Phenomenology and the Cognitive Sciences* 5:5–16.

Lomofsky, L. 2014. Instrumental enrichment. In L. Green (ed.), *Schools as thinking communities*. Pretoria, South Africa: Van Schaik.

Losada, M. 1999. The complex dynamics of high performance teams. *Mathematical and Computer-Modeling* 30:179–92.

Louis, K., and S. Kruse. 1995. *Professionalism and community: Perspectives on reforming urban schools*. Thousand Oaks, CA: Corwin Press.

Lozano, A. 2001. A survey of styles of thinking: From learning styles to thinking styles. In A. Costa (ed.), *Developing minds: A resource book for teaching thinking*. Alexandria, VA: Association for Supervision and Curriculum Development.

Lunsford, B. 1995, April. A league of our own. *Educational Leadership* 52(7): 59–61.

MacDonell, L., and D. Lindsey. 2011, February. The inside-out approach. *Journal of Staff Development* 32(1): 34–38.

Mansilla, V. B., and A. Jackson. 2011. *Educating for global competence: Preparing our youth to engage the world*. New York: Council of Chief State School Officers' Ed Steps Initiative and Asia Society Partnership for Global Learning.

Markus, H. R., and A. Conner. 2013. *Clash! 8 cultural conflict that make us who we are*. New York: Hudson Street Press.

Marzano, R. J. 1998. *A theory-based meta-analysis of research on instruction*. Aurora, CO: Mid Continent Regional Educational Laboratory.

———. 2001. *A new taxonomy of educational objectives*. In A. Costa (ed.), *Developing Minds: A resource book for teaching thinking*. Alexandria, VA: Association for Supervision and Curriculum Development.

———. 2007. *The art and science of teaching*. Alexandria, VA: Association for Supervision and Curriculum Development.

———. 2011, April. The inner world of teaching. *Educational Leadership* 68(7): 90–91.

Marzano, R., T. Frontier, and D. Livingston. 2011. *Effective supervision: Supporting the art and science of teaching*. Alexandria, VA: Association for Supervision and Curriculum Development.

Marzano, R. J., and J. S. Marzano. 2010. The inner game of teaching. In R. J. Marzano (ed.), *On excellence in teaching*, 345–67. Bloomington, IN: Solution Tree.

Marzano, R., D. Pickering, and J. Pollock. 2001. *Classroom instruction that works: Research-based strategies for increasing student achievement*. Alexandria, VA: Association for Supervision and Curriculum Development.

Marzano, R., and J. Simms. 2012. *Coaching classroom instruction*. Bloomington, IN: Marzano Research Laboratories.

Massachusetts Institute of Technology. 2008. Exploring the mechanics of judgment, beliefs: Technique images brain activity when we think of others. *ScienceDaily*. www.sciencedaily.com/releases/2008/05/080515212112.htm (accessed February 20, 2015).

McLaughlin, M. 1990, December. The RAND change agent study revisited: Macro perspectives and micro realities. *Educational Researcher* 19(9): 11–16.

McNeill, B. 1991. *Beyond sports: Imaging in daily life: An interview with Marilyn King*. Sausalito, CA: Institute of Noetic Science.

McNerney, R. F., and C. A. Carrier. 1981. *Teacher development*. New York: Macmillan.

Megginson, L.1963. Lessons from Europe for American business. *Southwestern Social Science Quarterly* 44(1): 3–13.

Merchant, K. 2012. How men and women differ: Gender differences in communication styles, influence tactics, and leadership styles. Senior thesis, Claremont College. http://scholarship.claremont.edu/cmc_theses/513.

Miller, G. A. 1963. The magical number seven, plus or minus two: Some limits on our capacity for processing information. *Psychological Review* 2:81–97.

Miyake, A., N. P. Friedman, M. J. Emerson, A. H. Witzki, A. Howerter, and T. Wagner. 2000. The unity and diversity of executive functions and their contributions to complex "frontal lobe" tasks: A latent variable analysis. *Cognitive Psychology* 41:49–100.

Moir, E., W. Barron, and C. Stobbe. 2001. Personal communication.

Moncada, G. 1998. *A model for conversation and change within interdisciplinary teams*. Unpublished PhD diss., University of Minnesota.

Moore, A., and P. Malinowski. 2009. Mediation, mindfulness, and cognitive flexibility. *Conscious Cognition* 18:176–86.

National Institute of Education. 1975. *Teaching as clinical information processing* (Panel 6, National Conference on Studies in Teaching). Washington, DC: National Institute of Education.

Neason, A. 2014, July 23. Half of teachers leave the job after five years: Here's what to do about it. *Huffington Post*. http://www.huffingtonpost.com/2014/07/23/teacher-turnover-rate_n_5614972.html?&ir=Education&ncid=tweetlnkushpmg00000023 (accessed 20 February 2015).

Newmann, F. M., H. M. Marks, and A. Gamoran. 1995. *Authentic pedagogy: Standards that boost student performance*, 1–11. Madison: Wisconsin Center for Educational Research.

Newmann, F. M., and G. G. Wehlage. 1995. *Successful school restructuring: A report to the public and educators by the Center on Organization and Restructuring of Schools*. Madison: Wisconsin Center for Educational Research.

Newton, J., and J. Davis. 2014, July 14. Three secrets of organizational effectiveness: How the practices of "pride builders" can help you build a high-performance culture. *Strategy + Business* 76.

Nummenmaa, L. 2014. Bodily maps of emotions. *Proceedings of the National Academy of Science* 111(2): 646–51. doi: 10.1073/pnas.1321664111.

Nummenmaa, L., E. Glerean, R. Hari, and J. Hietanen. 2013, November 27. Bodily maps of emotions. *Proceedings of the National Academy of Sciences of the United States*. http://www.pnas.org/content/111/2/646.full (accessed 20 February 2015).

O'Connor, J., and J. Seymour. 1990. *Introducing neuro-linguistic programming: The new psychology of personal excellence*. London: HarperCollins.

Office for Standards in Education of the UK. 2004. Why colleges succeed. http://www.ofsted.gov.uk/resources/why-colleges-succeed (accessed 1 August 2013).

Oja, S., and A. Reiman. 1998. Supervision for teacher develop-ment across the career span. In G. Firth and E. Pajak (eds.), *Handbook on research on school supervision*, 463–87. New York: Simon and Schuster/Macmillan.

Ornstein, R. 1991. *The evolution of consciousness: Of Darwin, Freud, and cranial fire—The origins of the way we think*. New York: Prentice-Hall Press.

Ouchi, W. 1981. *Theory Z: How American business can meet the Japanese challenge*. Menlo Park, CA: Addison-Wesley.

Pajak, E. 2000. *Approaches to clinical supervision*. Norwood, MA: Christopher-Gordon.

Pajares, M. F. 1992. Teachers' beliefs and educational research: Cleaning up a messy construct. *Review of Educational Research* 62(3).

Paulson, D. 1995, January/February. Finding the soul of work. *At Work: Stories of Tomorrow's Workplace* 4(1): 18–20.

Perkins, D. 1983. *The mind's best work: A new psychology of creative thinking*. Cambridge, MA: Harvard University Press.

Pert, C. 1997. *Molecules of emotions: Why you feel the way you feel*. New York: Scribner.

Peters, W. 1987. *A class divided: Then and now*. New Haven, CT: Yale University Press.

Peterson, P. 1988. Teachers' and students' conditional knowledge for classroom teaching and learning. *Educational Researcher* 17:5–14.

Piaget, J. 1950. *The psychology of intelligence*. New York: Routledge.

———. 1954. *The construction of reality in the child*. New York: Basic Books.

Pink, D. 2006. *A whole new mind: Why right-brainers will rule the future*. New York: Riverhead Books.

———. 2009. *Drive: The surprising truth about what motivates us.* New York: Penguin Books.

———. 2014, March. *Perfecting your power to move others.* Presentation at the Association for Supervision and Curriculum Development conference in Los Angeles, CA.

Pletzer, B., O. Petasis, and L. Cahill. 2014. Switching between forest and trees: Opposite relationship of progesterone and testosterone to global-local processing. *Hormone Behavior* 66:257–66.

Pogrebin, L. 1995. The same and different: Crossing boundries of color, culture, sexual preference, disability and age. In John Stewart (ed.), *Bridges not walls*, 6th ed., 445–59. New York. McGraw Hill.

Pollan, M. 2015, February 9. The trip treatment. Research into psychedelics, shut down for decades, is now yielding exciting results. *New Yorker*, 36-47.

Poole, M. G., and K. R. Okeafor. 1989, Winter. The effects of teacher efficacy and interactions among educators on curriculum implementation. *Journal of Curriculum and Supervision*, 146–61.

Popham, J. 2014, September. The right test for the wrong reason. *Phi Delta Kappan* 96(1): 47–52.

Powell, O. 2014. Personal correspondence.

Powell, W. 2000. Recruiting educators for an inclusive school. In W. Powell and O. Powell (eds.), *Count me in: Developing inclusive international schools*, 191–204. Washington DC: Overseas Advisory Council.

Powell, W., and O. K. Powell. 2015. Overcoming resistance to new ideas. *Phi Delta Kappan*. May 2015. 96: 66–69.

———. 2014, Autumn. Personalizing learning and teacher expectations. *Evidence-Based Education.* http://www.betterevidence.org/issue-16/personalizing-learning-and-teacher-expectations/ (accessed 20 February 2015).

Poyatos, F. 2002. *Nonverbal communication across disciplines.* Vol. 2: *Paralanguage, kinesics, silence, personal and environmental interaction.* Philadelphia: John Benjamins.

Provine, R. 2000. *Laughter: A scientific investigation.* New York. Viking Press.

Ramachandran, V. S. 2011. *The tell-tale brain: A neuroscientist's quest for what makes us human.* New York: W. W. Norton.

Raz, G. 2014a, November 14. What does it mean to be articulate? An interview with Jamila Lyiscott. NPR. *TED Radio Hour.*

———. 2014b, November 14. Why do we create stereotypes? An interview with Paul Bloom. NPR. *TED Radio Hour.*

Reeves, D. 2002. *The leader's guide to standards: A blueprint for educational excellence.* San Francisco: Jossey-Bass.

Richardson, J. 1998a. Putting student learning first put these schools ahead. *Journal of Staff Development* 18(2): 42–47.

———. 1998b. We're all here to learn. *Journal of Staff Development* 19(4): 49–55.

Riley, D. W., ed. 1991. *My soul looks back, 'less it forget: A collection of quotations by people of color.* New York: Harper-Collins.

Riley, S. 2000. Personal communication.

Rizzolatti, G., L. Fogassi, and V. Gallese. 2006. Mirrors in the mind. *Scientific American* 295(5): 54-61.

Rizzolatti, G., and C. Sinigaglia. 2008. *Mirrors in the brain: How we share our actions and emotions.* New York: Oxford University Press.

Robinson, H. L. 2009 *School leadership and student outcomes: Identifying what works and why. Best evidence synthesis iteration.* http://www.educationcounts.govt.nz/__data/assets/pdf_file/0015/60180/BES-Leadership-Web.pdf (accessed 26 March 2013).

Rock, D. 2008, June 15. SCARF: A brain-based model for collaborating with and influencing others. *Neuroleadership Journal* 1.

———. 2009. *Your brain at work: Strategies for overcoming distraction, regaining focus, and working smarter all day long.* New York. Harper Collins.

Rock, D., and L. Page. 2009 *Coaching with the brain in mind: Foundations for practice.* Hoboken, NJ: Wiley.

Rogowsky, M. 2014.Verizon, AT&T are at it again with more data for less money. *Forbes Insights.* http://www.forbes.com/sites/markrogowsky/2014/11/01/verizon-att-are-at-it-again-with-more-data-for-less-money/ (accessed 5 April 2015).

Rosenholtz, S. J. 1989. *Teachers' workplace: The social organization of schools.* New York: Longman.

Rosenthal, R., and L. Jacobson. 1992. *Pygmalion in the classroom.* New York: Holt, Rinehart, and Winston.

Rowe, M. B. 1974. Wait time and rewards as instructional variables: Their influence on language, logic and fate control. *Journal of Research in Science Teaching* 11:81–94.

———. 1986, January–February. Wait time: Slowing down may be a way of speeding up! *Journal of Teacher Education* 42–49.

———. 1996, September. Science, silence, and sanctions. *Science and Children*, 35–37.

Sadker, M., and D. Sadker. 1985. Sexism in the schoolroom of the '80s. *Psychology Today*, 54–57.

Sanford, C. 1995a. *Myths of organizational effectiveness at work.* Battleground, WA: Springhill.

———. 1995b. *Feedback and self-accountability: A collision course.* Battleground, WA: Springhill.

———. July 8, 2011.Why feedback is irresponsible and what to do instead: Part 2. www.carolsanford.com/blog/?p=818.

Saphier, J., M. A. Haley-Spica, and R. Gower. 2008. *The skillful teacher: Building your teaching skills.* Acton, MA: Research for Better Teaching.

Sarason, S. 1991. *The predictable failure of educational reform.* San Francisco: Jossey-Bass.

Schein, E. 2013. *Humble inquiry—the gentle art of asking instead of telling.* San Francisco: Berrett-Koehler.

Schmuck, R., and P. Runkel. 1985. *The handbook of organization development in schools*, 3rd ed. Prospect Heights, IL: Waveland Press.

Schon, D. 1983. *The reflective practitioner* New York: Basic Books.

Schwartz , A. 2000, Fall. The nature of spiritual transformation: A review of the literature. In *The Spiritual Transformation Scientific Research Program*, 4. Philadelphia, PA: Metanexus Institute on Religion and Science. http://www.metanexus.net/archive/spiritualtransformationresearch/research/pdf/STSRP-LiteratureReview2-7.PDF (accessed 20 February 2015).

Seagal, S., and D. Horn. 1997. *Human dynamics.* Cambridge, MA: Pegasus Communications.

Seashore-Lewis, K. S., H. M. Marks, and S. Kruse. 1996. Teachers' professional community in restructuring schools. *American Educational Research Journal* 33(4): 757–98.

Seligman, M. 1990. *Learned optimism.* New York: Alfred A. Knopf.

Senge, P. 1990. *The fifth discipline: The art and practice of the learning organization.* New York: Doubleday.

———. 2006. *The fifth discipline: The art and practice of the learning organization*, rev. ed. New York: Doubleday.

Sergiovanni, T. 1994. *Building community in schools.* San Francisco: Jossey-Bass.

Shavelson, R. 1976. Teacher decision-making. In *The psychology of teaching methods: 1976 yearbook of the National Society for the Study of Education, Part I.* Chicago: University of Chicago Press.

———. 1977. Teacher sensitivity to the reliability of information in making pedagogical decisions. *American Educational Research Journal* 14:144–51.

Sheridan, R. 2013. *Joy Inc.: How to build a workplace people love.* New York: Penguin.

Shermer, M. 2011. *The believing brain: From ghosts and gods to politics and conspiracies—How we construct beliefs and reinforce them as truths*. New York: Times Books.

Shettleworth, S. J. 1998. *Cognition, evolution, and behavior*. New York: Oxford Press.

Shonkoff, J., and D. Phillips, eds. 2000. *From neurons to neighborhoods: The science of early childhood development*. Washington, DC: National Academies Press. http://www.nap.edu/openbook .php?record_id=9824 (accessed 20 February 2015).

Shrestha, L., and E. Heisler. 2011. *The changing demographic profile of the United States*. Prepared for Members and Committees of Congress. Congressional Research Service. https://www.fas.org/sgp/ crs/misc/RL32701.pdf (accessed 20 February 2015).

Shulman, L. S. 1987. Knowledge and teaching: Foundations of the new reform. *Harvard Educational Review* 19(2): 4–14; *Educational Review* 57(1): 1–22.

Sigel, I. E. 1984, November. A constructivist perspective for teaching thinking. *Educational Leadership* 42(3): 18–22.

Silver, H., et al. 2014. *The theory of classroom effectiveness framework*. http://www.thoughtfulclassroom .com/.

Sinek, S. 2011. *Start with why: How great leaders inspire everyone to take action*. New York: Penguin.

Smith, E. R., and R. W. Tyler. 1945. *Appraising and recording student progress*. New York: Harper & Row.

Soodak, L. C., and D. M. Podell. 1995, April. *Teacher efficacy toward the understanding of a multifaceted construct*. Paper delivered at the annual conference of the American Educational Research Association, San Francisco.

Sparks, D. 2002. *An interview with Robert Kegan and Lisa Lahey: Inner conflicts, inner strengths*. National Staff Development Council.

Sparks-Langer, G. M., and A. Colton. 1991. Synthesis of research on teachers' reflective thinking. *Educational Leadership* 49(6): 37–44.

Sprenger, M. 1999. *Learning and memory*. Alexandria, VA: Association for Supervision and Curriculum Development.

Sprinthall, N., and L. Theis-Sprinthall. 1982. The teacher as an adult learner: Cognitive developmental view. In G. Griffin (ed.), *Staff development: 1982 yearbook of the National Society for the Study of Education, Part II*. Chicago: University of Chicago Press.

Squire, L., ed. 2009. *Encyclopedia of neuroscience*. Amsterdam: Elsevier.

Stacey, R. 1996. *Complexity and creativity in organizations*. San Francisco. Berrett-Koehler.

Sremac, S., and R. R. Ganzevoort. 2013. Addiction and spiritual transformation. *Archive for the Psychology of Religion* 35(3): 399–435.

Stern, D. 2004. *The present moment in psychotherapy and everyday life*. New York. Norton.

Sternberg, R., and J. Horvath. 1995, August–September. A prototype view of expert teaching. *Educational Researcher* 24(6): 9–17.

Stetler, R. 2014, March. Third-generation coaching: Reconstructing dialogues through collaborative practice and a focus on values. *International Coaching Psychology Review* 9(1): 51–52.

Suri, M. 2013, September 15. How to fall in love with math. *New York Times*. http://www.nytimes .com/2013/09/16/opinion/how-to-fall-in-love-with-math.html?_r=3& (accessed 20 February 2015).

Sylwester, R. 1995. *A celebration of neurons*. Alexandria, VA: Association for Supervision and Curriculum Development.

———. 2000. *A biological brain in a cultural classroom*. Thousand Oaks, CA: Corwin.

———. 2005. *How to explain a brain*. Thousand Oaks, CA: Corwin Press.

Taba, H. 1962. *Curriculum development: Theory and practice*. New York: Harcourt Brace and World.

———. 1965, May. The teaching of thinking. *Elementary English* 42:15.

Tannen, D. 1990. *You just don't understand: Women and men in conversation.* New York: Ballantine Books

———. 2014, March 11. "Bossy" is more than a word to women. *USA Today.*

Taylor, P. 1970. *How teachers plan their courses.* Slough, Berkshire, England: National Institute for Educational Research.

Thompson, M., L. Goe, P. Paek, and E. Ponte. 2004. *Study of the Impact of the California Formative Assessment and Support System for Teachers (CFASST).* Report 1. Princeton, NJ: Educational Testing Service/CFASST. http://www.ets.org/Media/Research/pdf/RR-04-30.pdf (accessed 20 February 2015).

Tice, L. 1997. *Personal coaching for results: How to mentor and inspire others to amazing growth.* Nashville: Pacific Institute.

Tognetti, S. S. 2002. "Bateson, Gregory." In Peter Timmerman (ed.), *Encyclopedia of Global Environmental Change,* 183–84. Chichester: Wiley.

Tokuhama-Espinosa, T. 2014. *Making classrooms better: 50 practical applications of mind, brain, and education science.* New York: W. W. Norton.

Tolson, J. 2000, July 3. Into the zone. *U.S. News & World Report,* 39–45.

Tomasello, M. 2009. *Why we cooperate.* Boston: MIT Press.

Tomlinson, C. A. 2014. The differentiated classroom: Responding to the needs of all learners. Alexandria, VA: Association for Supervision and Curriculum Development.

Tosey, P. 2006, May. *Bateson's Levels of Learning: A framework for transformative learning?* Paper presented at Universities' Forum for Human Resource Development Conference, University of Tilburg.

Tough, P. 2012. *How children succeed.* New York: Houghton Mifflin Harcourt.

Treadwell, M. 2014. *Learning: How the brain learns.* Whangamata, New Zealand: Author. www.Mark Treadwell.com/products (accessed 20 February 2015).

Tschannen-Moran, M. 2001. Collaboration and the need for trust. *Journal of Educational Administration* 39:308–31.

———. 2004. *Trust matters.* San Francisco: Jossey-Bass.

Tschannen-Moran, M., A. W. Hoy, and W. K. Hoy. 1998, Summer. Teacher efficacy: Its meaning and measure. *Review of Educational Research* 68(2): 202–48.

Tsui, A. 2003. *Understanding expertise in teaching: Case studies of second language teachers.* Cambridge: Cambridge University Press.

U.S. Census Bureau. 2011. American Community Survey. www.census.gov.

U.S. Department of Education, Office of Educational Technology. 2013, February 14. *Promoting grit, tenacity, and perseverance: Critical factors of success in the 21st century.* Washington, DC: Government Printing Office.

U.S. Population by Race, Census 2000, and Census 2010. N.d. http://www.infoplease.com/us/statistics/us-population-by-race.html#ixzz35bane0Qq (accessed 20 February 2015).

Ulich, E. 1967. Some experiments of the function of mental training in the acquisition of motor skills. *Ergonomics* 10:411–19.

Vaida, P. 2014, June 6. Do your filters stop you from hearing? *Management Issues.* http://www.management-issues.com/opinion/6917/do-your-filters-stop-you-from-hearing/ (accessed 20 February 2015).

Vescio, V., and D. Adams. 2006, January. *A review of research on professional learning communities: What do we know?* Paper presented at NSRF Research Forum. Galveston: University of Texas.

Vygotsky, L. S. 1978. *Mind in society: The development of higher psychological processes*. Cambridge, MA: Harvard University Press.

Wagner, T. 2008. *The global achievement gap: Why even our best schools don't teach the new survival skills our children need—And what we can do about it*. New York: Basic Books.

Ward, G., and K. Burns. 2007. *The war: An intimate history, 1941–1945*. New York: Knopf.

Watzlawick, P., J. Beavin, and D. Jackson. 1967. *Pragmatics of human communication*. New York: Norton.

Wellman, B., and D. Zimmerman. 2001. Personal communication.

Wheatley, M. 1992. *Leadership and the new science: Learning about organization from an orderly universe*. San Francisco, CA: Berrett-Kohler.

———. 2006. *Leadership and the new science: Discovering order in a chaotic world*, 3rd ed. San Francisco: Berrett-Khoeler.

Wheatley, M., and M. Kellner-Rogers. 1999. *A simpler way*. San Francisco: Barrett-Kohler.

Whetten, D. A., and K. S. Cameron. 2010. *Developing management skills*, 8th ed. Upper Saddle River, NJ: Prentice-Hall.

Whiteley, G. 2012. *Alaska administrator coaching project institute notebook*.

Wiggins, G. 2014, April 14. *What feedback is and isn't*. http://grantwiggins.wordpress.com/2014/04/15/what-feedback-is-and-isnt (accessed 20 February 2015).

Williams, S. 2012. *Leaders letter*. Dayton, OH: Raj Soin College of Business, Wright State University.

Willis, J. 2013. Cooperative learning: Accessing our highest human potential. In A. Costa and P. O'Leary (eds.), *The power of the social: Brain teaching, learning, and using interdependent thinking*, 119–29. New York: Teachers College Press.

Witherall, C. S., and V. L. Erickson. 1978, June. Teacher education as adult development. *Theory into Practice*, 17.

Witkin, H. A., P. K. Oldman, W. Cox, E. Erlichman, R. M. Hamm, and R. Ringler. 1973. *Field dependence–independence and psychological differentiation: A bibliography*. Princeton, NJ: Educational Testing Service.

Wolfe, P. 1987. *Facilitator's manual: Instructional decisions for long-term learning*. Alexandria, VA: Association for Supervision and Curriculum Development.

Wolfe, P., and R. Brandt. 1998, November. What do we know from brain research? *Educational Leadership*, 56(3).

Woods, F. 1997. Staff development: A process approach. In A. Costa and R. Liebmann (eds.), *The process-centered school: Sustaining a renaissance community*, 70–85. Thousand Oaks, CA: Corwin Press.

Xiao-Qin Mai, Jing Luo, Jian-Hui Wu, and Yue-Jia Luo. 2004. *"Aha!" effects in a guessing riddle task: An event-related potential study*. Key Laboratory of Mental Health, Institute of Psychology, Chinese Academy of Sciences, Beijing, People's Republic of China. *Human Brain Mapping* 22:261–70.

Yinger, R. J. 1977. *A study of teacher planning: Description and theory development using ethnographic and information processing methods*. Unpublished PhD diss., Michigan State University, East Lansing.

Yoon, K. S., T. Duncan, S.W.-Y. Lee, B. Scarloss, and K. Shapley. 2007. *Reviewing the evidence on how teacher professional development affects student achievement* (Issues & Answers Report, REL 2007–No. 033). Washington, DC: U.S. Department of Education http://ies.ed.gov/ncee/edlabs/regions/southwest/pdf/rel_2007033.pdf (accessed 20 February 2015).

York-Barr, J., W. Sommers, G. Ghere, and J. Montie. 2001. *Reflective practice to improve schools: An action guide for educators*. Thousand Oaks, CA: Corwin Press.

Zahorick, J. 1975. Teachers' planning models. *Educational Leadership* 33:134–39.

Zhang, L. F. 2000. Are thinking styles and personality types related? *Educational Psychology* 20:271–83.

Zimmer, C. 2014. Secrets of the brain. *National Geographic*. Feb.

Zinsser, N., L. Bunker, and J. M. Williams. 1998. Cognitive techniques for building confidence and enhancing performance. In J. M. Williams (ed.), *Applied sport psychology*, 270–95. Mountain View, CA: Mayfield.

Zoller, K. 2007. *Nonverbal patterns of teachers from five countries: Results from the TIMSS-R Viteo Study*. PhD diss., California State University, Fresno.

———. 2014. *Graphic credits*. Sacramento, CA: Sierra Consulting.

Zoller, K., and C. Landry. 2010. *The choreography of presenting: The 7 essential abilities of effective presenters*. Thousand Oaks, CA: Corwin Press.

Index

About the Authors

Arthur L. Costa, EdD, is an emeritus professor of education at California State University, Sacramento. He is cofounder of the Institute for Habits of Mind and cofounder of the Center for Cognitive Coaching. He served as a classroom teacher, a curriculum consultant, and an assistant superintendent for instruction, and as the director of educational programs for the National Aeronautics and Space Administration. He has made presentations and conducted workshops in all 50 states as well as on six of the seven continents

Art has written and edited numerous books, including *The School as a Home for the Mind*, *Techniques for Teaching Thinking* (with Larry Lowery), and *Cognitive Coaching* (with Bob Garmston). He is editor of *Developing Minds: A Resource Book for Teaching Thinking*, and co-editor (with Rosemarie Liebman) of the Process as Content Trilogy: *Habits of Mind*, *Learning and Leading with Habits of Mind*, and *Habits of Mind across the Curriculum* (both with Bena Kallick). He is co-author of *Assessment Strategies for Self-Directed Learning* (with Bena Kallick), *Cognitive Capital: Investing in Teacher Quality* (with Bob Garmston and Diane Zimmerman), *The Power of the Social Brain* (with Pat Wilson O'Leary), and *Dispositions: Reframing Teaching and Learning* (with Bena Kallick). His books have been translated into Arabic, Chinese, Italian, Spanish, and Dutch.

Active in many professional organizations, Art served as president of the California Association for Supervision and Curriculum Development and was the national president of the Association for Supervision and Curriculum Development from 1988 to 1989. He was the recipient of the prestigious Lifetime Achievement Award from the National Urban Alliance in 2010.

Robert J. Garmston, EdD, is an emeritus professor of education at California State University, Sacramento, and co-developer of Cognitive Coaching with Dr. Art Costa. Formerly a classroom teacher, principal, director of instruction, and acting superintendent, he works as an educational consultant and is director of Facilitation Associates, a consulting firm specializing in leadership, learning, and personal and organizational development. He is co-developer and founder of the Center for Adaptive Schools with Bruce Wellman, now organized under the heading of Thinking Collaborative (http://www.thinkingcollaborative.com), representing its relationship to Cognitive Coaching and Habits of Mind. The Center for Adaptive Schools develops organizational capacity for self-directed, sustainable improvement in student learning. He has made presentations and conducted workshops for teachers, administrators, and staff developers throughout the United States as well as in Canada, Africa, Asia, Australia, Europe, China, and the Middle East.

Bob has written and co-authored a number of books, including *Cognitive Coaching: A Foundation for Renaissance Schools* with Art Costa, *How to Make Presentations That Teach and Transform*, and *A Presenter's Fieldbook: A Practical Guide*. In 1999, the National Staff Development Council (NSDC) selected *The Adaptive School: A Sourcebook for Developing Collaborative Groups* as book of the year. In that same year, Bob was recognized by NSDC for his contributions to staff development. Second editions of the *Presenter's Fieldbook* and *Adaptive Schools* were published in 2005 and 2009. His recent publications include *I Don't Do That Any More: A Memoir of Awakening and Resilience* (2011), *Unlocking Group Potential for Improving Schools* (2012), and *Lemons to Lemonade: Resolving Problems in Meetings, Workshops and PLCs* (2013). His books have been translated into Arabic, Dutch, Hebrew, and Italian.

In addition to educational clients, he has worked with diverse groups including police officers, probation officers, court and justice systems, utilities districts, the U.S. Air Force, and the World Health Organization.